W9-BWF-021

The Electric Power Industry: A Nontechnical Guide

THE ELECTRIC POWER INDUSTRY

A NONTECHNICAL GUIDE

THOMAS O. MIESNER
A. ANDREW GALLO

PENNWELL
BOOKS

Disclaimer

The recommendations, advice, descriptions, and the methods in this book are presented solely for educational purposes. The author and publisher assume no liability whatsoever for any loss or damage that results from the use of any of the material in this book. Use of the material in this book is solely at the risk of the user.

PennWell Books, LLC
10050 E 52nd Street
Tulsa, OK 74146

866.777.1814
sales@pennwellbooks.com
www.pennwellbooks.com

Publisher: Matthew Dresher

Library of Congress Cataloging-in-Publication Data

Names: Miesner, Thomas O., author. | Gallo, Andrew, author.
Title: The electric power industry a nontechnical guide / Tom Miesner & Andrew Gallo.
Description: First edition. | Tulsa, OK : PennWell Books, [2022] | Includes bibliographical references and index.
Identifiers: LCCN 2022030785 (print) | LCCN 2022030786 (ebook) | ISBN 9781955578103 (hardcover) | ISBN 9781955578110 (epub)
Subjects: LCSH: Electric utilities. | Electric power production.
Classification: LCC HD9685.A2 G355 2022 (print) | LCC HD9685.A2 (ebook) | DDC 333.793/2—dc23/eng/20220817
LC record available at https://lccn.loc.gov/2022030785
LC ebook record available at https://lccn.loc.gov/2022030786

Printed in the United States of America

1 2 3 4 5 26 25 24 23 22

Contents

Chapter 3

Chapter 4

Chapter 5

Figures

Table

Dedication

This book is dedicated to all the people who work, or have worked, in the electrical power industry – around the clock, often in difficult conditions – so we can enjoy the security and comfort brought to us by that electricity.

Foreword

Clean energy – the phrase rolls pleasantly and hopefully off the tongue. Only "clean energy" does not really exist. No matter the energy type, all energy production and consumption has consequences, some readily apparent, others more subtle. The most apparent consequence of using energy – releasing heat into the surrounding environment – is so common it sinks into the background and is not much discussed or debated. The more energy we use, the hotter our world becomes, even if we achieve "carbon neutral" status. Cleaner (not clean) energy is a complex topic with known and unknown consequences.

Conserving and limiting our use of energy and improving energy efficiency in the quest to reduce energy use *per capita* must be part of the energy dialogue. Those goals serve as important cornerstones of the energy policy debate and are, arguably, as important, if not more important, than other parts of achieving "cleaner energy."

We hope this book provides insight into the electrical power industry and aids in the energy dialogue moving the world along its journey to cleaner energy.

Acknowledgments

First, we thank our families, and particularly our wives, for the respect, patience, understanding, encouragement, and love they provided throughout this effort. Those five gifts provided the support we needed to produce this book.

Next, it would not have been possible for an oil and gas pipeline executive and an electric industry lawyer to write this book without the help of numerous people who willingly reviewed chapters, answered questions, and provided ideas and updates. We appreciate the assistance they provided and the interest they took.

Special thanks go to Dr. John Miesner (engineer and inventor extraordinaire and Tom's brother), who patiently taught us about the difference between magnetic force and magnetic field, helped us understand other aspects of magnetism and electricity, and produced many of the figures contained in Chapter 1.

We sincerely thank our contributing authors: Dr. Paul F. Meier, energy consultant and author of *The Changing Energy Mix*; William T. Shaw, author of *Cybersecurity for SCADA Systems*; and Tom Ortman. Each of these three wrote the first draft of three different chapters and then critiqued our edits of their draft.

We are also grateful to our reviewers who pored over chapters, finding errors both large and small, offered suggestions for improvements, and answered a long list of questions. These reviewers include:

- Scott G. Gudeman, P.E. Chief Engineer – Distribution Engineering at TRC Companies, Inc., who reviewed Chapters 1 through 6
- Dan Smith, V.P. of Transmission System Operations for the Lower Colorado River Authority, who reviewed Chapter 4, helping us understand the difference between transmission and distribution
- Steven Dockery, P.E., who reviewed and provided comments to Chapter 5, including reviewing the figures for accuracy and completeness and helping us understand that the neutral wire might also be used as a shield wire
- Lisa Martin, Vice-President, Engineering & Technical Services for Austin Energy, who reviewed and provided comments on Chapter 10 and 12
- Erika Bierschbach, V.P. of Energy Market Operations and Resource Planning for Austin Energy, who reviewed and commented on Chapter 11

Finally, our thanks to William Leffler, who wrote the first book in the PennWell Books nontechnical series and patiently co-authored two editions of *Oil and Gas Pipelines in Nontechnical Language* with Tom.

To all those people, we express our thanks. Nonetheless, and as always, we had the last word, so we carry the burden of interpreting correctly what they offered.

We also want to acknowledge each other. While we have been friends for many years, working together on this text provided opportunities for us to review each other's work, challenge each other's understanding, and critique each other's writing style, all in the interest of providing the best work possible. Happily, our friendship survived the ordeal.

1

Electrical Energy Basics

There is no greater satisfaction for a just and well-meaning person than the knowledge that he has devoted his best energies to the service of the good cause.

—Albert Einstein (1879–1955)

"Grandpa Tom, it was exciting! The teacher told us about electricity and how electricity does not cause pollution or climate change," Luke exclaimed. But then he turned quizzical and asked, "If electricity does not cause pollution, why don't we just switch to using only electricity?"

Luke's grandfather explained that electricity does not occur naturally, or at least not in sustainable amounts. Rather, electricity is generated from other forms of energy. "The world is trying to switch from carbon-based fuels like crude oil, coal, and natural gas to renewable fuels – primarily wind and solar," Grandpa Tom said. "But wind and solar are not as dependable as those carbon-based fuels and require lots of storage for when they aren't generating," Grandpa Tom elaborated.

The grandfather went on to explain that one of the primary advantages of electricity produced from "fossil fuels" is fossil fuels contain chemical energy derived many years ago from the sun, and the chemical energy stored in fossil fuels is convertible to electrical energy nearly instantaneously and on demand, whereas wind and solar can only generate significant amounts of electricity when atmospheric conditions allow.

A pensive frown came onto Luke's face, and, after a brief pause, he said, "It sounds like maybe electricity is not as simple as I thought."

"And life in general," Grandpa Tom added.

Electricity and Magnetism

Magnetic fields generate most of the electricity consumed in the world. Conversely, electricity can generate magnetic fields. So, magnetic fields and electricity are causally, and directly, related – a recurrent theme throughout this text. Most people use the terms magnetic field and magnetic force interchangeably, but physicists consider that incorrect. They say magnetic fields acting on an object produce a magnetic *force*. This "nontechnical" book attempts to use the two concepts correctly but, like most people in the world, the authors may occasionally conflate them.

Electrical Energy

Electrical energy is a secondary form of energy, meaning it does not occur naturally in commercial quantities. Rather, primary energy sources produce electrical energy (commonly just called electricity, rather than electrical energy) – more on this in Chapter 3.

Electricity

The Merriam-Webster online dictionary offers the following definitions for electricity:

1. A fundamental form of energy observable in positive and negative forms that occurs naturally (as in lightning) or is produced (as in a generator) and that is expressed in terms of the movement and interaction of electrons.
2. Electric current or power.
3. A science that deals with the phenomena and laws of electricity.
4. Keen contagious excitement.[1]

The authors use definition 2 in this text and leave definitions 1 and 3 to physicists. The authors hope readers experience definition 4 as they read this text.

Dr. Christopher S. Baird, in his website "Science Questions with Surprising Answers," offers the following definition: "This word is very general and basically means, 'all things relating to electric charge.'"[2]

Combining the Merriam-Webster and Baird definitions, electricity means different things to different people. In real life, most people use the terms electricity, electrical energy, and electrical power interchangeably.

Magnetism

The same dictionary defines magnetism as:

1. A class of physical phenomena that include the attraction for iron observed in lodestone and a magnet
   a. are inseparably associated with moving electricity
   b. are exhibited by both magnets and electric currents
   c. are characterized by fields of force
2. A science that deals with magnetic phenomena.
3. An ability to attract or charm.[3]

Definitions 1.a. and 1.b. apply to magnetism as discussed in this text. Definitions 1.c. and 2 are left to physicists. The authors hope definition 3 applies to them.

Electrical Terms and Relationships

Five of the most common electrical terms – volt, current, amps, watts, and watt-hours are discussed in this section. Also included is an "extra credit" section discussing two less common terms – coulombs and joules. Equations showing the relationship between the various terms are also included. Feel free to skip or read the extra credit section.

Volt

Volt (V) – defined as one coulomb per joule – comes from the Italian physicist Alessandro Volta (1745–1827). Volts measure the difference in electrical potential between two points. One volt also means, "the electrical potential that drives one ampere of current against one ohm of resistance." Volts are defined in terms of current and resistance, so those terms are discussed next. Most residences in North America use 120 volts or 240 volts. Most other parts of the world use only 240 volts.

Current

Current (I) – the *transmission* of electrical energy caused by a difference in electrical potential between a source and a load. (Loads are devices using electricity.) Current is transmitted only when the load needs it and only in the amount needed by the load. Current is measured in amps.

Ampere

Ampere (A) – abbreviated "amp" and defined as one coulomb of charge flowing past a point in a circuit in a second – is named for André-Marie Ampère (1775–1836), a French mathematician and physicist. Amps express the *rate* of energy transmission. Residential circuits are rated in amps. A fifteen-amp circuit can safely handle a maximum current of fifteen amps.

Watt

Watt (W) – defined as one joule per second (and *volts* multiplied by *amps* measures power) – is named after James Watt (1736–1819), a Scottish instrument maker and inventor whose steam engine contributed substantially to the Industrial Revolution.

Watt-Hour

Watt-hour (Wh) – defined as one watt of power consumed over an hour. Watt-hours define energy consumed over time.

Ohm

Ohm (Ω or R), named after Georg Simon Ohm (1789–1854), a German physicist and mathematician, measures *resistance* to transmission and defined as the resistance in a circuit transmitting a current of one amp when subjected to a potential difference of one volt. Electrical power professionals commonly use the symbol R rather than the Greek letter omega (Ω).

Prefixes

One thousand volts equals one kilovolt (kV). One million watts equals one megawatt (MW). The prefixes for the terms included in this section carry on to Giga-, Tera-, Penta-, and so forth.

Ohm's Law

Dr. Ohm, through experiments with equipment he designed, discovered that current, voltage, and resistance relate to each other and vary in a predictable manner. From this finding, he developed Ohm's Law, one of electricity's fundamental laws:

$$V = I \times R$$ (where V is voltage, I is current, and R is resistance)

If two of the variables are known, the unknown variable can be determined. Electrical engineers and designers use Ohm's law to design systems.

Resistance in this equation is a property of the conductor carrying the current and does not change as voltage and current change. Resistance does change as the conductor material or cross-sectional area of the conductor changes.

Electrical Terms Extra Credit

Some people like to dig deeper into the relationships or terms and find equations helpful. This section is for them.

Coulomb

Coulomb (C), named for the French physicist Charles-Augustin de Coulomb (1736–1806), is defined as the amount of electrical charge contained in approximately 6,240,000,000,000,000,000 electrons or, more practically, the amount of electricity transported in one second by a current of one amp. The discerning reader will recognize amps and coulombs are defined in terms of each other.

Joule

Joule (J), named after the English physicist James Prescott Joule (1818–1889), is defined as the amount of work done by moving an electric charge of one coulomb through an electrical potential difference of one volt. It also means the work required to produce one watt of power for one second. Joules measure *work* and describe either the amount of *work performed* or *energy stored.*

Equations

The following equations demonstrate the various relationships:

1 Volt = 1 Joule/1 Coulomb
1 Amp = 1 Coulomb/1 second
1 Watt = 1 Joule/1 second
1 Watt = 1 Volt × 1 Amp
1 Watt-hour = 1 Watt × 1 hour
1 Coulomb = 1 Amp × 1 second
1 Joule = 1 Coulomb × 1 Volt
1 Joule = 1 Watt/1 second
1 Ohm = 1 Volt/1 Amp

Engineers know equations do not apply to every case and often need extra factors in certain situations. For example, one of the equations for watts works for direct current (DC) but not alternating current (AC). Therefore, readers should take these as general guidelines and not absolutes.

Relationship Summary

The following summarizes the essential electrical terms and what they measure:

- volt – electrical potential
- amp – current
- watt – electrical power
- watt-hours – electrical energy
- ohm – resistance to transmission

Electrical Potential

In this text, the term *electricity* generally means the *transmission* of electrical energy. Electrical energy is transmitted in a circuit from areas of higher electrical potential to areas of lower electrical potential, much like water flows from areas of higher pressure (water towers or pumps) to areas of lower pressure (faucets in houses). The term *volt* (V) represents the amount of electrical potential of the circuit.

As with water, which sits waiting in the faucet at a particular pressure and flows only when someone opens the faucet (valve), electricity waits at the switch at a particular voltage (usually 120V or 240V, depending on the country) and is *transmitted* only when someone (or some device like a thermostat) closes the switch, completing the circuit.

Electrical potential increases electrical energy to a system. For example, electrical generators add electrical energy, which is transmitted from the generating station to an end-use electrical device such as a light, air conditioner, stove, oven, or motor.

Flowing water may be a useful analogy, but it is not an accurate one. Electricity does not flow; it is transmitted from one electron to the other, as explained later in the text.

Electrical Transmission

Conductors – in this case, metal wires – consist of metal atoms. Each metal atom has layers of electrons orbiting the nucleus (center) in successive layers. The atom's nucleus holds electrons closer to the center tightly and electrons on the outside layer loosely – in fact, so loose that they move freely from one atom to another. These loosely held electrons are called "valence electrons" or "free electrons."

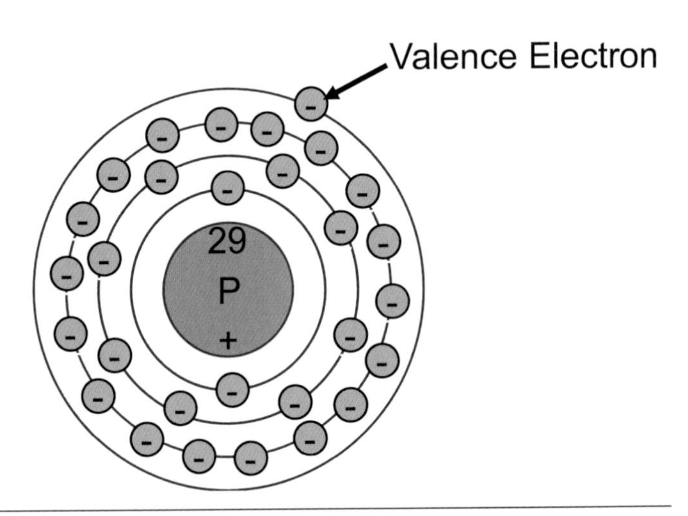

Figure 1-1. Two-dimensional Representation of the Atomic Structure of a Copper Atom. While this representation is two-dimensional, the electrons move in three dimensions.

Valence Electrons

Valence electrons move extremely fast, in constantly changing directions, pushing on each other as they go. Physicists say all these free electrons look like a swarm of bees as they buzz around the outside of metal objects (unlike water, where the molecules in the middle of a pipe move more freely than those along the pipe wall because of friction). Some people say electricity is moving electrons, but that is not quite right. Electrons do move and it is this moving that enables transmission of electricity but the moving electrons are not themselves electricity.

Transmission

When both ends of a wire are at the same voltage (electrical potential), the electrons all buzz around and push on each other, but all the pushing balances out and no energy is transmitted into or out of the wire.

When a light switch is open (not turned on), the voltage on the switch's supply side is 110V and 0V on the load side. Even though the voltage at the load side of the switch (and at the light bulb it controls) are both 0V, the wire between them still has electrons swarming around pushing on each other. Because each location has the same voltage, no net energy transfer occurs – in other words, no electricity is transmitted.

> **Load**
>
> In electrical parlance, loads are the devices or equipment receiving or using electricity – for example, lights, air conditioner motors, stove heating elements, and so forth.

When the switch is closed (turned on), the light bulb instantly illuminates because the voltage differential across the switch (110V versus 0V) causes immediate transmission of electrical energy across the switch through a wire and to the light bulb.

One way to visualize electrical transmission is that the electrons at 110V are pushing on the electrons at 0V harder than those electrons push back. This pushing happens very quickly as each electron pushes on its neighbor, transmitting electricity to the light bulb. Not only does the electrical energy transmit across the switch and to the light bulb, but it also transmits to the switch from the supply lines and from the generating station to the supply line, with each electron pushing harder on its downstream neighbor than that neighbor pushes back. This imbalance in pushing moves electrical energy from the generation site (source) to the end-use device (load). The light bulb converts the electrical energy into light and heat. But a complete circuit back to the source must also exist. That is why home wiring has at least two wires and not just one.

> **Circuit**
>
> Before electricity will transmit, it must have a complete path from the source to the load and back to the source (called a *circuit*).

The term "very quickly" in the previous paragraph is an understatement because electricity is transmitted at *almost* the speed of light – a little over 670,000,000 miles per hour. Because the earth's circumference at the equator is a little less than 25,000 miles, electricity will travel completely around the earth's circumference *seven times in one second.*

Electrical Source

In the example above, the electricity must come from somewhere. It starts as an electrical force produced by a generator pushing on the electrons closest to it, which, in turn, push on the next, and so forth. Some texts incorrectly refer to electricity as electrons flowing. For clarity, electrons do flow, and physicists call the flow "electron drift," but electron drift is much slower than electricity is transmitted. Think of electricity as continuous waves of energy produced at the generator and traveling to the load at almost the speed of light, rather than as electrons flowing.

Water Flow as an Analogy for Electricity Transmission

Imagine a water hose connected to a house. If the hose is full of water but the faucet (source) is turned off (the valve is closed), no difference in pressure exists between the faucet and the end of the hose and water does not flow. When the faucet at the house is turned on (by opening the valve), a pressure wave starts at the faucet and quickly flows to the hose's end where it pushes water out. The wave of pressure that traveled quickly from the house pushed the water out. In the same way, a wave of electrical force travels from the generator to the load.

A rock tossed into a pond provides another water example. When the rock hits the water, it generates a series of small waves. Many people call these waves ripples. The ripples travel from the place of impact to the shore, but the individual molecules do not move from the point of impact to the shore, rather they push their neighbor a little bit. In the same way, electrical transmission is a continuous wave (or waves) of electrical energy. Rocks are like voltage. Larger rocks make larger waves and higher voltage pushes more current.

Resistance to Flow

Another water analogy: water pressure is measured in pounds per square inch (psi), and the electrical potential is measured in volts. Larger pipes carry more water at the same pressure over the same distance versus smaller pipes. Higher pressures move more water than lower pressures in the same size pipes over the same distance.

In other words, more pressure is required to push the same amount of water through smaller pipes than larger pipes – smaller pipes have higher resistance to flow. Figure 1-2 shows the difference in resistance to flow caused by size and distance.

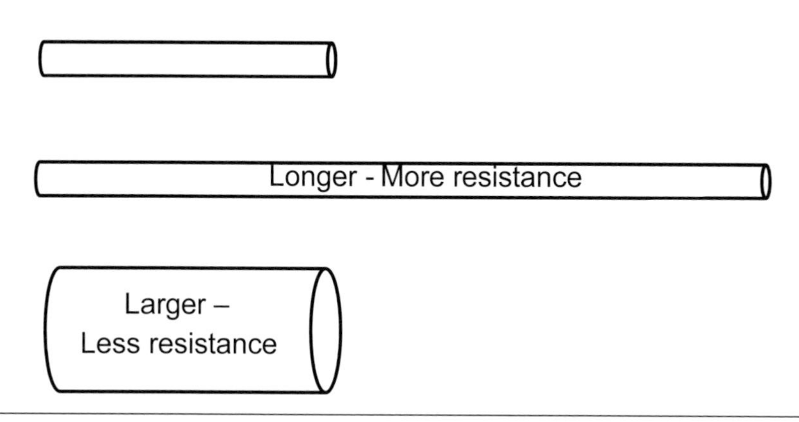

Figure 1-2. Resistance to Water Flow.

In the same way, conductors having greater cross-sectional areas carry more electricity than conductors having smaller cross-sectional areas at the same voltage (V) over the same distance. Higher voltages transmit more electricity than lower voltages in the same cross-sectional area conductors over the same distance. In other words, it takes more voltage differential to "push" the same amount of electricity through smaller wires than larger wires – smaller wires have higher resistance to transmission.

Continuing with the water analogies, when users turn on more faucets, more water flows. The flow rate (gallons per minute) depends on the

- pressure differential pushing the water;
- size and internal roughness of the pipe the water passes through; and
- number of devices using the water at any one time.

Fortunately, faucets, dishwashers, ice makers, and other water-using devices do not care about how many gallons per minute they receive and, instead, care more about the total number of gallons. Low-flow rates are a nuisance but not a catastrophe.

Electrical transmission on the other hand, is quite different. Electric devices are designed to use a set amount of power. The dishwasher may not care how fast the water comes in but does care how fast the electrical energy comes in. If the dishwasher tries to draw more current than the system can safely supply, a circuit breaker opens removing the dishwasher from the system.

One difference between water and electricity is water molecules move from one point to another and exit the system, but electricity, because the electrons do not exit the system, requires a circuit with a return line, as shown in Figure 1-3, so the electrons can drift.

Not to confuse the issue, but the electrical return path need not be a wire. Fiber-optic transmission lines under the ocean have only a high-voltage DC supply wire to power the optical amplifiers. The return goes through the ocean water – somewhat like water supply where the return comes through clouds and rain.

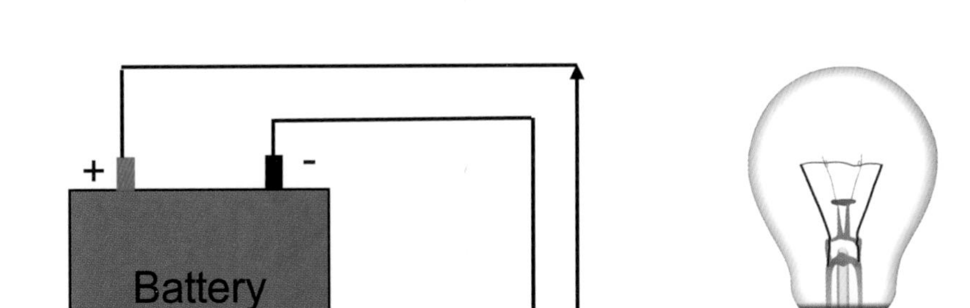

Figure 1-3. Simple Electrical Circuit. When the switch is closed, the difference in electrical potential between the positive and negative terminals causes electrical transmission from the battery to the light.

Electrical Power and Electrical Energy

Understanding the difference between two often conflated terms, power and energy, can help when learning about electricity.

Power

Power depends on:

- volts driving the circuit; and
- combined load of the devices connected to that circuit.

As an example, light bulbs do not require as much power as air conditioners. Thus, the power transmitted to a light bulb is less than the power transmitted to an air conditioner. In this example, the light bulb and the air conditioner receive electricity at 110V, but the air conditioner requires more power (watts), which means it requires a higher current (I) than the light bulb.

Power Extra Credit

Understanding this section is not necessary to understand the electrical power industry but provides additional insight.

Power depends on volts and amps and is measured in terms of volt-amperes (VA). One VA equals one watt of direct current (DC) power but alternating Current (AC) introduces other factors called real, reactive, and apparent power that usually mean one watt requires more than one VA of energy. A discussion of real, reactive, and apparent power appears later in this chapter and again in Chapter 4.

Electrical Energy

Continuing with the light example, if the light stays on for an hour and consumes electricity at the rate of 10 watts during that hour, then the energy consumed over that hour equals 10 watt-hours (Wh). Watt-hours measure the energy used as opposed to just watts, which is simply the power the load requires.

Peak Demand and Average Demand

Electrical demand varies during the day and over the seasons, with more electrical energy required on hot or cold days than on temperate days.

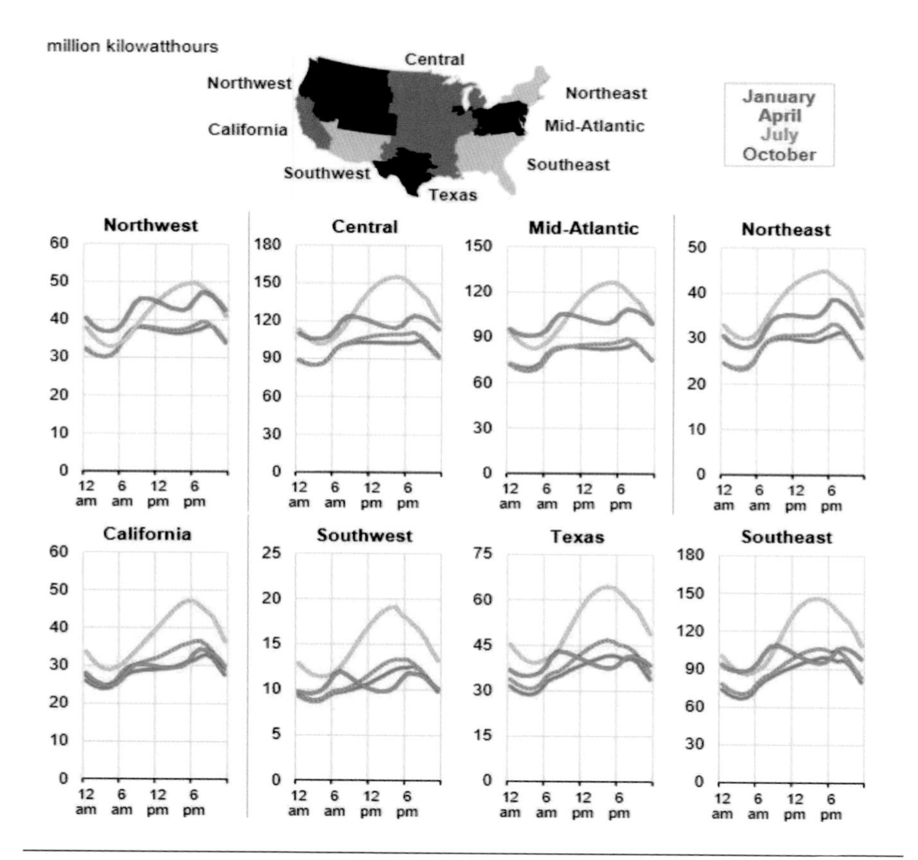

Figure 1-4. Average Hourly Electrical Load for Selected Regions and Months.

Between October (red line) at 3:00 a.m. and July (yellow line) at 6:00 p.m., the average hourly demand varies by about a factor of two. In other words, load demand is twice as much on summer afternoons as on fall mornings. This wide demand swing means electric utilities must cycle generation and storage throughout the day and, in some cases, several times per day.

Electrical devices operate at specific voltages, alternating at specific frequencies. The residential voltage standard varies by country but is generally either 110V or 220V, and 60 or 50 hertz. Higher or lower voltages and frequencies than the standard cause safety or operating problems. To make sure enough voltage exists and to keep frequencies constant, electric utilities build facilities (generation plants, storage, transmission, and distribution lines) to meet the *maximum* load.

> **Frequency**
>
> Perhaps even more important than keeping voltage constant is maintaining the grid's *frequency*. In the United States, the grid frequency is 60 hertz. When frequency drops below 60 Hz, electric devices run more slowly (e.g., an electric clock would run slow) and if frequency goes above 60Hz, electric devices run fast.

When the cumulative energy demanded by users on the electric grid (load) falls, generators adjust to produce less electricity or get removed from the grid supply. When load increases – for example in the afternoon of a hot summer day – the system operator adjusts generators to produce more electricity and/or connects more generators to the grid. These activities help maintain the voltage and frequency of the electrical system.

Figure 1-4 shows only average hourly demand. It does not show maximum instantaneous demand, which is even higher. Extreme heat and extreme cold call for more generators connected to the grid at the same time, and any interruption in generation capacity can result in local or regional power loss.

The second draft of this chapter, coincidentally, was delayed for several days because one of the authors lost power to his office for more than two days in February 2021 when Texas experienced unprecedented cold weather that drove up demand. Compounding the problem, the cold weather affected natural gas supply, causing some generators to lose natural gas supply and shut down, and freezing temperatures caused some wind turbines to stop working. The generators remaining online tried to work harder to make up for the increased demand and reduced supply but, eventually, load outstripped available generation and the grid operator had to order rotating load outages.

Just like cars slow down going up hills (at a constant fuel flow) because going uphill requires more power than driving on level ground, when the grid draws more current, generators start slowing which also slows the frequency of the electricity they generate.

Multiple news sources reported the electrical grid in Texas, operated by Electric Reliability Council of Texas, Inc. (ERCOT), came within four minutes of crashing. To avoid frequency dropping to dangerous levels and causing safety devices to automatically remove generators, ERCOT ordered electrical distribution companies to disconnect customers (load) from the system to balance demand with supply.

Resistance

Transmission is not 100% efficient. Electricity must overcome resistance in the conductor and other electrical equipment, including transformers, regulators, and switches, meaning less electricity arrives at the destination than started at the generation plant. This lost electricity manifests itself as heat along the conductors. The amount of heat generated depends on the conductor's properties. Some materials, like copper, have low resistance to transmission and are called *conductors.* Other materials like plastics and ceramics have high resistance to transmission and are called *insulators.* Conductors used in electrical service are generally copper or aluminum. Many people use the terms conductors and wires synonymously – a practice that shocks physicists and electrical power professionals (pun intended).

Resistance, measured in Ohms, comes in two forms:

- Inherent resistance – primarily wire size and construction material
- Resistance intentionally added to the system using resistors

Resistors make it harder for electrical transmission. They are not much used in the electrical power industry but are used extensively in appliances such as electric heaters, electric ovens, and toasters, to turn current into heat. The electronics and semiconductor industries also use resistors.

Static Electricity

Some electricity is called static electricity. When people walk across carpet wearing woolen socks, their feet rub on the carpet generating an electrical charge. Most of this electrical charge enters the people walking, increasing the electrical potential of their bodies. When they touch a metal object, like a doorknob, the electrical potential between their body and the doorknob equalizes, causing a spark. This simple example illustrates energy transmits between objects at different electrical potentials

> **Static Electricity**
>
> Static electricity is the result of an imbalance between negative and positive charges in an object. These charges can build up on the surface of an object until they find a way to release or discharge.

Benjamin Franklin and his kite experiment provide another example of static electricity. The wind passing over the kite built up a static charge on the kite that was transmitted down the kite string to the key. Because the key became more and more charged, it reached the point when the difference in charge between the key and Franklin's hand increased enough to create a spark between the hand and key. Static electricity gave early physicists clues about electricity, but its usefulness ends at that point, so the rest of this text leaves it behind.

Energy Sources

As mentioned previously, electricity is a *secondary* energy source generated from *primary* energy sources. Most people say primary energy sources are "original sources" and secondary energy sources are "produced from primary energy sources."

Primary Energy Sources

Primary energy sources exist in roughly three categories:

- Fossil fuels (coal, natural gas, and oil)
- Nuclear
- Renewable (wind, water, biomass, geothermal, and sunlight)

The exact grouping and definition of each energy source can get a little fuzzy. Some people argue water is not *renewable* energy because gravity is the real primary energy. After all, the force of gravity makes the water flow through the turbine driving the generator. Therefore, gravity–not water– should get the credit. These same people might argue wind is not a primary source because atmospheric conditions cause wind (it does not exist on its own). In both cases, the important point is not whether the source is primary or secondary but how the generated electricity impacts the environment.

Hydrostatic Head

Hydropower depends on hydrostatic "head"—the distance a given water source falls as it generates power. Gravity provides the force for hydropower so a hydro-electricplant with a tall/high head can produce more energy per hour than a similar plant with a short/low head.

Secondary Energy Sources

Secondary sources do not occur naturally and must be produced from primary sources.

Some of the most common secondary energy sources include:

- Electricity
- Gasoline
- Liquid Fuel Oil
- Biofuels
- Hydrogen

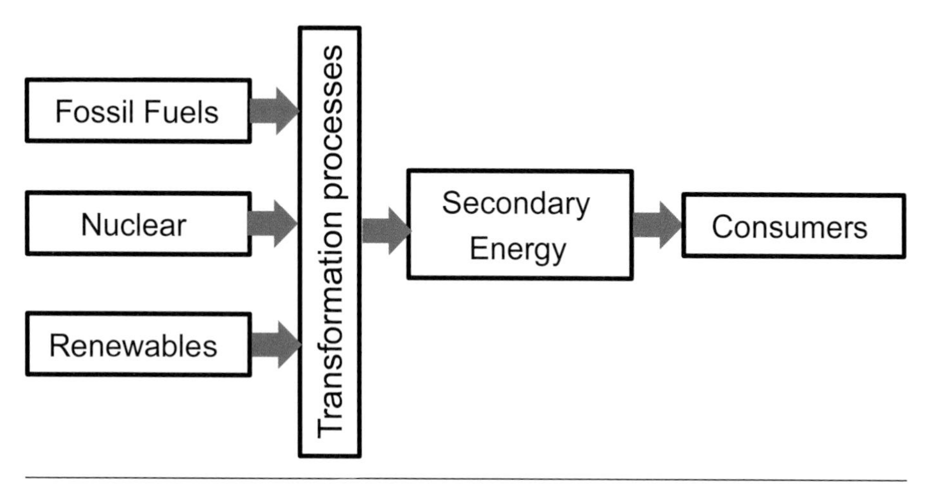

Figure 1-5. Primary Energy Transmission to Consumers.

Transformation Process

Transforming primary energy into secondary energy, generating electricity, refining crude oil into fuels (gasoline, diesel, and jet), converting corn to ethanol, converting wind and solar into electricity, and extracting hydrogen from fossil fuels or water, have various consequences.

Some of these consequences are well known, for example, fossil fuels releasing CO_2 into the air, solar panel manufacturing producing toxic by-products, biofuels reducing farmland for raising crops, reducing global food supplies, and raising food prices.

Wind turbines, in addition to creating visual and noise pollution, cause local climate change by altering the atmospheric boundary layer.[4] Other consequences are not well known or have not yet surfaced. This text does not take a position on the "best" type of energy but does attempt to deal factually with all forms of energy by noting their positive and negative aspects.

Energy Consumption

Figure 1-6 shows the history of primary energy sources consumed by the world starting in 2010 and the forecast for use out to 2050. The graph shows renewables increasing by a factor of about two and one-half over approximately thirty years.

Most people can agree fossil fuels have greatly improved the world's standard of living over the past 200 years. Most people also agree fossil fuels caused a significant increase in the level of pollutants over that same time. Despite the forecast growth in renewables, the U.S. Energy Information Administration (EIA) forecasts that, by 2050, fossil fuels will still comprise nearly 70% of primary energy used in the world.

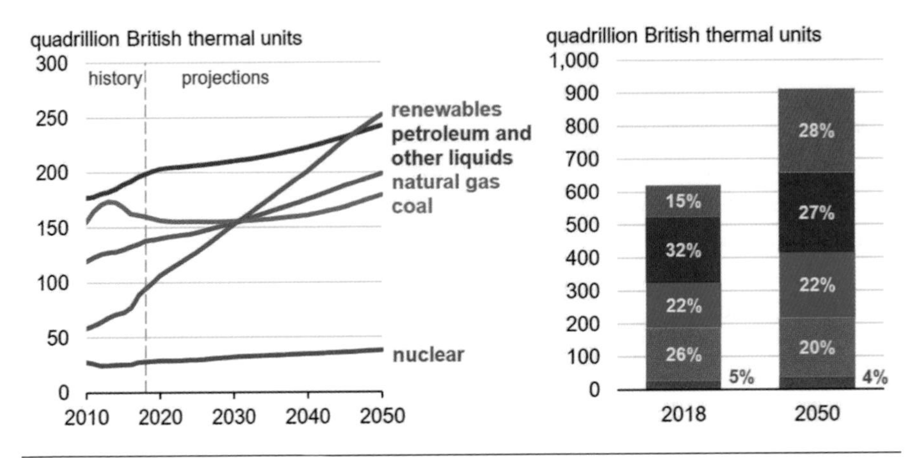

Figure 1-6. World Consumption of Primary Energy Sources.

Electrical Generation

In 1831, Michael Faraday, an English scientist, built the first version of an electromagnetic generator (the Faraday Disk), which used the force of magnetism to generate electrical energy and remains the basic technology for electromagnetic power generation to this day. Many of the practices and features of electricity everyone takes for granted, like alternating current, developed because of the characteristics of the power electromagnetic generators create as they generate electricity. This chapter discusses only electromagnetic generation, the source of electricity in most of the world, and Chapter 3 deals with other forms of generation such as solar voltaic and fuel cells.

Magnets

Magnets produce magnetic fields that exert forces on metallic objects. Magnets can be either *permanent* or *temporary*. Some metals possess magnetic properties naturally due to a phenomenon called ferromagnetism, which results from the orientation of various atomic particles relative to each other. These magnets are called *permanent* magnets. Electromagnets are produced when metals are exposed to electrical current, temporarily magnetizing them.

Magnetic Fields

Permanent magnets and electromagnets have two ends with magnetism (magnetic field lines) surrounding them. We refer to the two ends as *poles* – the north pole and the south pole. The ends of the magnets have no special qualities, and the magnetic fields do not start or finish at the end. Rather, the fields extend through the magnet's entire length.

The field lines have a magnitude and direction. The field is stronger near the magnet and weaker farther away from it. Figure 1-7 shows a permanent magnet creating its own persistent magnetic field. Nonpermanent magnets (electromagnets) can be produced by wrapping a conductor (wire) around a core of steel and sending an electrical current through the wire. The current temporarily magnetizes the steel.

Poles of a Magnet

The earth is a huge magnet, which is why compasses work. One end of the compass needle points to the North Pole and the other to the South Pole, which is why one end of a magnet is called the north pole and the other is called the south pole.

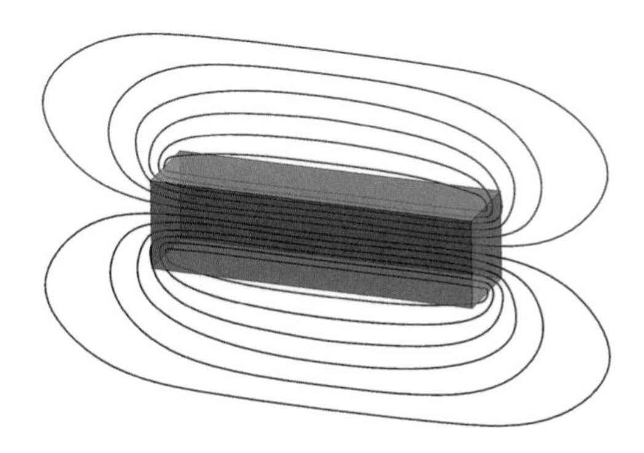

Figure 1-7. Magnetic Field Lines of a Permanent Magnet.

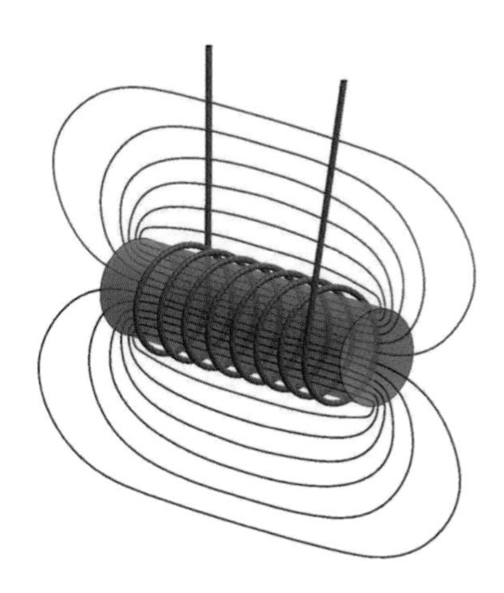

Figure 1-8. Magnetic Field Lines of an Electromagnet.

Magnetic Flux

The magnetic field lines passing through an area are called magnetic flux and measure the flux density (amount) in terms of the "Tesla," after Nikola Tesla, a Serbian-American inventor, electrical engineer, mechanical engineer, and futurist, best known for his contributions to the design of the modern alternating current electricity supply system.

Electromagnetism

The interaction between magnetism and electricity is called electromagnetism. The movement of a magnet can generate electricity; in turn, electricity can generate a magnetic field.

Electromagnetic Generators

Holding a magnet steady next to a conductor makes the magnet's fields remain constant as it tries to attract the metal but does not generate electricity. If, however, the magnet moves past the conductor, the magnitude and direction of the magnetic field relative to the conductor change, generating electricity. In other words, magnetic fields induce an electromagnetic force in a conductor as their position changes relative to each other.

Figure 1-9 shows a magnet rotating in a conductor loop. As the magnet rotates, each pole gets closer and then farther from the conductor bars, so the magnetic flux acting on the conductor bars oscillates from strong to weak.

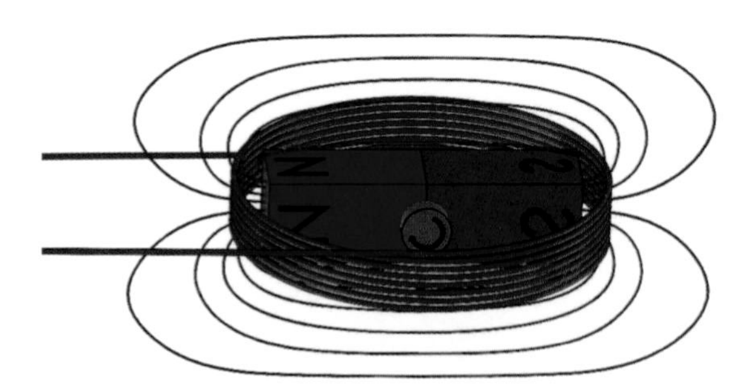

Figure 1-9. Schematic of a Conductor Loop.

As the flux varies, it pulls and then pushes the metal conductor's electrons. This pulling and pushing excites (adds energy to) the conductor's electrons. The added energy increases the difference in potential energy between the generator and the circuit to which it connects, causing electrical energy transmission. For simplicity, Figure 1-9 shows only one conductor loop, commonly called a *coil* or *bar*. Commercial generators contain a whole series of coils or bars, each connected in a specific pattern to capture and direct electricity transmission.

Alternators and Dynamos

Two broad types of electromagnetic generators exist: the alternator (which produces alternating current or AC) and the dynamo (which produces direct current or DC). Both the alternator and dynamo generate alternating current, but dynamos also have a device called a commutator that collects the current from the dynamo's rotating parts (the rotating part of the dynamo is called an armature). Commutators are switches connecting and disconnecting as the shaft turns. Switching as the armature rotates changes the output's polarity current so it always remains the same. Without a commutator, the dynamos produce alternating current like an alternator.

> **Direct Current**
>
> Other common uses of DC are battery-operated devices like radios and smart phones. LED lights also use DC.

More than a century ago, Thomas Edison built dynamos (containing commutators) and George Westinghouse built alternators (which did not). AC and DC each had unique advantages and disadvantages, and Edison and Westinghouse entered a great competition with Edison promoting DC and Westinghouse AC. The story of this competition between the two electricity tycoons to determine whose electrical system would power the modern world is now chronicled in the movie, *The Current War*.

Besides Edison and Westinghouse, the third main character in this movie is Nikola Tesla, who worked for Edison. Tesla left Edison to start his own company where he invented and patented the AC induction motor (among other things). Prior to that time, only DC could power motors.

AC wins over DC

In the great battle between currents, the main advantage of AC was the ability of transformers to increase its voltage, making it more efficient to transmit. Before the invention of the induction motor, however, AC could not power motors. Some say Tesla's invention of the AC induction motor essentially drove the final nail in the DC coffin.

> **Power Transmission**
>
> Because $V = I \times R$, higher voltage means more current for the same wire size or smaller wire size for the same power moved.

Alternators

Alternators consist of two primary parts: a stationary stator, which does not turn, and a rotor, which turns inside the stator. Either the rotor or the stator (but normally the rotor) has a magnetic field, and the other component (normally the stator) has conductors wound around a metal core (windings), called coils or bars. Some generators use permanent magnets to produce the magnetic field while others use electromagnets. More on alternators in Chapter 3.

Alternating Current

Figure 1-10 shows the rotor with the north pole to the right (in the left drawing) and, one-half revolution later, the rotor with the south pole to the right (in the right drawing). This movement pushes the electrons in the stator one way and then the other in an alternating and predictable fashion causing the current to alternate and giving rise to the term "alternating current" (AC).

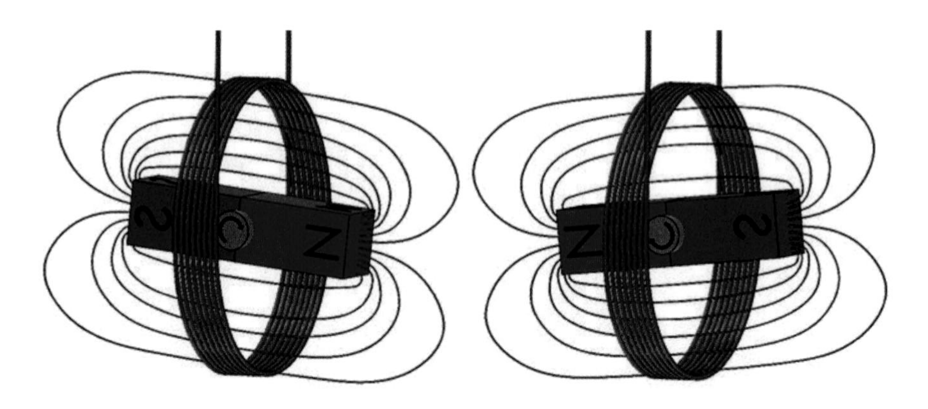

Figure 1-10. Rotor at a Point in Time and then One-Half Revolution Later.

When the north and south poles of the magnet in Figure 1-10 point directly at the windings (the twelve o'clock and six o'clock positions), the north pole's electromotive force on the windings next to it and the south pole's force on the windings next to it are the strongest. As the magnet rotates and the north pole reaches three o'clock and the south pole reaches nine o'clock, the forces acting on the windings drop to zero.

As the magnet continues rotating, the forces on the windings rise, becoming strongest in the opposite direction at twelve o'clock

> **The Hertz**
>
> Hertz is named after Heinrich Rudolf Hertz, the first person to provide conclusive proof of the existence of waves and measure how quickly AC alternates.

and six o'clock. During the first half-revolution, the forces on each winding go from maximum from the north pole, to zero, to maximum from the south pole. During the second half-rotation, the forces go from maximum from the south pole, to zero, to maximum from the north pole, inducing current in the windings, first in one direction and then in the other. The intensity and direction of voltage and current generated over time varies depending on the position of the rotor to the stator.

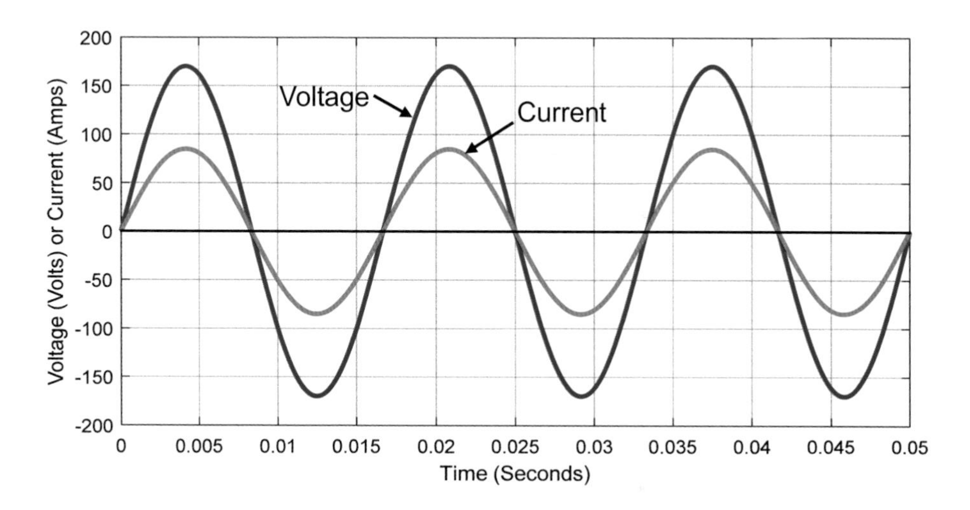

Figure 1-11. Graph of Alternating Voltage and Current Resulting from 60 Hz, 120 Volt AC Applied to a 2-Ohm Resistor.

The vertical axis of the graph shows volts and amps, and the horizontal axis shows time (in seconds). The current values are exactly one-half of the voltage values because, as Mr. Ohm said, V = I × R. In this graph, resistance is two, meaning current (in amps) is one-half the voltage.

The blue line alternates between about positive 170 volts and negative 170 volts on a frequency of 60 hertz. The DC equivalent value of the voltage during each cycle is about 110 volts, giving rise to the 110-volt convention for AC, even though voltage is not actually 110 volts for very long during each cycle. The same type of graph could show electrical systems operating at 220 volts and 50 hertz.

The frequency of 60 hertz means voltage and current make one complete cycle every one-sixtieth of a second (or every 0.166 of a second). During this 0.166 of a second, the voltage and current each drop to zero twice, change from positive to negative twice, and reach maximum twice.

When volts and amps alternate in synchronization with each other, they are both either positive or negative at the same time. Because positive multiplied by positive is positive, and negative multiplied by negative is also positive, power is always positive – at least as long as voltage and current are synchronized.

Figure 1-12 combines the voltage and current from Figure 1-11 to graph the power they produce.

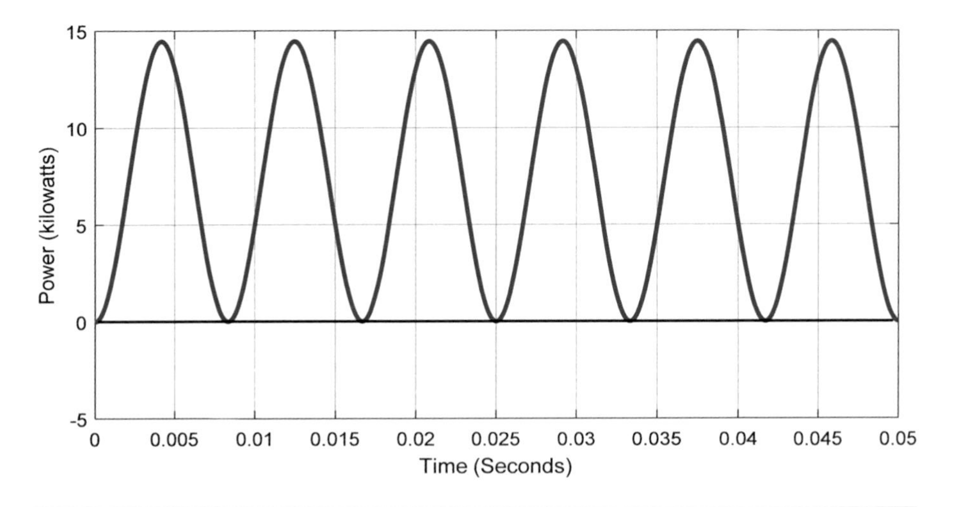

Figure 1-12. Graph of Power Resulting from 60 Hz, 120 Volt AC Applied to a 2-Ohm Resistor. Voltage and current are in phase, so power is always positive, indicating energy transfer from the generator is to a load resistor.

AC can be thought of as electrical energy waves alternating quickly and traveling fast.

Figure 1-12 shows power driving a resistive load. This text will discuss in a later section the concept of reactive loads and how those loads impact volts, current, and power.

Converting AC to DC

Many electronic devices operate on DC current, meaning AC must be converted to DC for them to work properly. Devices called *rectifiers,* which allow current transmission in only one direction, commonly accomplish that conversion. Figure 1-13 shows what happens if the alternating current from Figure 1-11 passes through a *half-wave* rectifier.

Figure 1-14 shows what happens if the alternating current from Figure 1-11 passes through a *full-wave* rectifier.

From these two figures, it is easy to guess where each got its name.

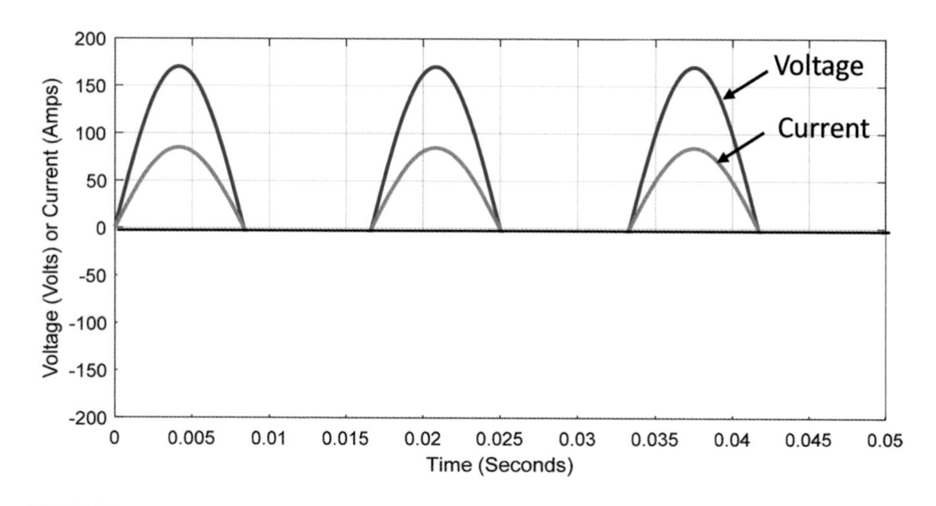

Figure 1-13. Graph of Direct Current after the Alternating Voltage and Current from Figure 1-11 Pass through a Half-Wave Rectifier.

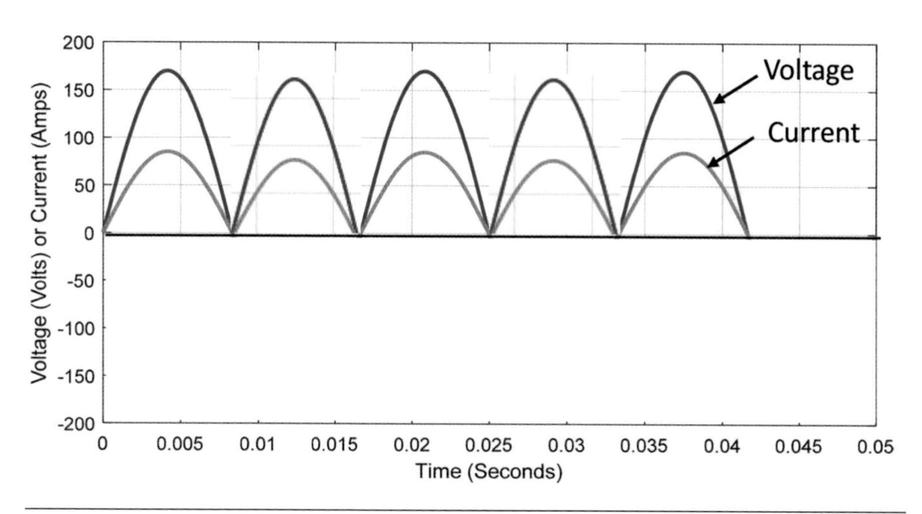

Figure 1-14. Graph of Direct Current after the Alternating Voltage and Current from Figure 1-11 pass through a Full-Wave Rectifier.

Rectifiers

Rectifiers come in various types and technologies but use the same principle of operation – diodes made of semiconductors connected in different patterns produce half- or whole-wave outputs. In Figures 1-13 and 1-14, the current, converted from AC to DC, still cycles between zero volts and about 170 volts. Some DC devices can use that approach, but many need a more constant (smoother) power transmission. In that case, the rectified current passes through devices called *capacitors* or *chokes,* or through sets of capacitors, chokes, and resistors, possibly followed by a voltage regulator to produce steady voltage.

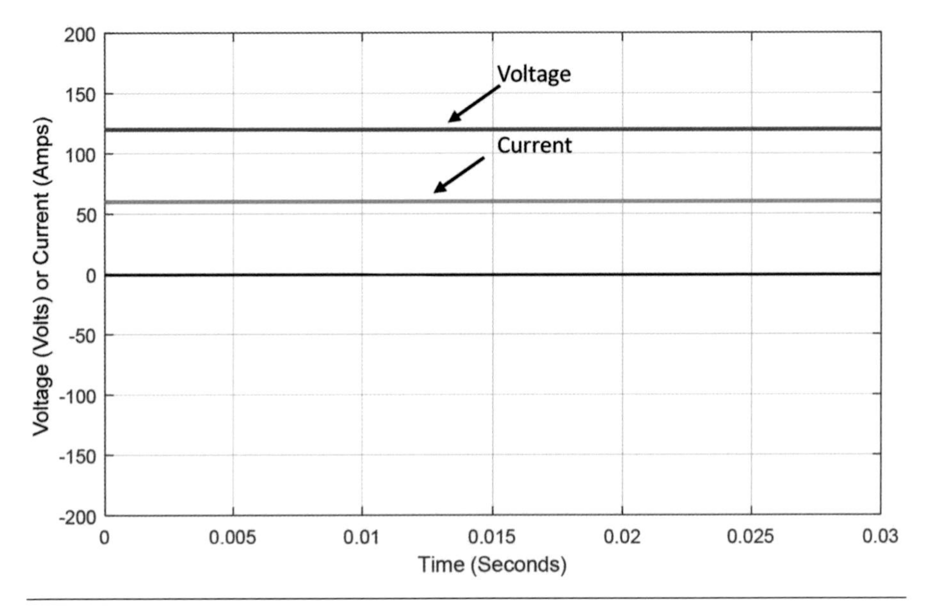

Figure 1-15. Graph of Direct Current after the Voltage and Current from Figure 1-14 pass through a Set of Capacitors and Chokes and then a Voltage Regulator.

Capacitors

Like rectifiers, capacitors come in various types and technologies, but use the same basis of operation. They contain two or more conductors separated by an insulator. The insulator prevents current from transmitting and causing charge to build. The higher the voltage difference between the two conductors, the higher the stored charge. If a capacitor connects to the terminals of a battery, energy transmits across the plates until the voltage difference between the plates equalizes to the full battery voltage.

Capacitors store energy in an electric field. The higher the voltage, the more energy stored. They store energy when voltage is high and give it back when voltage drops, thus smoothing the output. Capacitance is measured in *Farads.*

Inductors

Inductors store energy in the magnetic field caused by the current (rather than the voltage). The higher the current, the more energy stored in an inductor.

Inductors smooth the current, while capacitors smooth the voltage. Inductance is measured in *Henrys.*

Looking back at Figure 1-15, the DC voltage and current appear nice and smooth. But that DC voltage and current started out lumpy, as shown in Figure 1-14. Capacitors smoothed the voltage, and inductors smoothed the current.

Converting DC to AC

Inverters convert DC to AC. For example, they take DC from home solar systems and produce the correct voltage and frequency to feed that DC power into the AC grid.

Transformers

Transformers use alternating electrical current to generate a changing magnetic field on one side of the transformer. The changing magnetic field interacts with a conductor on the other side, generating electricity on that side.

Figure 1-16. Schematic of a Transformer.

The voltage on the primary core, combined with the ratio of the number of windings on the primary core to the number of windings on the secondary core, determine the voltage transferred to the secondary core. Voltage induced into the secondary by the primary core can be calculated as:

$$V_2 = V_1 \times C_2W/C_1M$$

V_2 – Voltage produced in secondary core

V_1 – Voltage of primary core

C_2W – Number of windings in secondary core

C_1W – Number of windings in primary core

Figure 1-16 shows twice as many windings on the secondary core as on the primary core. If the primary core has 110 volts, the secondary core outputs 220 volts. Because power must remain constant, the current output by the secondary core is one-half the amps supplied into the primary core. The power into a transformer equals the power out of the transformer, except for losses caused by the transformer process.

The alternating (cyclical) nature of AC makes transformers possible for AC but not for DC, because the magnetic force intensity must change to induce a current in a nearby conductor, which occurs for AC but not DC.

Power generators generate electrical energy at convenient voltages and then *step-up* (transform) the generation voltage to higher voltages for long-distance transmission. As electricity leaves long-distance transmission lines, transformers *step-down* the voltage. Not surprisingly, the transformers are called step-up and step-down transformers.

Frequency

The number of times AC cycles per second depends on the rotor's rotational speed and its number of poles.

$$\text{Frequency} = \text{Rotational speed} \times \text{number of poles} / 120$$

Most electric grids contain multiple generators. At the connection point of two power systems, the frequency and voltage must be equal and in phase.

As end users turn on more electrical devices, the overall grid draws more current. Voltage regulators controlling generators respond by increasing current to the rotor's field coils, causing the generator to produce more electricity, thereby keeping voltage output and frequency constant. Increasing the current field makes the generators work harder. When all generators are fully loaded, more generation must be added to the grid or frequency may drop to dangerous levels triggering automatic shutdown devices.[5]

A practical example of this phenomenon comes from one author's experience working at Seattle City Light (SCL). Aircraft maker Boeing, has a large facility attached to SCL's power grid. At that facility, Boeing used a large wind tunnel for testing aircraft designs. The motors turning the fans for the wind tunnel use *a lot* of electricity so, before Boeing started using the wind tunnel, they would contact SCL's control center to warn the system operators because the instantaneous load increase would cause a voltage and frequency "sag." Thus, the SCL system operators had to prepare to increase generation at the precise moment the wind tunnel fans started up to avoid having breakers activate to disconnect the grid from the generators.

Three-Phase Generation

As the electrical industry developed, engineers found they could use more than one set of windings on generators. By using three sets of windings, they could generate more than one electricity phase.

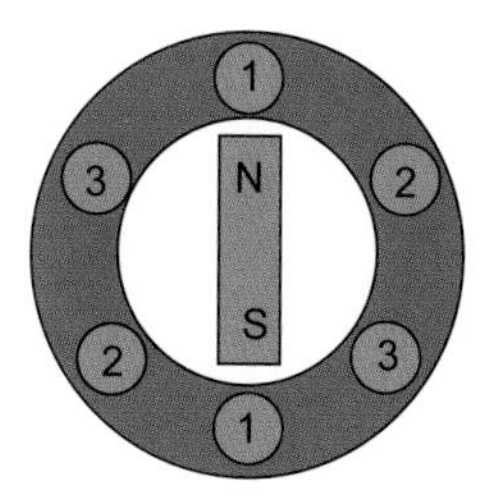

Figure 1-17. Cross Section of a Simple Two-Pole Generator with Three Windings.

Figure 1-17 contains a simple schematic depicting the concept of three-phase generation and shows one set of poles and three sets of windings. Modern generators are more complex than the simple schematic in Figure 1-17 (more about generators in Chapter 3).

The plot in Figure 1-18 shows how the three phases alternate by taking the plot in Figure 1-11 and overlapping the phases. Imagine the rotor going around in a circle of 360 degrees with the phases offset by 120 degrees. Figure 1-18 omits the current to avoid showing even more lines.

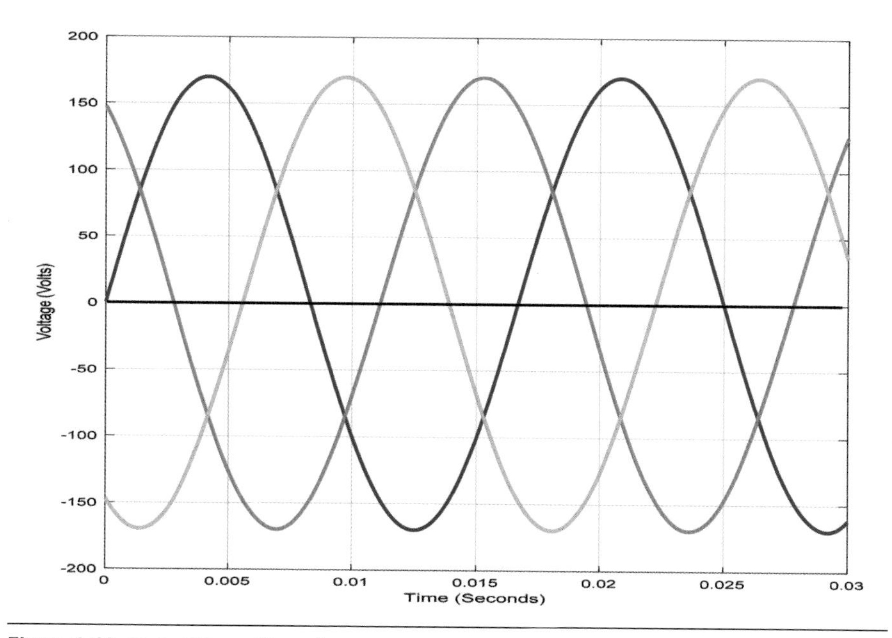

Figure 1-18. Plot of Three-Phase Power.

There is no magic to three phases versus two phases or four phases. Over the years, electrical engineers experimented with different numbers of phases and finally settled on three phases as the most efficient and cost-effective.

Most houses have two electrical wires connected and use only one of the three phases. While it is easy to assume that, if one phase requires two wires, three phases must have six wires, but that is not the case. Most three-phase lines have only three wires carrying electricity. Because the three phases are out of phase from each other (shown on Figure 1-18), they can connect in ways that create a neutral wire, allowing a complete circuit from the generator to the load and back.

The details of these connections, called *delta* (or star) and *wye*, are discussed later in this text. The pattern in which three-phase loads connect to three-phase supplies determines how the load operates and the rotational direction of electrical motors.

Single-Phase Power

All modern AC power has three phases, but most homeowners need only a single phase of power, which is normally obtained by connecting the load to one of the three phases. Distribution systems must spread out these connections across three phases to keep them in balance.

Alternating Current Extra Credit

This section introduces some interesting electrical concepts beyond the scope of a nontechnical book. Accordingly, readers should feel free to skip this section.

Electrical Loads

Loads use electrical power and convert it into another form of energy. Electrical loads come in two broad categories:

- Resistive
- Reactive

The essential difference between resistive loads and reactive loads is that reactive loads store part of the energy rather than just dissipating it, like resistive loads. Loads may not be purely resistive or purely reactive. LED lights, for example, have characteristics of both.

Resistive load

Resistive loads make it harder for electricity to transmit through them – they resist transmission and convert electrical energy into light and heat. Examples of resistive loads include incandescent lights, toasters, ovens, space heaters and coffee makers. The amount of power used by resistive loads is simply voltage times current, measured in watts, or kilowatts, megawatts, gigawatts, terawatts, and so forth.

Reactive loads

Reactive loads are not as straightforward as resistive loads. They *react* with alternating current transmission, causing the voltage and current to go out of phase with each other. Figure 1-19 shows the impact on voltage and current synchronization of a reactive load.

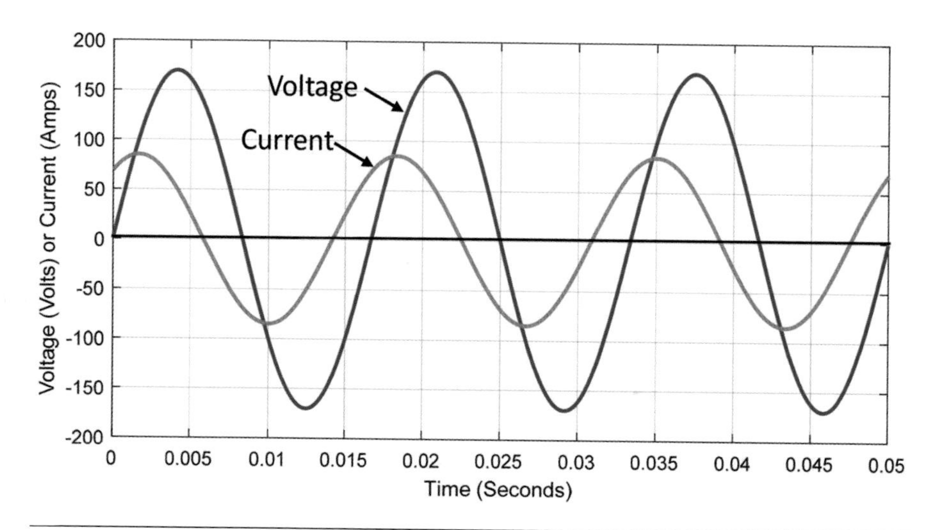

Figure 1-19. Voltage, Current, and Power Resulting from 60 Hz, 120 Volt AC Applied to an Induction Motor.

There are two types of reactive loads, *capacitive* and *inductive*. Capacitive loads store energy in an electric field caused by the voltage, and inductive loads store energy in a magnetic field caused by the current. The stored energy returns to the generator in part of the cycle, which means power is transmitted in both directions.

When voltage and current are not in phase with each other, electrical engineers use mathematical formulas to calculate real power (power supplied to the motor), measured in kilowatts (kW); reactive power (power pushed by the motor back towards the generator), measured in terms of volt amperes reactive (VAR); and apparent power (the power a line must be able to handle), measured in volt amperes (VA). The graph in Figure 1-20 shows how power fluctuates. Real power is above the line and consumed by the load. Reactive power is below the line and is pushed back towards the generator.

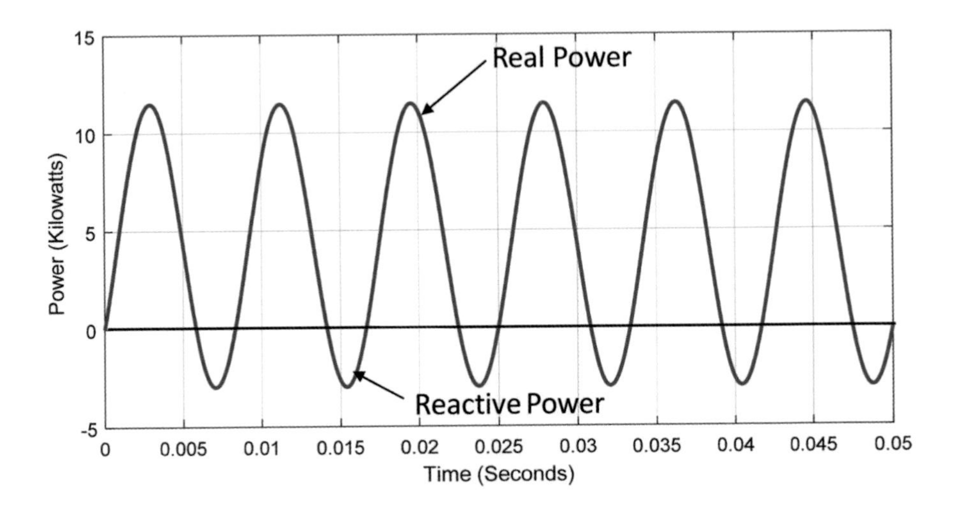

Figure 1-20. Real Power and Reactive Power. Positive power indicates real power, and negative power reactive power.

Reactive power means transmission and distribution lines must handle more than just real power. Reactive power reduces the amount of real power lines can carry. Apparent, real, and reactive power fluctuate as system load and power change.

Power Factor

Power factors are the amount of real power delivered versus the amount that would have been delivered if the voltage and current were in phase with each other, expressed as a number between 0.0 and 1.0.

Managing Power Factors

Because capacitive loads are essentially the opposite of inductive loads, engineers often use capacitors to offset (or manage) the effects of inductive loads and improve the power factor.

This extra credit section demonstrates some of the complexities of electrical power. Chapter 4 contains more information about real, reactive, and apparent power and their importance to electrical power transmission and distribution.

The Electrical Value Chain

The value chain, first described by Michael Porter in his 1985 bestseller, *Competitive Advantage: Creating and Sustaining Superior Performance*, consists of the set of activities industries perform as they deliver a product or service. Companies may specialize in parts of the value chain or may have different business units to handle parts of the value chain. Figure 1-21 shows the electric industry value chain. Traditionally, the steps in the value chain happened sequentially with little, if any, storage. As wind and solar generation increase, so does the need for the storage shown in Figure 1-21, stretching across the rest of the value chain. This text contains one chapter for each part of the electric industry value chain, including storage.

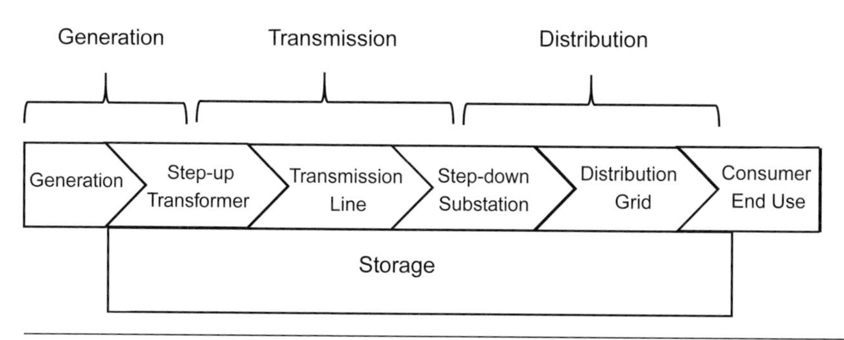

Figure 1-21. Electrical Industry Value Chain.

Electric Generation from Primary Energy Sources

Figure 1-22 shows the percentage of worldwide electricity generated from primary energy sources.

At the global level, coal dominates power generation. Its share, however, fell 1.5% (to 36.4%) in 2019. The shares of both natural gas and renewables rose to record levels in 2019 (23.3% and 10.4%, respectively), and renewables surpassed nuclear for the first time.[6] In 2019, coal (36.4%) and natural gas (23.3%) together accounted for almost 60% of world electrical supply, with a little over 10% generated from renewable fuels.

Figure 1-23 shows the primary energy sources currently used for electrical generation in the United States. While the U.S. uses about the same percentage of fossil fuels as the world in total, the percentages for coal and natural gas are 23% and 38%, respectively.

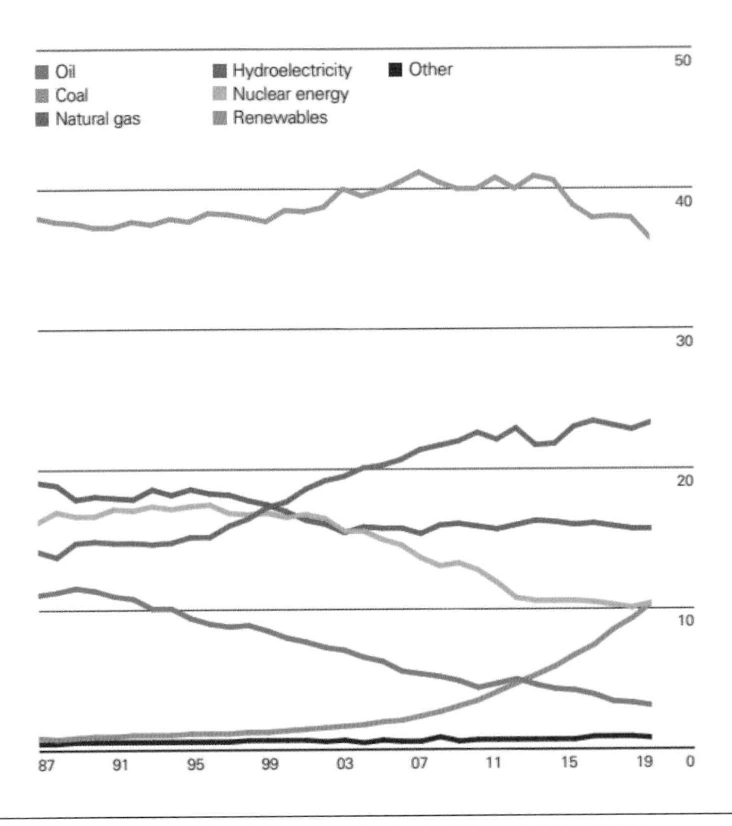

Figure 1-22. Share of Global Electricity Generation by Fuel from 1987 to 2019.

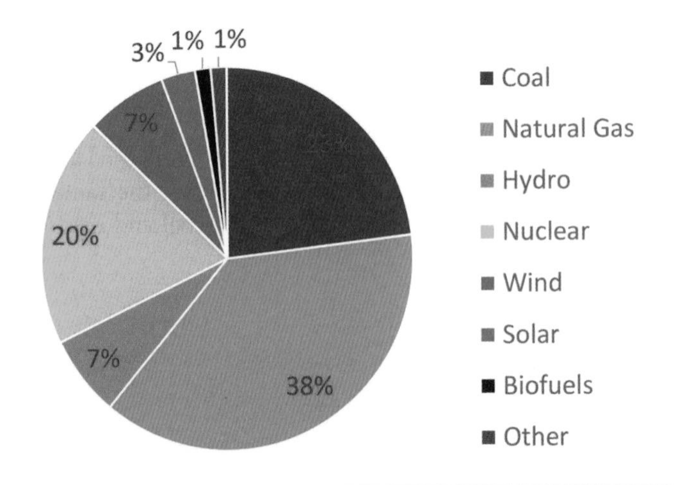

Figure 1-23. U.S. Primary Energy Sources for Electrical Generation.

Electricity Usage

One measure of a country's or region's standard of living is how much energy it consumes. Figure 1-24 shows current levels of electric energy consumption for the top ten consuming nations.[7]

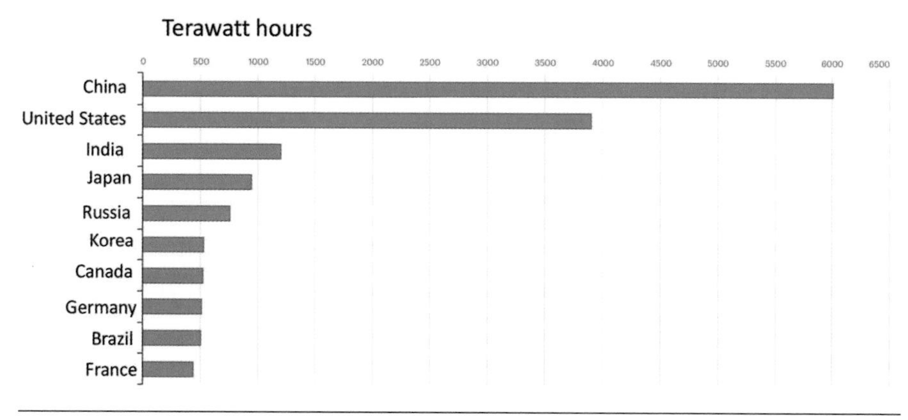

Figure 1-24. Top Ten Electricity-Consuming Countries.

Electric demand by China and the United States dwarfs that of all other countries, with India a distant third. Turning to demand growth, Figure 1-25 shows the forecast of electrical energy demand growth for selected countries and regions from 2019 to 2030. Electrical demand for India, Southeast Asia, Africa, and China is forecast to grow much faster than the U.S. and European Union demand during that same time period.

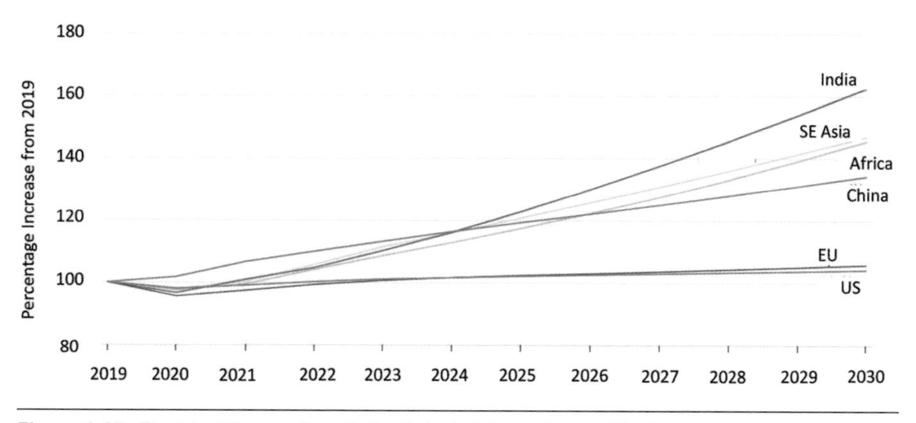

Figure 1-25. Electrical Energy Growth for Selected Countries and Regions.

Starting with present-day usage (shown in Figure 1-24), and extrapolating the growth predicted in Figure 1-24 means a huge growth in total electrical energy usage for China and India. While the growth rates forecast for Africa and Southeastern Asia are higher than for China, 12 to 13% of the world's population (~840 to 940 million people) do not have access to electricity, and approximately 63% of those people reside in sub-Saharan Africa and 25% in South Asia.[8] In other words, the growth rate for those two areas is forecast to be high but they start with a smaller base, meaning those regions will see a relatively small amount (rather than rate) of growth.

Summary

- Electrical energy is transmitted from one point to another when the difference in electrical potential (measured in volts) between the points exceeds the resistance keeping it from transmitting.
- Primary sources of energy are original sources and secondary energy sources are produced from primary energy sources.
- Electricity is a secondary energy source.
- Electricity and magnetism are closely related.
- A difference in electrical potential (measured in volts) causes current.
- Current equals current required by the load.
- Amps express the rate of energy transmission.
- Watts are a unit of power.
- Watt-hours are a unit of energy.
- Electrons interact with each other, causing electrical transmission.
- Electricity is transmitted at almost the speed of light.
- Electrical current always follows a path called a circuit.
- Nearly all electrical power used worldwide comes from alternators rotating a magnetic field past a metal conductor.
- Alternators consist of two primary parts: a stationary stator and a rotor turning inside the stator.
- Alternating current (AC) prevailed over direct current (DC) because its alternating nature allows for transformers, and the induction motor was invented at about the same time as Edison and Westinghouse fought the "great current war."
- Generating electricity and storing it results in a net loss of electrical energy manifesting itself as heat.
- China and India will consume vastly more electrical energy in 2030 than they do now.

- Fossil fuel-generated electricity has given the world the highest standard of living of all time.
- Every form of generating electrical power comes with its own set of challenges.

2

History of Electricity

If I have seen further, it is by standing on the shoulders of Giants.

—Isaac Newton (1642–1726)

"Grandpa Tom," said Luke. "Today in school we learned that Benjamin Franklin invented electricity."

"I'm not sure that's correct. I'm pretty sure electricity exists naturally," Luke's grandfather replied. "Let's use your laptop for some research," he added.

Luke entered the search phrase, "who invented electricity", into the search box. *Universe Today* was the first site given. It contained the following:

> *Electricity is a form of energy and it occurs in nature, so it was not "invented." As to who discovered it, many misconceptions abound. Some give credit to Benjamin Franklin for discovering electricity, but his experiments only helped establish the connection between lightning and electricity, nothing more.*
>
> *The truth about the discovery of electricity is a bit more complex than a man flying his kite. It actually goes back more than two thousand years.*
>
> *In about 600 B.C., the ancient Greeks discovered that rubbing fur on amber (fossilized tree resin) caused an attraction between the two – and so what the Greeks discovered was actually static electricity. Additionally, researchers and archeologists in the 1930s discovered pots with sheets of copper inside that they believe may have been ancient batteries meant to produce light at ancient Roman sites. Similar devices were found in archeological digs near Baghdad meaning ancient Persians may have also used an early form of batteries.*[9]

After reading from the site, Luke said, "So, no one 'invented' electricity. It appears in nature. Right?"

"Yes." said Grandpa Tom, "It was 'discovered.'"

In the modern world, we use electricity for almost everything, every day – appliances, heating, cooling, smartphones, entertainment, lighting, and a myriad of other purposes. Exactly who first discovered electricity is lost in a combination of fact, fiction, and lore.

Before 1600

Thales of Miletus

In 600 B.C., Thales of Miletus, one of the seven sages of Greece, a mathematician, astronomer, and philosopher from Miletus in Ionia, Asia Minor, wrote about the charging of amber (fossilized tree resin) by rubbing it on fur – what we now refer to as static electricity. His work is the earliest known investigation into the properties of electricity.[10]

Earlier Lore

Some sources say ancient Egyptians (circa 2750 B.C.) witnessed shocks from electric fish. Many years later, ancient Greek, Roman, and Arabic naturalists reported electric fish experiences, described the numbing effect of electric shocks from fish. They seemed to understand shocks could travel along conductors.

Interestingly, the Parthians (247 B.C.–224 A.D.) may have known about electroplating, based on a 1936 discovery of the *Baghdad Battery*, which resembles a galvanic cell. The earliest and most proximate discovery of the identity of lightning and electricity from any other source is credited to the Arabs, who before the fourteen hundreds had a word for lightning.

Some even believe ancient Egyptians harnessed the power of electricity to light their underground tombs and monuments, based on a series of carvings in the walls of a crypt in the temple of Hathor at Dendera (125 B.C.). A Norwegian engineer visiting the temple believed he identified a light bulb (known as the *Dendera Light*) in one of the carvings.

Other researchers investigated the matter and agreed the carving seems to depict a light bulb. After further study, researchers hypothesize the ancient Egyptians may have had light bulbs using so-called *Crookes tubes*. A Crookes tube is an experimental electrical discharge tube invented by English physicist William Crookes, *et al.* in approximately 1869–1875. In the temple carving, a snake represents the electron beam of a Crookes tube. The snake's tail begins where a cable from a power box enters the tube and the snake's head touches the opposite end. In Egyptian art, a serpent typically symbolized divine energy.

Supporters of the hypothesis point out that the ancient Egyptian temples and crypts have elaborate designs of sculptures, reliefs, and murals in areas with no light. Historically, researchers believed the Egyptians relied on lanterns and lamps. Researchers have, however, found no traces of soot to support that theory.

1600–1700: Early Discoveries

The sixteen hundreds saw the birth of the word *electric* and led to early understanding of electricity and magnetism and the beginning of investigation into how those two forces relate to each other.

Electricity

Around 1600 A.D., an Englishman named William Gilbert wrote the book *De magnete, Magneticisique Corporibus* (*On the Magnet*), detailing the attractive nature of amber. He used the Latin word *Electricus* to describe the force and coined the term *electricity* from the Greek word *elektron* (the word for amber). Mr. Gilbert included the electrification of many substances and used the terms *electric force, magnetic pole,* and *electric attraction.*[11]

Subsequently, Sir Thomas Browne (another Englishman) wrote physics books using the word *electricity* to describe his investigations based on Gilbert's work. Thus, Gilbert and Browne are considered the first scientists to use the term *electricity*.

Electricity as a Force

In 1660, German scientist Otto von Guericke described and demonstrated how to produce a vacuum and invented the first machine to produce static electricity.

In 1675, an Irishman, Robert Boyle, discovered he could transmit an electric force through a vacuum and observed the forces of attraction and repulsion being transmitted through a vacuum.

Robert Boyle also founded the Royal Society and was elected a member in 1663; he is probably more famous for "Boyle's law," dealing with the compression and expansion of a gas at constant temperature.

1700–1800: Building on Previous Discoveries

The seventeen hundreds built on the discoveries of the previous century and continued improving knowledge about conductors, insulators, and electrical charges. Scientists even developed a few equations during that time to explain the interrelationship of electrical terms.

Conductors, Insulators, and Charges

In 1729, Stephen Gray, an Englishman, became the first person to systematically experiment with electrical conduction. Previous work focused on the generation of static charges and investigated the static phenomena such as electric shocks and plasma glows.

Shortly thereafter, in 1733, Charles Francois Dufay, a French chemist, discovered two types of electricity he named *vitreous* and *resinous.* Mr. Dufay also identified the distinction between *conductors* and *insulators,* which he

Vitreous (+) and *resinous* (-) were subsequently named *positive* and *negative* charges by Ben Franklin and Ebenezer Kinnersley.

referred to as *electrics* and *nonelectrics,* and he found like-charged objects repel and unlike-charged objects attract.

In his later work, Dufay disproved several electrical theories such as those of Stephen Gray, who postulated electrical properties depended on a material's color. Mr. Dufay reported his observations about charge and color in a paper in December 1733.

> C.F. Dufay became a member of the French Academy of Sciences in 1723 and died of smallpox in 1739.

Electrical Storage

In 1745, Ewald Georg von Kleist of Germany invented the "Leyden jar," an electric circuit element to store electricity and early form of a *capacitor.* Pieter van Musschenbroek, from Leyden in the Netherlands, discovered the capacitor independently, but seems to have gotten most of the credit since the device is named for his city and not that of Mr. von Kleist.

In its original form, the Leyden jar was a glass vial filled partly with water. The jar's top was closed by a cork penetrated by a wire or nail extending into the water.

Figure 2-1. Discovery of the Leyden Jar in Musschenbroek's Lab. The chain conducted the static electricity produced by the rotating glass sphere electrostatic generator, through the suspended bar to the water in the glass held by assistant Andreas Cunaeus. A large charge accumulated in the water and an opposite charge in Cunaeus's hand on the glass. When he touched the wire dipping in the water, he received a powerful shock.

In its present-day form, the inner and outer surfaces of an insulating container are coated with metal. The outer coating connects to ground and the inner coating connects through a central brass rod projecting from the jar's mouth. The Leyden jar became important as a prototype for modern-day capacitors widely used in radios, television sets, and other electronic equipment.

Capacitors hold charge. A practical example is a computer where the power light glows for a few seconds after you unplug the device from the wall.

Electrical Transmission

In the mid-seventeen hundreds, Benjamin Franklin of the United States took an interest in electricity. Before that time, scholars knew about and experimented with only static electricity by generating and then discharging it. Mr. Franklin, however, believed electricity had positive and negative elements and charges flowed between those elements. He suspected lightning was a form of flowing electricity and experimented with charges, postulating the existence of an electrical fluid composed of particles.

In 1752, Mr. Franklin conducted his now-famous experiment of flying a kite with a metal key tied to the string during a lightning storm. The key provided a path to conduct the electricity from the string. As Mr. Franklin expected, electricity was transmitted from the lightning to the kite and down the string, shocking him. In the same year, Franklin suggested an experiment using conductive rods to attract lightning to a Leyden jar.

Around the same time, William Watson of England discharged a Leyden jar through a circuit, which began the understanding of current and circuits. Henry Cavendish of England also started measuring the conductivity of different materials at that time.

In 1767, Joseph Priestley of England observed the force of attraction or repulsion between two charges depends on the magnitude of those charges, and the force is inversely proportional to the distance between them. Some people speculate Mr. Priestley used Isaac Newton's gravitational inverse-square law to help him ponder invisible forces and how they work.

In 1785, Charles-Augustin de Coulomb, a French military engineer, relying on the work of Mr. Priestley and others, devised "Coulomb's law," which identifies the amount of force between two stationary, electrically charged particles. The electric force between those charged particles is called *electrostatic force* or *Coulomb force*. The law holds that the magnitude of the electrostatic force of attraction or repulsion between two-point charges is directly proportional to the product of the magnitudes of charges and inversely proportional to the square of the distance between them. Coulomb's law led to the theory of electromagnetism because it made it possible to examine the quantity of electric charge in a meaningful way.

Figure 2-2 shows two charges, one positive and one negative. The top charges are one unit apart, and the bottom charges of the same magnitude are two units apart, meaning the attraction between the bottom charges is one-fourth the attraction of the top charges.

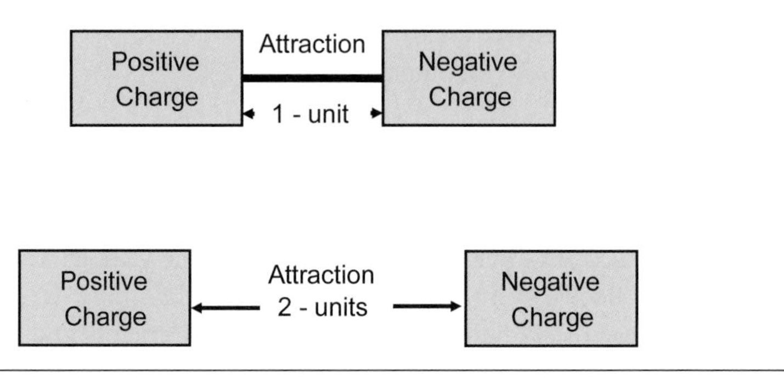

Figure 2-2. Opposite Charges and their Force of Attraction.

Animal Electricity

In the 1780s, Italian scientist Luigi Galvani and his wife, Lucia Galeazzi, made frog muscles spasm by jolting them with sparks from an electrostatic device. They described the nerve impulses to trigger muscle movement as *animal electricity.* That work laid the foundation for the subsequent knowledge of the nervous system's electrophysiology.

1800–1900: Discoveries Lead to Generation and Lighting

Physicists of the nineteenth century built on the knowledge of charges, forces, conductors, capacitors, Leyden jars and a variety of other bits of knowledge developed in the previous century as they worked to understand electricity and magnetism more fully.

Current and Batteries

Alessandro Volta, an Italian physicist, was the first person to discover transmission of electrical charge, and is most famous for inventing the electric battery (known at the time as the *voltaic pile*). Mr. Volta studied Luigi Galvani's observations about the effect of electricity on frog muscles and concluded an electrical potential exists between two metals which transmits an electrical charge through the frog's leg causing it to twitch. Mr. Volta discovered that, in the presence of electrical potential, an electrical charge could pass through a metal wire – like water flowing through a pipe – and he used that work to develop electric batteries. Dependable batteries finally allowed scientists to produce steady electrical current, unleashing a torrent of new discoveries and technologies. Volta also:

- discovered *contact electricity* resulting from contact between different metals;
- recognized two types of electric conduction;
- wrote the first electromotive series, which showed, from maximum to minimum, the voltages different metals will produce in a battery; and
- discovered electric *potential* in a capacitor directly relates to electric charge.

> To honor Volta's contributions to science, the unit of electric potential is called the *volt.*
>
> Different types of batteries (e.g., AAA (1.5V), car (12V), transistor (9V)) have different *voltages* or electrical *potentials.*

After Mr. Volta, many scientists advanced the theory of static and moving charges and their connection to magnetism. After this period, developments in the field began to come fast and furiously.

Electrical Lighting

In 1807, Sir Humphrey Davy of England invented the first successful *arc lamp,* which produced light by sustaining an electric arc across a gap between two conductors. In an arc lamp, light comes from the heated ends of the conductors (typically carbon rods) and from the arc itself.

> **Arc Lamps**
>
> Arc lamps prove useful for applications needing a great deal of brightness (searchlights, large film projectors, and floodlights) but are too bright for home use.

Sir Davy used a battery of 2,000 cells to create a four-inch arc between two charcoal sticks. When electric generators became available in the late 1870s, arc lights became practical. The city of Paris and other European cities used the Yablochkov candle (an arc lamp invented by the Russian engineer Paul Yablochkov) for street lighting after 1878.

In 1878, manufactured gas was used for street lighting, and the invention of the Yablochkov candle started the move to replace gas lighting with electric lighting.

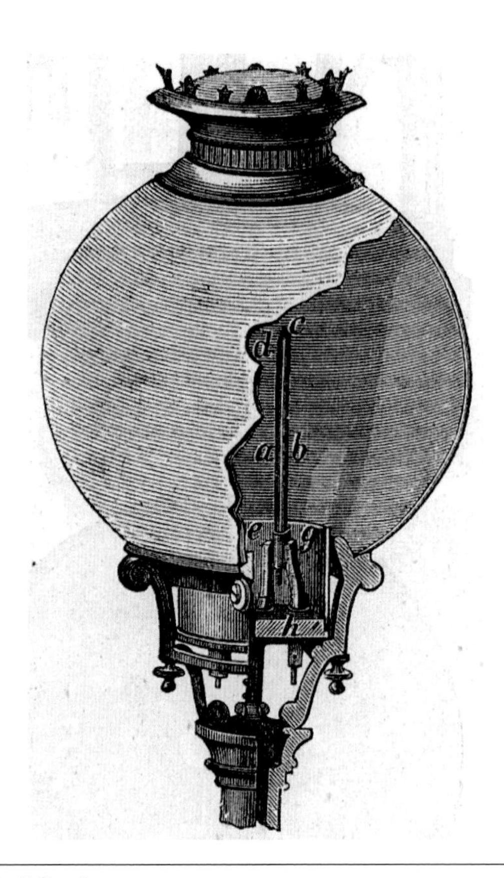

Figure 2-3. Jablochkoff Candle.

Electricity and Magnetism

In 1820, Hans Christian Øersted of Denmark confirmed the relationship between electricity and magnetism when he observed electrical currents affecting the needle of a compass. From his observations, Mr. Øersted developed "Øersted's law," which states an electric current creates a magnetic field.

François Arago, a French physicist, discovered the principle of producing magnetism by rotating a nonmagnetic conductor. Mr. Arago showed that passing an electric current through a cylinder-shaped spiral of copper wire made it attract iron shavings like a magnet. The shavings dropped off when he turned off the current. In 1824, in another demonstration of the connection between electricity and magnetism, he showed a rotating copper disk would make a nearby magnetic needle rotate.

In September 1820, Mr. Arago demonstrated Øersted's electromagnetic effect to the French Academy in Paris. Coincidently, André-Marie Ampère attended the presentation and became interested in the topic. Mr. Ampère began his study of electricity by replicating Ørsted's work, and he discovered electrical current transmitted in the same direction in two nearby parallel wires caused the wires to attract each other. He also found if electric currents are transmitted in opposite directions the wires repel each other. Mr. Ampère dubbed this new field of study *electrodynamics*.

Mr. Ampère then devised an equation ("Ampère's circuital law") linking the size of a magnetic field to the electric current producing it. Later, he proposed the existence of a particle now recognized as the *electron*.

Michael Faraday of England became interested in the subjects of electricity and magnetism after attending chemical lectures by Sir Humphry Davy (inventor of the arc lamp) in London. Faraday watched the lectures with great interest and took extensive notes. He sent a bound copy of his notes to Sir Davy with a letter seeking employment and Sir Davy hired him.

> James Clerk Maxwell, a Scottish physicist, stated, "The experimental investigation by which Ampère established the law of the mechanical action between electric currents is one of the most brilliant achievements in science.... It is perfect in form, and unassailable in accuracy, and...must always remain the cardinal formula of electrodynamics." (*Electricity and Magnetism,* Vol. 2, Chapter 3, 1873)

After Ørsted announced the flow (his term) of an electric current through a wire created a magnetic field around the wire, Ampère also demonstrated the magnetic force was circular, producing a sort of *tube* of magnetism around the wire – a phenomenon no one had previously observed.

Mr. Faraday, understanding the implications of that discovery, posited that, if a magnetic pole could be isolated, it would move in a circle around a wire containing current. He constructed a device to confirm this hypothesis. The device successfully converted electrical energy into mechanical energy and became the first electric motor. Faraday considered electricity a *vibration* or *force* somehow transmitted through tension in a conductor.

In 1831, Faraday started working with Sir Charles Wheatstone on another vibrational phenomenon – sound. He focused on patterns in powder spread on iron plates when a violin bow made them vibrate. That phenomenon demonstrated a dynamic cause (sound) could create a static effect. He believed the same thing happened in wires carrying current, which led to his next significant experiment.

In August 1831, Faraday wound a thick iron ring with insulated wire, connected it to a battery on one side and connected it to a galvanometer on the other side. He anticipated the production of a *wave* when he

> Michael Faraday also discovered that every material has a specific inductive capacity.

closed the battery circuit. When he closed the primary circuit, he saw the galvanometer needle jump, thereby demonstrating a current in the primary coil induced a current in the secondary coil. Surprisingly, though, he saw the galvanometer jump in the opposite direction when he opened the circuit. Apparently, turning off the current created in the secondary circuit an induced current equal and opposite to the original current. This observation led Faraday to propose an *electrotonic* state of particles in wire he considered a state of tension. He believed a current was responsible for setting up or collapsing that state.

In 1831, an American named Joseph Henry also found the induction law independent of Faraday and assembled a magnetic rocker generally accepted as one of the first electric motors.

Later in 1831, Faraday tried to determine how to induce this current. His previous experiment involved a strong electromagnet created by the primary coil's winding. This time, he attempted to create a current using a permanent magnet. The experiment showed, when he moved a permanent magnet into and out of a coil of wire, it induced a current in the coil. He knew magnets had forces around them and could see that fact by placing iron filings on a surface above the magnet.

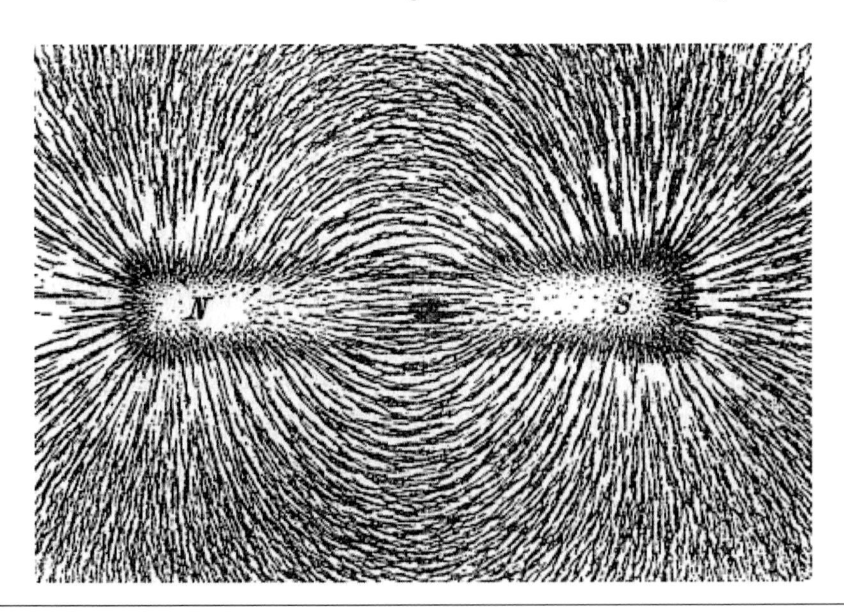

Figure 2-4. The Direction of Magnetic Field Lines.

He saw those forces as *tension* in the air around the magnet. Ultimately, this led to the law regarding how magnets create electric current – a current's *magnitude* depends on how many lines of force the conductor creates during a specific time.

Mr. Faraday then concluded he could create a continuous current by spinning a copper disk between the poles of a strong magnet and remove the current from leads attached to the disk's rim and center. He postulated the disk's exterior would cut more lines of force than the interior, creating a continuous current in the circuit between the rim and center. This device is now known as Faraday's Disc.

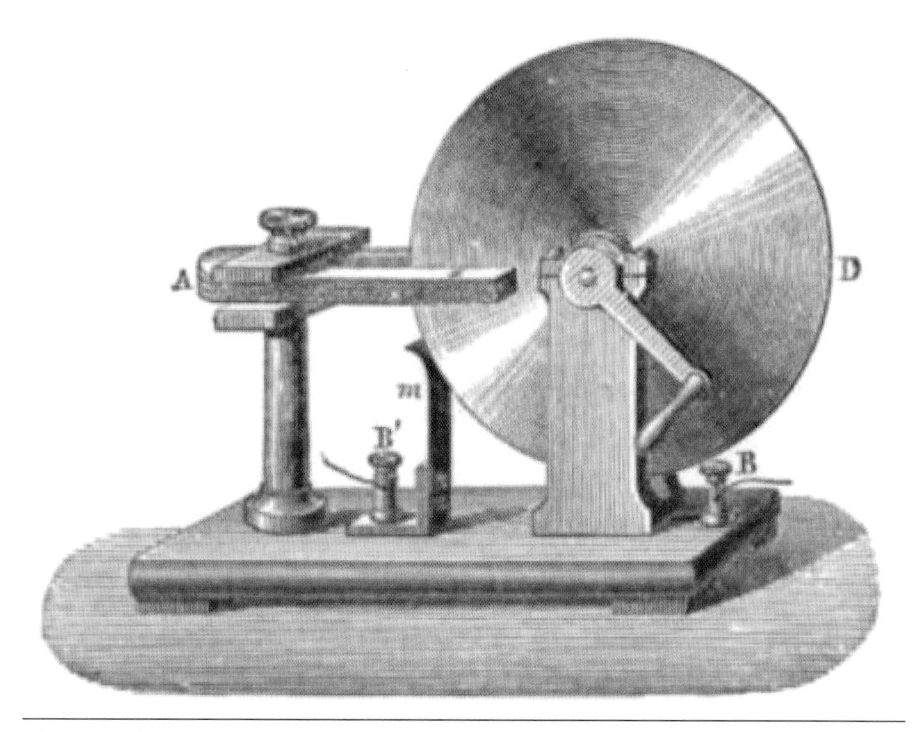

Figure 2-5. Faraday's Disc.

Mr. Faraday's device became the first *dynamo* (which, in turn, became the basis of electric generators). It also served as the precursor to the electric motor because reversing the arrangement would supply electricity to the disk, causing it to rotate.

In 1826, around the same time Faraday did his work, Georg Simon Ohm of Germany performed experiments allowing him to define the relationship between power, voltage, current, and resistance (Ohm's law). Ohm's law states current transmitted through a conductor is directly proportional to the potential difference (voltage) and inversely proportional to the resistance.

Ohm was a professor of mathematics and set forth the most important parts of Ohm's law in a pamphlet in 1827: *Die galvanischeKette, mathematischbearbeitet* (*The Galvanic Circuit Investigated Mathematically*). His work influenced theory and applications of current electricity but was poorly received, causing Ohm to resign his post and accept a position at the Polytechnic School of Nürnberg in 1833. Ultimately, his work was recognized and, in 1841, he received the Copley Medal of the Royal Society of London, which made him a foreign member the following year.

In 1832, following Hans Christian Oersted's discoveries, Hippolyte Pixii of France built a hand-cranked machine that rotated a magnet past a bar of iron wrapped with wire. This motion created alternating current, which Pixii converted to direct current. Although not as efficient as subsequent devices using electromagnets, the invention is generally recognized as the first true electrical generator.

Joseph Henry also served as the first Secretary of the Smithsonian Institute.

Joseph Henry (who previously discovered the induction law independent of Faraday), in 1835, invented the electric relay, which allowed electric current to travel over long distances. This relay became the basis for the telegraph years later.

Davenport used his electric motor to build an electric car in 1834.

Thomas Davenport, an American, purchased Mr. Henry's electromagnet system and started his studies on electromagnetism in 1833. He built a device that used four electromagnets – two electromagnets pivoted and two remained stationary. Davenport connected a switching device called a commutator to a battery and, when activated, the device spun, creating the first real electric motor. The interaction between two of the magnets made the rotor turn one-half a revolution. By reversing the wires to one magnet, he made the rotor complete the other half-turn.

In 1839, Sir William Robert Grove of England devised a two-fluid electric cell (the "Grove battery") consisting of amalgamated zinc in sulfuric acid and a platinum cathode in concentrated nitric acid. A porous container separated the liquids. Grove used his platinum-zinc batteries to create electric light for his college lectures. Later, in 1842, he designed the *gas battery* (the first fuel cell) in which hydrogen and oxygen gas form water and generate an electric current.

James Joule, an Englishman, grew up fascinated with electricity. In early adulthood, while managing his family's brewery. Mr. Joule investigated replacing the brewery's steam engines with electric motors, leading him to explore the relationship between heat and mechanical work.

In 1841, Mr. Joule discovered "Joule's first law," which states the power of heating generated by an electrical conductor is proportional to the product of its resistance and the square of the current. His work ultimately led to the law of conservation of energy, which, in turn, led to the first law of thermodynamics. The generally accepted unit of energy, the joule, is named after him.

Generation and Lighting

Following Hippolyte Pixii's hand-cranked machine, various others made improvements to this concept until finally Antonio Pacinotti built a generator that provided continuous, direct-current power around 1860.

In 1867, Werner von Siemens, Charles Wheatstone, and S.A. Varley almost simultaneously devised the "self-exciting dynamo-electric generator." Then, in 1870, a Belgian inventor, Zenobe Gramme, devised a dynamo that produced a steady direct current, well-suited to powering motors—a discovery that generated a burst of enthusiasm about electricity's potential to light and power the world.

By 1877, arc lighting lit the streets of many cities across the world. Arc lights were too bright for lighting houses, which limited their use to large outdoor areas. Thomas Edison invented a less powerful incandescent lamp in 1879 and, in 1881, purchased property on Pearl Street in Manhattan to house a generator to power a one-square-mile light system (known as the Pearl Street Station). On September 4, 1882, the staff energized the small grid and Edison connected the electric lighting to it.

Figure 2-6. Edison's Pearl Street Generating Plant.

For many years after the Pearl Street Station was commissioned, companies built and operated small generators relatively close to the end users. After Westinghouse won the great current war, electric service became more widespread, as transporting it longer distances became easier because transformers allowed stepping up the voltage for transmission and stepping it down for distribution. Utilities built generation stations further from customers and constructed transmission lines to get the power from the point of generation to the point of use.

Other Global Developments

The late nineteenth century was a time when electrical power came to countries around the world. Three more examples of global progress follow.

In 1879, Shanghai Public Concession installed 10 hp DC generating units that started the electric light age in China and, in 1882, the first public electricity company in China – Shanghai Electric Company – opened for business. That company subsequently built a power plant equipped with 16 hp steam engine generating units. It also constructed poles and lines from Shanghai Waitan to China Investment Bureau and installed fifteen arc lights.[12]

During the mid-1890s, South Africa's first public power plant was constructed. Located in Brakpan, the plant supplied electricity to the gold mines on the Witwatersrand in the Transvaal and Johannesburg, forty kilometers away.[13]

In late 1881, electric lights came to the streets of the Surrey town of Godalming in the U.K. A waterwheel on the River Wey drove a Siemens alternator that supplied a number of arc lamps in the town and electricity to a number of shops and premises.[14]

Steam

In 1884, Sir Charles Algernon Parsons invented the steam turbine (as opposed to the steam engine, generally attributed to James Watt). Parsons, born in London, attended Trinity College, Dublin, and St. John's College, Cambridge, where he received a first-class honors degree. He began his career at a ship-engine manufacturer at Clarke, Chapman, and Co., where he became head of electrical-equipment development. After Sir Parsons built his steam turbine, he quickly realized it was an ideal driver for anything that rotated, including electrical generators, which he then designed and put into use. Figure 2-7 is a drawing of his steam turbine and generator.

> Parson also worked on rocket-powered torpedoes.

In 1911, Parsons reminisced about his work, stating:

> It seemed to me that moderate surface velocities and speeds of rotation were essential if the turbine motor was to receive general acceptance as a prime mover. I therefore decided to split up the fall in pressure of the steam into small fractional expansions over a large number of turbines in series, so that the velocity of the steam nowhere should be great...I was also anxious to avoid the well-known cutting action on metal of steam at high velocity.[15]

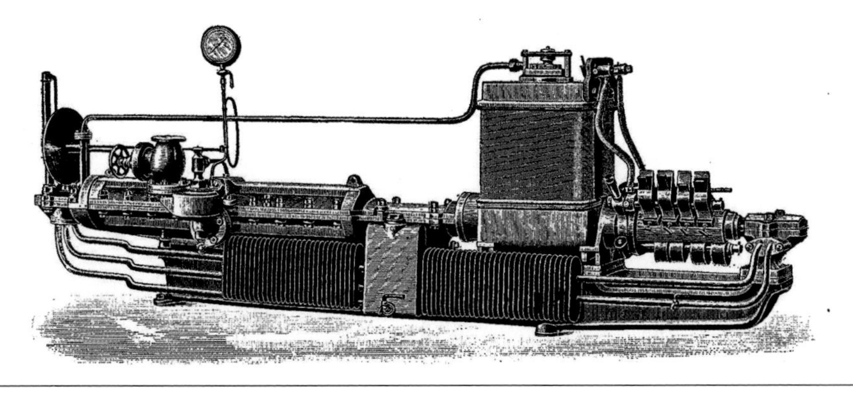

Figure 2-7. Parson's Compound Steam Turbine and Generator.

It took a relatively long time to perfect steam turbines because they rotated at high speeds in high-pressure, high-temperature environments, with many moving parts. Sir Parson's first machines produced very little electricity (1 kV). Over time, however, designs improved dramatically and, eventually, made cheap and easy electricity possible.

Waterpower

Some of the first innovations in using water for power occurred in China during the Han Dynasty between 202 B.C. and 9 A.D. where they used trip hammers powered by a vertical-set waterwheel to pound and hull grain, break ore, and make paper.[16] Waterwheels drove mills for many years before French engineer Benoit Fourneyron, in 1827, developed a turbine capable of producing around six horsepower – the earliest version of the Fourneyron reaction turbine.

A little less than twenty-five years later, in 1849, British-American engineer James Francis developed the first modern water turbine – the Francis turbine – which remains the most widely used water turbine in the world today.[17] According to www.hydropower.com, the world's first hydroelectric project happened in 1878 – but powered only a single lamp in Northumberland, England.[18] The first *commercial* hydroelectric project came online the evening of July 24, 1880, when Grand Rapids, Michigan, used DC hydroelectric power to light sixteen brush-arc lamps.

Shortly thereafter, in 1881, the city of Niagara Falls powered its streetlights with DC hydroelectric power, and by 1888, approximately 200 electric utilities relied on hydropower for their operations.

In 1889, the world's first AC hydroelectric plant went into operation. The Willamette Falls Electric Company (later Portland General Electric) built a hydroelectric facility at Willamette Falls in Oregon with four turbine dynamos and a fourteen-mile transmission line to Portland – the first "long-distance" transmission of AC energy in the U.S.[19]

Figure 2-8. First Hydroelectric Generating Plant. Grand Rapids Electric Light & Power Company buildings (left to right): William T. Powers' Sawmill built in 1868 and used from September 1880 to November 1881; G. R. E. L. & P. First Hydroelectric Power Plant with steam-engine backup system (Winter 1880–81); City Lighting Plant (Nov 1881); north brick building possibly late 1880s. Photo circa the late 1890s.

The 1891 International Electro-technical Exhibition showed off the world's first three-phase hydroelectric plant, located at Lauffen am Necka, Germany, with transmission lines going 109 miles north to Frankfurt am Main, Germany. The generator spun at 150 revolutions per minute (RPM), and the field magnet rotated with 32 poles. The plant and transmission line demonstrated the effectiveness of three-phase power and effectively ended the current war between AC and DC.[20]

Just four years later, the Niagara Falls Power Company completed the Edward Dean Adams Power Plant. Westinghouse Electric subcontracted to build 5,000 hp (3,700 kW) 25 Hz AC generators, based on the work of Nikola Tesla and Benjamin G. Lamme. The I.P. Morris Company of Philadelphia built the turbines based on the design of the Swiss company Faesch and Piccard. The Adams site consisted of several buildings southeast of Niagara Falls. General Electric provided the transmission and sent the electricity twenty-five miles to the booming industrial city of Buffalo, New York. An enormous worldwide publicity campaign accompanied the Adams power station. This long-lasting publicity forms part of the reason many people falsely believe the Niagara Falls power plant was the first AC hydroelectric power plant. In fact, AC hydroelectric power dates to the 1870s in the Austro-Hungarian Empire. The Adams Power Station stands as a monument to the first large-scale application of AC power in North America.

The Adams plant was a key start to modern electrical utility operations, and the Niagara Falls Power Company operated it until 1961,[21] at which point the Robert Moses Niagara Hydroelectric Power Station (owned and operated by the New York Power Authority and located in Lewiston, NY) replaced it.[22]

Figure 2-9. Adams Power Plant Transformer House. The only remaining building (still standing) of the old Niagara Hydroelectric power plant which Tesla and Westinghouse built in 1895.

Figure 2-10. Robert Moses Niagara Power Plant.

Wind Power

The 1880s also saw the rapid development of wind turbines to convert wind energy to mechanical energy. In 1882 and 1883, while working for the United States.

Wind Engine and Pump Company, Thomas Perry conducted a series of experiments on windmill rotors and rotor blades, leading to a windmill design 87% more efficient than other windmills at the time. Perry used concave windmill blades made from steel instead of flat, wooden blades. In 1888, he and LaVerne Noyes founded the Aermotor Windmill Company to build and sell windmills to pump water. Those windmills became common through the U.S. Midwest. In the 1880s, Charles F. Brush, an American engineer and inventor, made a fortune improving the arc lamp and other electrical devices. He was fascinated by electricity and wind power and, in 1886, built a home in Cleveland employing his inventions, powered by what is believed to be the world's first automatically operated wind turbine generator. It operated for twenty years.

Figure 2-11. Charles F. Brush's 60-foot, 80,000-pound Turbine. The turbine supplied 12kW of power to 350 incandescent lights, 2 arc lights, and a number of motors at his home for twenty years. The dynamo driven by the device turned fifty times for every revolution of the blades and charged a dozen batteries each with thirty-four cells.

As the world entered the twentieth century, electricity was well on its way to being a major energy source. But before leaving the nineteenth century, it is important to mention that regulation was just around the corner. In fact, the earliest legislation in the U.S. addressing the emerging electric industry took place in 1899 when the U.S. Congress passed the Rivers and Harbors Appropriation Act, making it illegal to dam navigable streams for any reason (including hydroelectric generation) without permission from Congress.

1900–1950: Electricity Comes of Age

As the world entered the twentieth century, generation technologies grew and matured, business models evolved, and regulations and government involvement increased, as the electrical power industry experienced significant growth.

Generation Technologies

Coal and water were the two primary energy sources used to drive electrical generators at the turn of the century, and dynamos (DC) and alternators (AC) generated the lion's share of the electricity.

Steam Power Surpasses Hydropower

By 1907, Charles Parsons' new company, C.A. Parsons & Co., manufactured steam turbines producing five megawatts (MWs) of electricity. Only five years later, in 1912, that number had increased five-fold to 25 MWs. Leaving the nineteenth and entering the twentieth century, Westinghouse and General Electric continued producing steam turbines and, by the start of World War I, steam turbines were the main driver for electric generation, a fact that continues to this day.

Geothermal Power

The first steam geothermal power plant in the world went into operation in Larderello (Tuscany), Italy, thanks to Prince Piero Ginori Conti. Prince Conti worked in the boric acid industry and eventually found his way into geothermal energy with the creation of the first geothermal energy generator in 1904. It was powered by reciprocating engines and not steam turbines. Based at the Larderello dry stream field, Conti's generator was able to produce 10 kW of energy and power five light bulbs.[23]

In an area near Larderello, the world's first real (not reciprocating engine) geothermal power plant, Larderello 1, was completed in 1913. It had a capacity of 250 kW and was able to produce 2750 kW of electricity to power the Italian railway system and the nearby villages of Larderello and Volterra. The area now produces approximately 10% of the world's supply of geothermal electricity (4,800 GWh per year), providing power to approximately one million Italian households.[24] It took until 1958 for the construction of the world's second geothermal power plant in Wairakei, New Zealand.[25]

Wind Power

At the same time, wind turbines continued to grow in popularity. In the 1920s, they consisted of blades like those of airplane propellers and could produce approximately 3 kW of electricity. Farmers in America's heartland built many such systems. In 1941, Vermont saw the most potent wind turbine built to that time. Constructed mostly of steel and having a 175-foot-diameter rotor blade, it generated 1.25 MW. This "Smith-Putnam turbine" did not last long, though, because of its large size and weight. After World War II, wind power grew dramatically in northern Europe (particularly in Germany and Denmark) as a response to the rising cost of fossil fuels.

Regulation

As the century turned, the growing electrical business attracted the attention of regulators who have remained interested ever since.

In 1907, the Reclamation Service became an organization within the Department of the Interior and was renamed the U.S. Bureau of Reclamation.

In 1977, the FPC was renamed the Federal Energy Regulatory Commission (FERC).

United States: In 1902, the U.S. Congress passed the Reclamation Act, creating the U.S. Reclamation Service in the U.S. Geological Survey to manage water resources and build hydroelectric power plants.

In 1920, the U.S. Congress passed the Federal Power Act which, among other things, required licenses for hydropower facilities on public lands and created the Federal Power Commission (FPC) to license hydropower facilities. In 1935, Congress extended the jurisdiction of the FPC to all hydroelectric projects involved in interstate commerce.

The federal government increased its activity in the electric industry during the Roosevelt administration when Congress passed the Public Utility Holding Company Act in 1935 and established the Securities and Exchange Commission in 1934. Congress also passed the Rural Electrification Act in 1936.

Worldwide: Some of the major regulatory bodies around the world include: State Electricity Regulatory Commission (SERC) – communist China's administrative and regulatory body (excluding nuclear power); Central Electricity Regulatory Commission (CERC) – the regulator for electric power utilities in India (excluding nuclear power); Canadian Energy Regulator (CER) – the government agency regulating inter-provincial/territorial electric utilities (excluding nuclear power); and Council of European Energy Regulators (CEER) – a nonprofit organization of Europe's national electricity regulators that voluntarily cooperates on electricity matters in Europe (CEER has twenty-nine members consisting of the energy regulators of the European Union (EU) Member States, plus Iceland and Norway).

The following EU countries have deregulated electricity markets:

- Austria
- Belgium
- Bulgaria
- Croatia
- Republic of Cyprus
- Czech Republic
- Denmark
- Estonia
- Finland
- France
- Germany
- Greece
- Hungary
- Ireland
- Italy
- Latvia
- Lithuania
- Luxembourg
- Malta
- Netherlands
- Poland
- Portugal
- Romania
- Slovakia
- Slovenia
- Spain
- Sweden
- United Kingdom

Australia has a power grid covering *most* of the country (only Western Australia and the Northern Territory have their own grid). Australia has the National Electricity Market (NEM) that purchases electricity from generators and delivers it to consumers. Retailers (electricity providers) manage the supply side part of the business. The development of the national grid led to energy deregulation that allows companies to compete for business. Distributors can sell to consumers from South Australia to Queensland.

Japan has recently begun deregulating its energy markets. Large commercial and industrial companies can choose their supplier, but smaller commercial companies and residential consumers do not have that option and must stay with their local utility.

Energy-deregulated countries in the Asia-Pacific area include Thailand, Vietnam, Mongolia, Russia, Pakistan, Nepal, and Sri Lanka.

City Involvement

As the twentieth century progressed, government entities became more involved in the new electrical power generation industry, as cities began to establish municipal electric utilities.

United States: In 1893, Tacoma City Light was created. Its first independent power generation was the construction of LaGrande Dam on the Nisqually River in 1912. The city of Austin, Texas, started its electric utility department in 1895. The Salt River Project, founded in 1903, serves the Phoenix metropolitan area. Seattle City Light was formed in 1905, as was the Burlington Electric Department of Burlington, Vermont. The Sacramento Municipal Utility District (SMUD) was founded in 1923, and CPS Energy in San Antonio was formed in 1942.

Worldwide: Tokyo Electric Power Co. Holdings, Inc. (founded in 1951) generates and distributes electric power in and around Tokyo. The company's *fuel* segment generates electricity using thermal power generation. Its *power grid* segment deals with power generation and supply, hydroelectric power generation, and power transmission and distribution.

Kansai Electric Power Co. (also founded in 1951) provides electricity to customers in Osaka, Japan, and the surrounding region. It builds and maintains electric power facilities and uses thermal, geothermal, nuclear, and hydroelectric power sources.

Companhia Energética de São Paulo (CESP) provides electricity to São Paulo, Brazil, and was founded in 1966.

Toronto Hydro (founded in 1998 by combining six municipal electric utilities into one) owns and operates an electricity distribution system for the city of Toronto in the Canadian province of Ontario.

Électricité de Strasbourg S.A. (founded in 1955) provides electricity in the Strasbourg and Bas-Rhin regions of France.

Orkuveitu Reykjavíkur provides electricity to the city of Reykjavik, Iceland. At the end of 1919, the City Council of Reykjavík decided to build a 1000-horsepower power plant, and construction ended in the spring of 1921. The plant began operations in June 1921, sending electricity above-ground to a reservoir located on the outskirts of the city (at that time). From there, the utility distributed electricity to eight transformer stations in the city.

Electric supply to the city of Mumbai, India (formerly known as Bombay) and its southern suburbs comes from Bombay Electric Supply and Transport. Originally, Brihanmumbai Electricity Supply and Transport was set up in 1873 as a tramway company called "Bombay Tramway Company Limited." In 1905, the company set up a thermal power station to generate electricity for its trams and to supply electricity to the city. It then rebranded itself to Bombay Electric Supply & Tramways (BEST) Company. In 1947, the Municipal Corporation took over BEST and rebranded it Bombay Electric Supply & Transport. In 1995, it was renamed Brihanmumbai Electric Supply & Transport.

Electrical Cooperatives

The Rural Electrification Act, passed by the U.S. Congress in 1936, provided federal loans for the installation of electrical distribution systems to serve isolated rural areas of the United States. The funding was channeled through cooperative electric power companies, hundreds of which still exist in the U.S today. By 1942, almost one-half of the farms in the U.S. had electricity due to the Rural Electrification Administration. By 1950, virtually all farms in the U.S. had electricity.

Two examples of a rural cooperative are Pedernales Electric Cooperative, headquartered in Johnson City, Texas, which was organized in 1938, and the San Bernard Electric (Co-op), which built eighty-nine miles of power lines in 1940. It initially served 141 members in the rural areas of Colorado and Austin Counties.

U. S. Federal Involvement

The U.S. Federal Government also became involved in electricity generation and transmission through various entities including the Boulder (later Hoover) Dam, beginning in 1931 (completed in 1936); the Tennessee Valley Authority (1933); Bonneville Power Administration (1937); the Grand Coulee Dam (between 1933 and 1942); Southwestern Power Administration (1943); and Southeastern Power Administration (1950).

1950–2000: Nuclear, Geothermal, Solar, Wind, Storage, and DC

The post-war years saw global economic growth driven by mounting consumer expectations for reliable and inexpensive energy and fueled by cheap energy of all kinds, including electricity produced from traditional fossil fuels. The world entered the last half of the twentieth century hungry for energy and exited the century concerned about the high environmental cost of meeting their energy demands.

Nuclear

The world's first nuclear power plant (EBR-I in Idaho) went into operation in 1951 as a liquid metal-cooled fast reactor, created to prove it could create more fuel than it consumed. ERB-1 became the first power plant to produce usable electricity through atomic fission. In 1954, the U.S.S.R.'s Obninsk APS-1 nuclear facility generated 5 milliwatts of electricity.[26] In Idaho in 1955, Borax-III (a boiling water reactor plant) created enough electricity to power the town of Arco.

> The Obninsk plant ceased operating in 2002.

In 1954, the U.S. Congress passed the Atomic Energy Act, which allowed private ownership of nuclear reactors. In 1957, the Shippingport Atomic Power Station (about twenty-five miles from Pittsburgh) became the world's first full-scale atomic electric power plant devoted to peacetime use. It was owned by the Duquesne Light Company and began producing electricity in December 1957. It continued in operation until 1982.

As of April 2020, there are 440 operable power reactors in the world, with a combined electrical capacity of 390 GW. Additionally, there are 55 reactors under construction and 109 reactors planned, with a combined capacity of 63 GW and 118 GW, respectively, while 329 more reactors are proposed.[27]

Geothermal

In 1960, Pacific Gas and Electric began operating its 11 MW geothermal electric plant at the Geysers in California, the largest geothermal field in the world (covering approximately thirty square miles). Power from The Geysers provides electricity to Sonoma, Lake, Mendocino, Marin, and Napa counties. Unlike most geothermal sources, The Geysers is a dry steam field producing superheated steam.

In 1984, the Blundell Geothermal Power Plant in the Roosevelt Hot Springs Geothermal Area (46.7 square miles in Utah) began generating electricity. Geothermal brines from the Geothermal Area provide fuel for the plant, which has a nominal capacity of 37 MW.

In 2018, twenty-seven countries, including the United States, generated a total of about 83 billion kWh of electricity from geothermal energy. Indonesia was the second-largest geothermal electricity producer after the United States, at nearly 14 billion kWh of electricity (equal to about 5% of Indonesia's total electricity generation). Kenya was the eighth-largest geothermal electricity producer at about 5 billion kWh, but had the largest share of its total annual electricity generation from geothermal energy at 46%.[28]

Solar

Solar power is as old as the earth. Generating electricity from the sun began in the early 1880s when, in 1839, Edmond Becquerel of France discovered the photovoltaic (PV) effect – the basics for modern photovoltaic cells. Another enterprising Frenchman Augustin Mouchot registered patents for solar-powered engines as early as the 1860s. New York inventor Charles Fritts created the first solar cell in 1883 by coating selenium with a thin layer of gold, and, in 1888, Aleksandr Stoletov (a Russian scientist) created a solar cell based on the photoelectric effect. Many more people aided in developing solar technologies over the years, but they were novelties rather than feasible sources of electrical power until the mid-1950s.

Solar began in earnest in the 1950s, when scientists at Bell Labs developed a PV cell made from silicon which, for the first time, provided sufficient electricity to run electrical equipment from solar power.

During the 1970s, the U.S. government sponsored a program to help develop PV cells. In 1978, the Energy Tax Act provided a 10% investment tax credit for PV equipment, and the Solar Photovoltaic Energy Research, Development, and Demonstration Act of 1978 authorized research and development of PV systems using solar energy or sunlight to generate electricity. The Act also promotes energy conservation by encouraging displacement of conventional energy systems using fossil fuels.

Wind

By the 1970s, wind turbines began to develop in a significant way to generate electricity and, by 1979, in a project sponsored by NASA, the MOD-1 wind turbine became the first of its kind rated at more than 1 megawatt. Later, Boeing designed, built, and installed several 2.5-megawatt wind turbines at Goodnoe Hills, Washington. In May 1981, three turbines at the site formed the first "wind farm" in the world.

Lithium-Ion Battery

Many sources credit John Goodenough, professor at the University of Texas in Austin, with the invention of the lithium-ion battery. Other people, however, contributed to this discovery and then to the development of a commercial version of the battery in the 1970s and 1980s as well. In 2019, the Nobel Prize in Chemistry was awarded to John Goodenough, Stanley Whittingham, and Akira Yoshino for the development of lithium-ion batteries. All three contributed to the evolution of the kind of lightweight, rechargeable batteries that power today's mobile phones and other portable electronic devices.

Direct Current

The twentieth century is the century of alternating current, but with the close of the century, direct current (DC) began to slowly displace AC, driven by the growth in DC demand for use in electronics, LED lighting, and vehicles powered by DC current. On the supply side, solar voltaic cells generate DC current, and the emergence of High Voltage Direct Current (HVDC) electric power transmission systems, enabled by inverters, allow long-distance transmission of electrical power.

The Environment

The second half of the twentieth century saw a growing concern for air quality – first, in the form of acid rain, and then in the form of greenhouse gas emissions. The topic of the environment is discussed in more detail in the final chapter of this text.

Legislation and Regulation

The laws of physics apply globally but laws passed by governments vary. Attempting to cover a history of a myriad of laws and regulations promulgated during the last half of the twentieth century is left to others.

2000–2025: A Rapidly Evolving Industry

This section is not so much about names and dates as it is about trends and how the electrical power industry seems to be evolving.

From its inception, the electric industry evolved through technological advancements, economies of scale, effective financial/regulatory structures encouraging capital investment, and the continuing demand for more energy. Since the mid- to late-1800s, changes in generation, transmission, distribution, and market design and regulation of the electricity industry have influenced the industry. While electricity will continue as an integral form of energy and large-scale centrally located electricity generation will remain essential, the electrical power industry is rapidly evolving.

This evolution is driven by development of new and more cost-efficient generation options for centralized generation (wind farms, utility-scale solar and combined cycle power plants using natural gas) and the emergence of microgrids – self-sufficient energy systems that serve a discrete geographic footprint, such as a college campus, hospital complex, business center, or neighborhood.

Within microgrids, one or more kinds of distributed energy (solar panels, wind turbines, combined heat and power) generators produce power for those connected to the grid. These new primary energy generation technologies are challenging fossil fuel and nuclear primary energy generation technologies. The newer technologies tend to be modular (built smaller and increased in size as needed) and have lower emissions, shorter development times and/or low (or no) fuel costs (e.g., renewables).

More on these trends and challenges in the final chapter, "Challenges for the Future."

Summary

- In the modern world, we use electricity for almost everything, every day – appliances, heating, cooling, smartphones, entertainment, lighting, and a myriad of other purposes.
- The 1600s saw the birth of the word *electric* and early understanding of electricity and magnetism, and investigation began into how those two forces relate to each other.
- The 1700s built on the discoveries of the previous century with increased knowledge about conductors, insulators, and electrical charges.
- Physicists of the nineteenth century built on the knowledge of charges, forces, conductors, capacitors, Leyden jars, and a variety of other bits of knowledge developed in the previous century, as our understanding of electricity and magnetism increased.
- As the world entered the twentieth century, generation technologies grew and matured, business models evolved, and regulations and government involvement increased as the electrical power industry went into exponential growth mode.
- By the start of World War I, steam turbines dominated electrical power generation, a fact that continues to this day.
- The post-World War II years saw global economic growth, driven by mounting consumer expectations for reliable and inexpensive energy, and cheap energy of all kinds, including electricity produced from traditional fossil fuels.
- The world entered the last half of the twentieth century hungry for energy and left the last half of the twentieth century concerned about the high environmental cost of meeting energy demands.
- From its inception, the electric industry evolved through technological advancements, economies of scale, effective financial/regulatory structures encouraging capital investment, and the continuing demand for more energy.
- Since the mid- to late-1800s, changes in generation, transmission, distribution, and market design and regulation of the electricity industry have influenced the industry.
- While electricity will continue as an integral form of energy and large-scale centrally located electricity generation will remain essential, the electrical power industry is rapidly evolving.

3

Electrical Generation

Energy cannot be created or destroyed; it can only be changed from one form to another.

—Albert Einstein (1879–1955)

"Grandpa Tom, the other day you told me electricity does not happen naturally but instead must be generated from other forms of energy. But how does that happen?" asked Luke.

"Most of the electricity the world uses today is generated at large generation facilities by devices called electromagnetic or electromechanical generators," Grandpa Tom explained. "These generators have two main parts, both containing magnets," Grandpa Tom added. "One of those parts is connected to a power source which converts primary energy into rotary motion. The magnet on the rotating part travels past the magnet on the stationary part, producing electricity," Grandpa Tom further explained.

"Wait," exclaimed Luke. "Is that how those windmills work? I can see one part is holding still while the blades are turning."

"That is right," declared Grandpa Tom. "The same thing happens at coal, natural gas, hydroelectric, and nuclear generating facilities."

"What about solar?" questioned Luke. "The solar panels on your house are just sitting there, so how do they generate electricity?"

"They capture energy released by the sun and pass the energy through various devices, converting it to electricity," Grandpa Tom explained.

"Very cool," Luke said with a smile.

"Well, sometimes cool and sometimes hot," laughed Grandpa Tom.

Introduction

Electricity does not occur naturally in commercial quantities. Because electricity must be produced from other forms of energy, it is commonly called "secondary energy," as opposed to "primary energy," the energy sources that produce electricity. Converting primary energy to secondary energy is normally referred to as *generating electricity*.

Generation Fuel Sources

Currently, about 60% of the world's electricity comes from converting chemical energy contained in fossil fuels (coal and natural gas) into electrical energy. Gravity pulling water downhill produces about 16% of electrical energy; converting atomic power into electricity yields another approximately 10%; a variety of sources, most commonly called *renewables*, produce around 12%; leaving approximately 3% for all other fuels. The numbers are rounded which is why they add to 101% rather than perfectly adding to 100%.

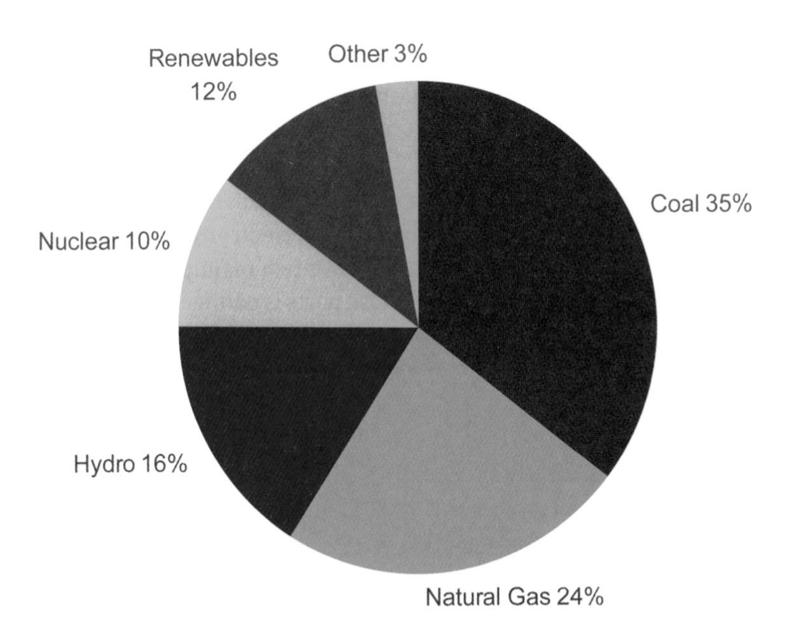

Figure 3-1. World Electrical Generation by Energy Source.

Of the renewables portion, a little over half is generated by wind, with 27% generated with biomass, geothermal, and a few other sources.

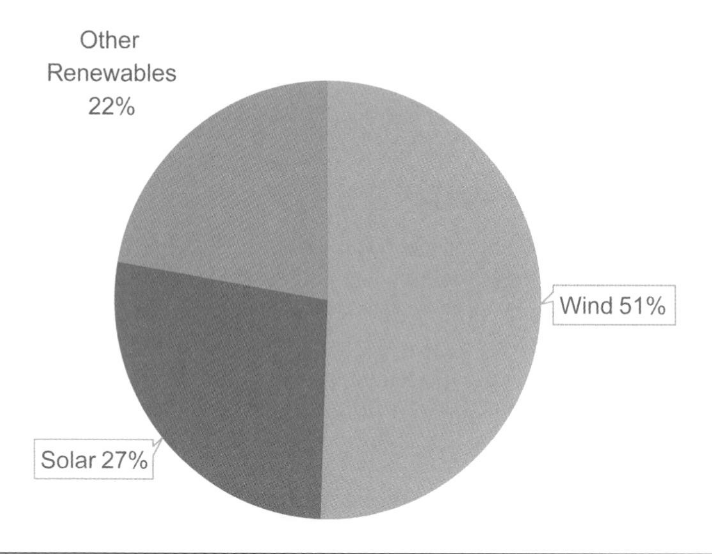

Figure 3-2. World Renewable Electrical Generation by Source.

Generation Types

Coal, natural gas, hydro, nuclear, and wind – all drive electromechanical generators. Figures 3-1 and 3-2 demonstrate that electromechanical generators account for over 90% of all electrical power generation. This chapter discusses electromagnetic and other types of generators and the primary fuel sources driving them.

Electromechanical or Electromagnetic?

Many people use the terms *electromechanical* generators and *electromagnetic* generators synonymously, with no ill effect. As nearly as the authors can deduce from the various definitions that abound, the term electromechanical refers to a machine that converts mechanical energy into electricity, and the term electromagnetic refers to a machine that converts mechanical energy into electricity using an electromagnet rather than a permanent magnet. The authors choose to use the term electromagnetic in this text rather than electromechanical.

Three important terms apply to generating electricity:

- Plant capacity denotes the maximum *rate* of electric power a plant can supply, expressed in megawatts (MW) or gigawatts (GW).
- Generation output is the *amount* of electrical energy produced over time, expressed in megawatt-hours (MWh) or gigawatt-hours (GWh).
- Sales is the *amount* of electricity sold to customers over a period and, like generation, is expressed in MWh or GWh (although residential-level sales are measured in kilowatt-hours (KWh)).

It seems the amount generated should equal the amount sold, but that is not quite right for two reasons. First, transmitting and distributing electricity from the power plant to the end user causes some energy loss (as heat). Second, storing electricity and releasing it for later use also causes energy loss.

European or North American?

Early in the development of electrical systems, North America adopted one combination of distribution voltage and electrical generation frequency, and Europe developed another. Endless debates ensued trying to prove one was "better" than the other. The truth is each system has its advantages and disadvantages. Once established, it was hard to change, so they continued on. Both follow the same laws of physics and behave largely the same.

Electromagnetic Generators

As shown in Figure 3-2, over 90% of the electricity the world uses today comes from electromagnetic generators that operate on the principle of electrical induction purportedly discovered by Michael Faraday.

As already stated several times in this text, electricity and magnetism are directly and causally related. Mr. Faraday discovered that changing the magnetic field around a coil of wire *induces* an electromagnetic force (emf) in that wire, measured in *volts*. Magnetic force can be changed by:

- changing the magnetic field strength;
- moving the magnet closer to or farther from the coil;
- moving the coil closer to or farther from the magnet;
- rotating the coil relative to the magnet or the magnet relative to the coil.

The magnet can be either a *permanent* magnet or an *electro*magnet.

Most generators today rotate the magnet relative to the coil and use electromagnets rather than permanent magnets.

Electromagnetic generators consist of three main parts:

- Rotor
- Stator
- Exciter

Winding vs. Coil

A coil is a bundle of wires produced by winding one wire in a circle or other geometric shape. Coils are sometimes generically called windings.

Electric Motors

Essentially the opposite of generators, motors have a rotor and stator. In the motor, the current drives the magnetic field making the rotor turn.

Not surprisingly, the rotor rotates, and the stator remains stationary. Either the rotor or the stator produces the magnetic field. The other has conductors – commonly comprised of coils of wire, giving rise to the term *coils*. As the coils on the one component pass through the magnetic field on the other component, the field induces an electrical current (electricity) into the coils.

While either the rotor or the stator can carry either the magnetic field or the coils, most commonly, the rotor carries the magnetic field, and the stator contains the coils. To make things a little more confusing, another term – armature – is sometimes used with generators. The armature creates the magnetic field. So, either the rotor or the stator serves as the armature depending on whether it carries the magnet. Consistent with the common practice of the rotor carrying the magnetic field, the rotor usually serves as the armature.

Rotors and Exciters

In nearly all power plant generators, electromagnets, and the shaft to which they connect, form the rotor. These electromagnets consist of coils of wire (more often now, bars of metal) that become electromagnets when powered with direct current from a small generator connected to the same shaft. The small generator energizing the electromagnets is called an *exciter.* The current supplied to the exciter is direct and not alternating (because, if alternating, the polarity of the magnets would switch as the polarity of the current alternated).

Coils or Bars?

Historically, the conductors were coils of wire. They are now often simply bars of metal, but some people still refer to them as coils.

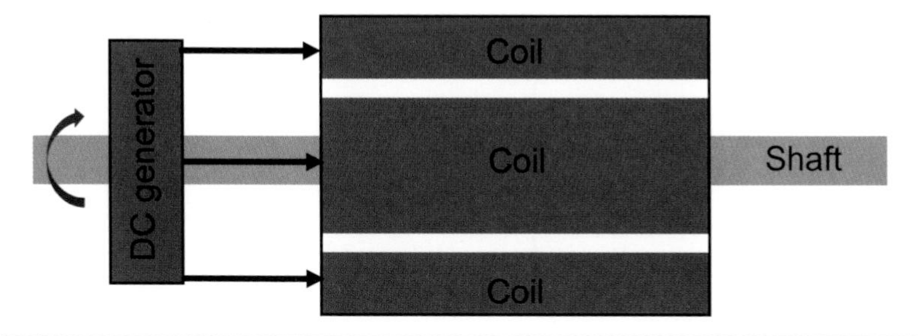

Figure 3-3. Drawing of a Rotor. The arrows show DC current transmission from the DC generator to the rotor coils.

Figure 3-4 depicts the rotor of a large generator. The cutaway reveals the metal bars attached to the shaft comprising the rotor's *poles*.

Figure 3-4. Rotor of an Industrial-Size Generator.

Stators

Stators remain stationary and, like rotors, contain coils. As the rotor rotates, its magnetic field changes relative to the stator coils, creating an electromagnetic force (emf) in the stator coils. The stator coils have leads connected to each end that transmit the voltage induced in the coil from the generator.

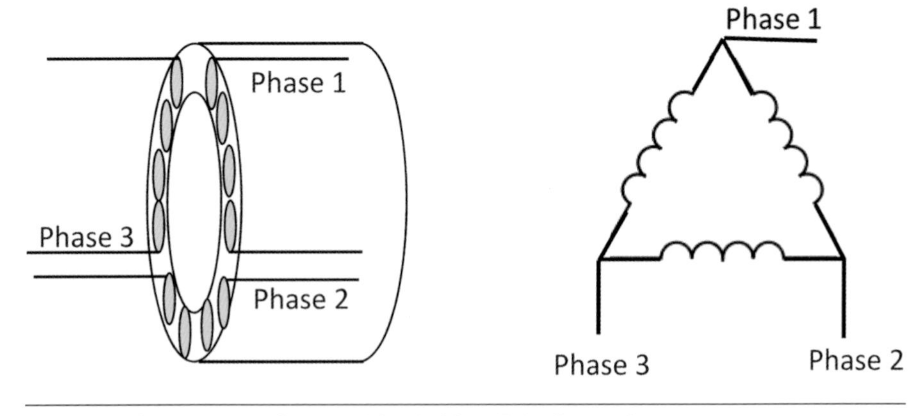

Figure 3-5. Schematic of a Stator and Symbol for a Delta Connection.

The schematic on the left shows three groups of coils generating one phase of alternating current each. It has six wires, two for each phase. The phase electricity could be transmitted with six conductors, one carrying the electricity and the other providing the return for each phase.

But, early, electrical engineers realized that because since the phases alternate, one lead from each coil could be connected to the next coil in what is called a delta connection. This connection pattern means the phases serve as a circuit return path for each other. Accordingly, three-phase transmission lines have three conductors (plus one or more shield wires) and not six.

Figure 3-6 shows the stator of a large generator.

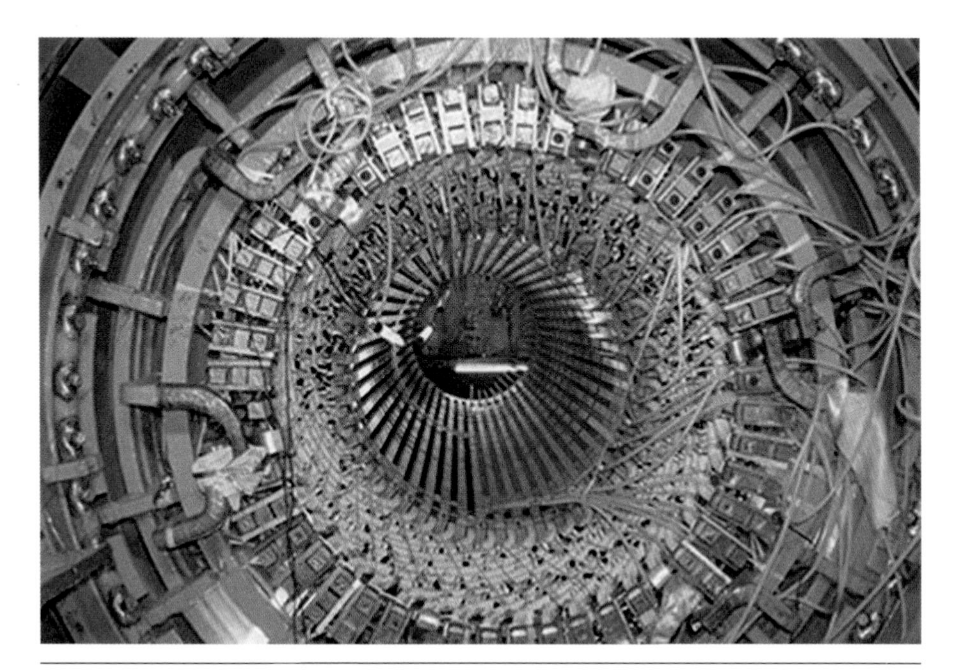

Figure 3-6. Stator of an Industrial-Size Generator.

Frequency

Frequency, discussed in Chapter 1, deserves a little more coverage in this chapter. The number and configuration of the rotor and stator windings, combined with the rotor's rotational speed, determine the frequency of the current produced.

The drivers discussed in the next section turn at different and potentially varying speeds. The generator's configuration must match the driver, meaning generators driven by water turbines versus those driven by wind, steam, and gas turbines, while similar, may vary somewhat. Controlling generator rotational speed controls frequency and helps in managing the grid.

Synchronization

When multiple generators connect to the same grid, they must synchronize their frequencies to keep the entire grid operating smoothly. Accordingly, one of the goals of the system operator's control room is keeping the voltage and current from each generation plant synchronized with every other generation plant connected to the grid. More about frequency and synchronization later in this chapter.

Drivers

By themselves, electromagnetic generators do not generate electricity. Rather, machines commonly called drivers or prime movers power them (i.e., cause them to rotate). Drivers start with a primary energy form and convert it into rotating motion (mechanical energy) to drive generators that convert mechanical energy into electrical energy.

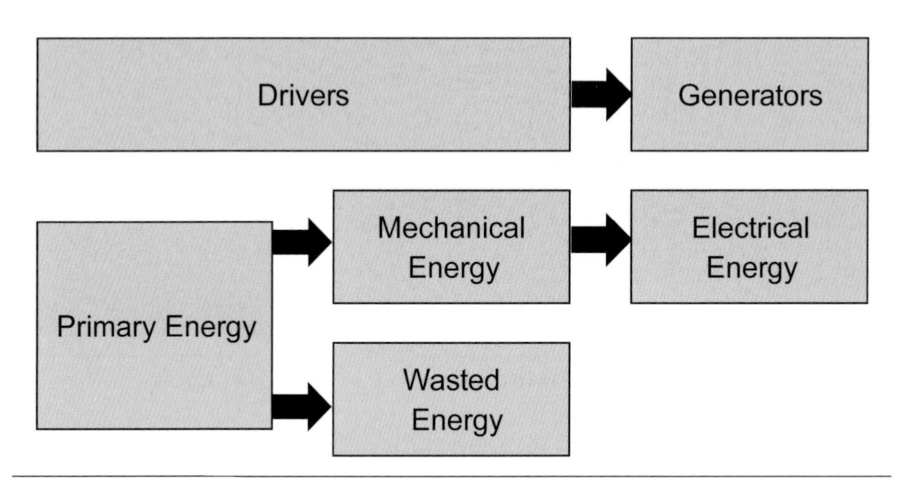

Figure 3-7. Primary Energy to Electrical Energy Flow Chart.

Figure 3-7 shows primary energy converted into mechanical energy doing useful work and also shows waste energy (heat). As anyone who has cooked on a stove knows, not all energy goes into the pot; part of it escapes into the surrounding environment. The same thing happens in generating plants.

Engineers say the amount of energy converted to useful work compared to the amount of energy used is the system's *efficiency*. The higher the efficiency, the less waste heat created.

Turbines of various designs and complexities provide the mechanical force shown in Figure 3-7. Wind turbines – propellers that take energy from the wind (rather than adding energy into it like airplane propellers) are simple turbines.

Figure 3-8. Four Wind Turbines

Turbines though, do not create energy; rather they convert kinetic energy contained in one of four flowing fluids into rotational motion which turns the rotor. The four flowing fluids are:

- Steam (water in vapor form)
- Gases (air and other substances in gaseous form)
- Hydro (water)
- Wind

"Heat engines" of various types (discussed in the next section) produce steam and gases.

Heat Engines

Heat engine is a generic term for a machine or device that produces heat from some primary energy source, uses part of the heat to create mechanical work, and discharges the remaining heat. In the case of electrical power generation, the heat comes from fossil fuel, nuclear fuel, and concentrated solar-thermal power.

The laws of physics dictate that the greater the temperature drop from the inlet to the outlet of the heat engine, the greater the amount of useable energy and the less wasted energy. The same laws of physics place practical limitations on the maximum temperature, based on the melting points of metals in the engine. The point is, the greater the temperature drop, the more efficient the heat engine.

Steam Turbines

Steam turbines are one of the more common methods of converting heat into useful work. Steam turbines obtain heat from a heat engine and use it to convert water into steam. Coal, nuclear, and concentrated solar-thermal (and parts of some natural gas) power plants use steam turbines as drives. Figure 3-9 shows a steam turbine rotating element.

Figure 3-9. Steam Turbine Rotating Element.

Gas Turbines

Adding heat to water converts it into steam driving steam turbines and combusting natural gas in the presence of compressed air creates expanding gases driving gas turbines.

Single Cycle and Combined Cycle

Figure 3-7 shows the process of converting primary energy into electrical energy and releasing wasted heat, that is, heat not used to produce electricity. Figure 3-10 shows the same concept with a few additional boxes. In this figure, a second generation cycle appears in green.

During the first cycle, expanding gases drive a gas turbine. The second cycle captures the exhaust gases from the first cycle and uses some of the heat remaining in the gases to generate steam which drives a steam turbine. Combined-cycle processes are more efficient than single-cycle processes because they waste less heat. More on combined cycle later in this text.

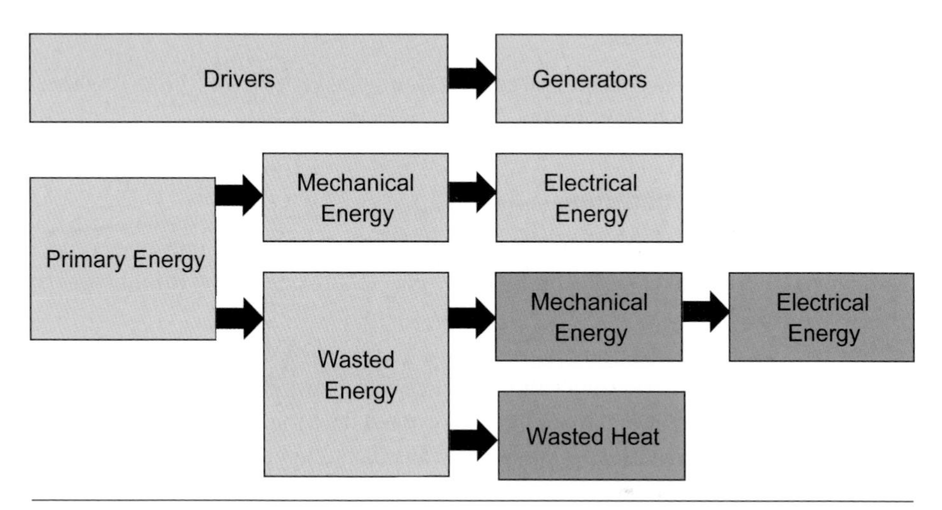

Figure 3-10. Wasted Heat from the First Cycle Converted into Electricity in the Second Cycle.

Coal Power Plants

Figure 3-11 shows a schematic for a single-cycle coal power plant that burns coal to produce steam in a boiler. As the water turns into steam, it expands, impacting the blades of the turbine, making it spin. The rotating shaft attached to the blades drives the generator producing electricity.

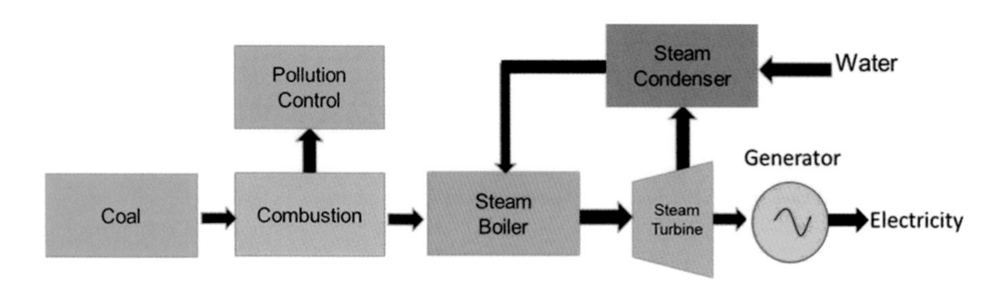

Figure 3-11. Schematic of a Single-Cycle Pulverized Coal Plant.

Figure 3-11 does not show the steps for coal preparation and the details for emission controls capturing particulates, sulfur oxides (SO_x), and nitrogen oxides (NO_x).

Three different types of coal combustion can support electricity generation:

- Stoker-fired
- Pulverized coal
- Fluidized bed

Stoker-Fired Combustion

This process crushes coal into large lumps and burns them on a grate to generate combustion heat to make steam.

Pulverized Coal Combustion

This method uses smaller particles (less than 200 microns) to improve contact with air relative to stoker-fired combustion. Two (some people say three) types of pulverized coal combustion exist.

The first, subcritical combustion, uses coal that is crushed but not so small, so it does not continue to behave as a solid. The second, supercritical, uses coal crushed into a powder so small it behaves as a supercritical fluid – a material that, above a certain temperature and pressure, no longer has distinct phases of liquid and gas.

The two types of supercritical combustion are supercritical and ultra-supercritical – the difference is the size of the powder, but both operate on the same principle. By operating above the critical point, the generator can have higher combustion temperatures, yielding better efficiency.

Fluidized Bed

This type of combustion uses extremely small coal particles (1–10 mm) and occurs at lower temperature than pulverized coal combustion, thereby reducing the amount of nitrogen oxides produced.

Table 3-1 shows typical efficiencies and demonstrates how increasing the steam temperature increases efficiency for pulverized coal combustion. Nevertheless, the best efficiencies of coal plants remain less than 50%.

Table 3-1. Coal Plant Efficiencies[29 30]

Coal Plant	Steam Temperature, °C	% Efficiency
Pulverized Combustion		
Subcritical	Up to 565	33-39
Supercritical	540-580	38-42
Ultra-supercritical	>580	>42
Fluidized Bed		35

Harkening back to Figure 3-7 and the wasted energy makes it easy to surmise power plants need a lot of cooling, so generation companies often build them near water sources, or they build water sources as part of the plant project. Figure 3-12 shows the Sam Seymour Power Plant with three units with a total nameplate capacity of 1,615 MW. The plant sits on the shores of Fayette County Reservoir, a 2,400-acre freshwater reservoir also used for recreation.

Figure 3-12. Sam Seymour Power Plant.

One more type of coal plant – combined cycle – is discussed later in this chapter.

Nuclear Power Plants

Just like coal plants, nuclear plants drive generators with steam turbines. Figure 3-13 shows the schematic for a nuclear power plant.

Uranium is currently the fuel of choice, but a lot of research and potential exists for using thorium in the future. Unlike coal or natural gas, nuclear power plants require greater safety measures to contain the nuclear reaction. Nuclear plants use two basic approaches.

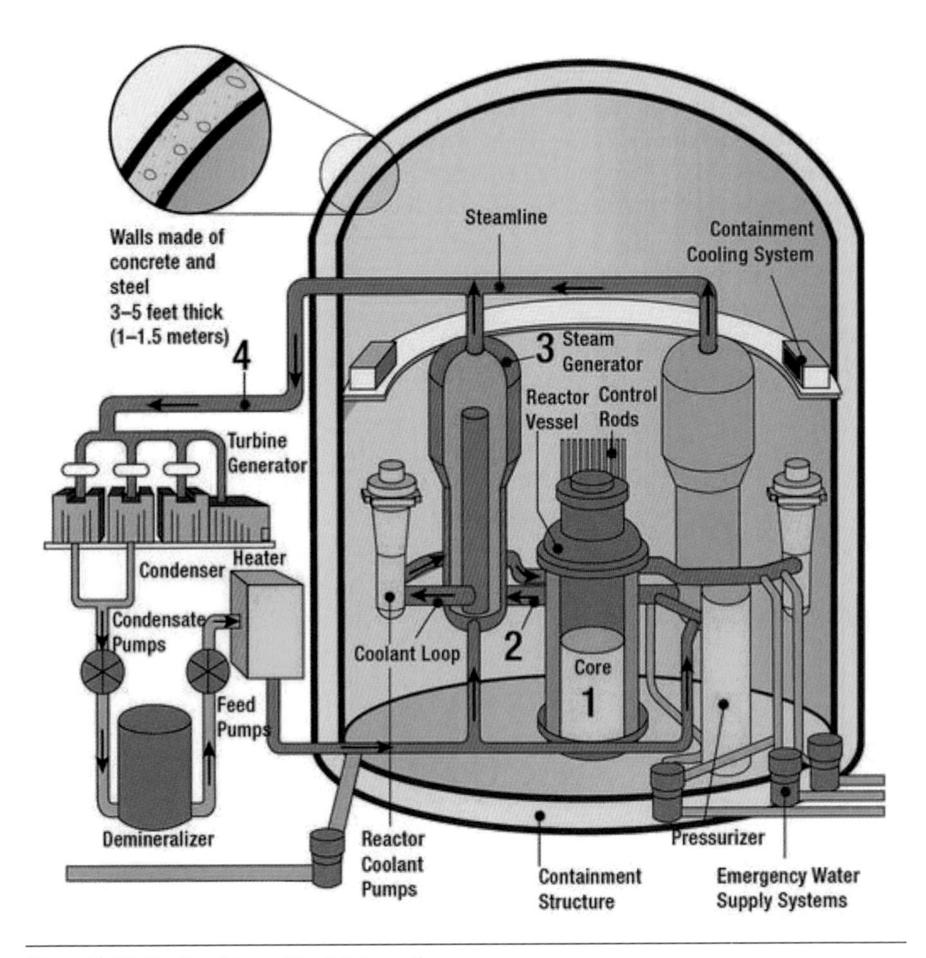

Figure 3-13. Nuclear Power Plant Schematic.

Pressurized Water Reactor

The most common approach is a pressurized water reactor (PWR). This process pumps water through the reactor in a pressurized loop for heating. The heat transfers to the boiler, producing steam to power the turbine and drive the generator. As a safety precaution, the water goes through an isolated loop and never directly contacts the turbine to prevent the potential for radioactive contamination of the turbine and condenser.

Figure 3-14 shows a pressurized water nuclear plant with a nameplate capacity of 2,560 MW.

Figure 3-14. South Texas Nuclear Plant.

Boiling Water Reactor

The second approach uses a boiling water reactor (BWR) and does not isolate the turbine and compressor from the reactor, meaning a fuel leak makes the water radioactive and contaminates the turbine and condenser.

Temperature Limits and Efficiency

Nuclear plants tend to have lower thermal efficiencies than coal plants because nuclear plants cannot produce steam at temperatures as high as fossil fuel plants due to temperature limits imposed by the nuclear reactor containing the fuel rods. According to the Energy Information Administration (EIA), nuclear power plant thermal efficiencies in the U.S. have averaged 33% for the last ten years.[31]

Natural Gas Power Plants

Natural gas power plants are either single-cycle steam or combined cycle. Combined-cycle plants are more efficient than single-cycle steam plants.

Natural Gas Steam Generation

In this approach, the plant burns natural gas, producing heat to make steam in a boiler. As with coal plants, the force of the expanding steam rotates the blades of a steam turbine that, in turn, rotates a shaft of the turbine delivering mechanical energy to the generator. A schematic of this approach looks just like Figure 3-11 except the box containing the word "coal" would say "natural gas."

Steam turbine natural gas plants tend to be coal plants converted to natural gas by removing the coal combustion equipment and replacing it with natural gas combustion equipment.

Natural Gas Combustion Generation

The second approach injects compressed air into natural gas combustion turbines, thereby increasing the combustion temperature. The combustion gases enter a gas turbine driving the generator. Gas turbine plants are less complex than many other drivers and relatively easy to start up and shut down. They are often referred to as quick start units, and utilities often use them as "peaker units" because they use them only when electricity demand peaks.

Both natural gas steam plants and natural gas combustion plants run at poor efficiencies (typically 20–35%[32]), meaning the high-end matches that of pulverized and fluidized bed coal combustion. Gas plants are, however, smaller, lighter, easier to permit and more environmentally friendly than coal plants.

Combined-Cycle Power Plants

As previously mentioned, combined-cycle plants use heat from the gas turbine to produce steam and drive a steam turbine. Doing so decreases the amount of heat wasted and increases the overall plant efficiency. A combined-cycle system combines two thermodynamic processes to generate electricity.

Combined-Cycle Natural Gas Plants

Natural gas combined-cycle (NGCC) systems capture the hot exhaust gases from the gas turbine and pass them through a heat recovery steam generator (HRSG) to power a steam turbine. Because two turbines generate electricity, they use the heat energy more effectively and increase thermal efficiency to 50-60%.[33] A schematic of the process appears in Figure 3-15. Basically, compressed air and natural gas combust in the gas turbine and turn the blades to generate electricity. The HRSG recovers heat from the exhaust gases and produces steam to rotate the blades of the steam turbine and generates additional electricity.

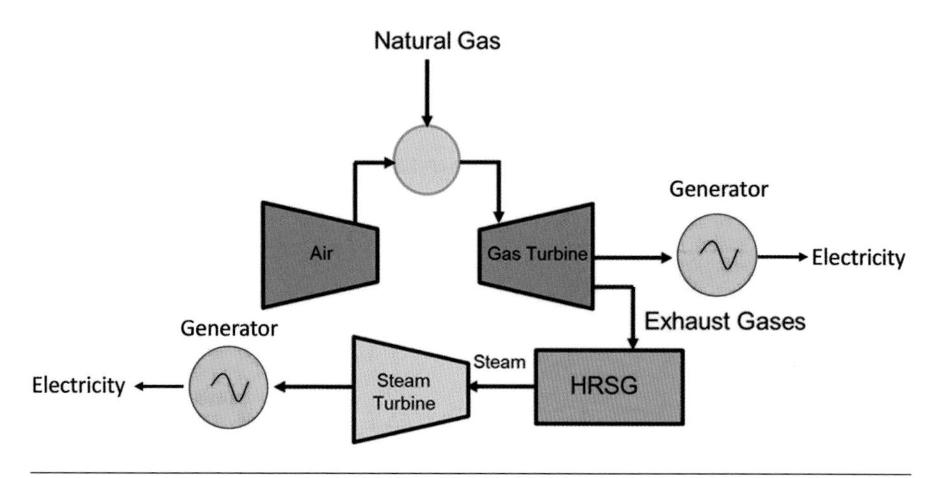

Figure 3-15. Schematic of a Natural Gas Combined-Cycle (NGCC) Plant.

Combined-Cycle Coal Plants

Synthetic gas produced from coal (rather than natural gas extracted from the earth) powered the first gas streetlights. In the same way, combined-cycle coal plants gasify the coal by heating it and using oxygen or air to produce a mixture of carbon monoxide (CO) and hydrogen (H_2). Unlike natural gas, coal contains heavy metals, such as mercury, selenium, and cadmium, as well as sulfur and nitrogen oxides, all of which require isolation and proper storage. After producing and purifying the synthesis gas, it goes to a gas turbine where, with the addition of air, combustion occurs. These combustion gases drive the turbine and generate electricity.

The hot exhaust from the gas turbine goes to a HRSG to recover heat. This heat then produces steam to turn the steam turbine and generate additional electricity. The industry refers to this process as integrated gasification combined cycle (IGCC).

Combined Heat and Power Plants

Combined heat and power (CHP) plants, also known as cogeneration (cogen) plants, produce electricity and thermal energy. Similar in concept to combined-cycle plants, CHP plants capture the waste heat from the generation process and, instead of using it to generate electricity, use it to heat a building. Two common CHP configurations exist: gas turbine/reciprocating engine and steam boiler.

Gas Turbine/Reciprocating Engine

This configuration burns gas or liquid fuel in a prime mover, such as a gas turbine or reciprocating engine. The prime mover connects to a generator that produces electricity. Heat contained in the prime mover's exhaust is captured and used to heat buildings.

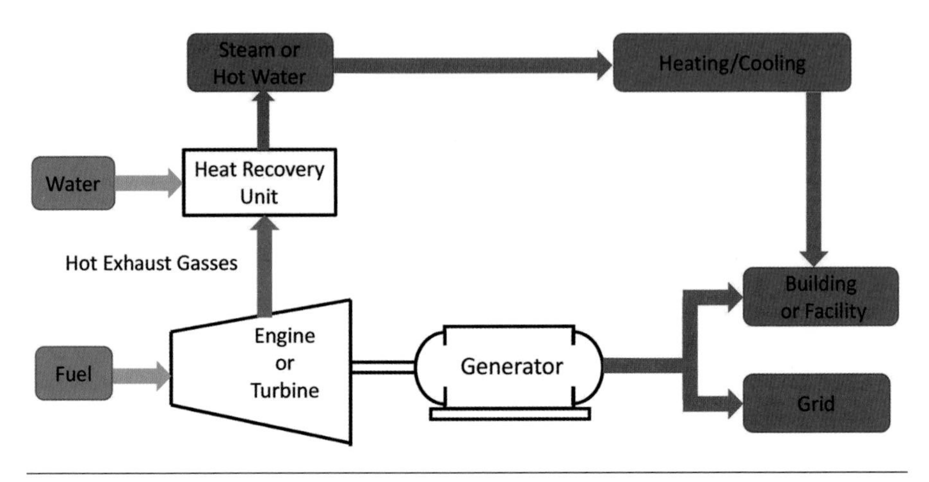

Figure 3-16. Schematic of a Combined Heat and Power Plant

Steam Boiler

This configuration is similar to the gas turbine/reciprocating engine but, instead of using a prime mover, it produces steam in a boiler that drives a steam turbine connected to a generator. The system then heats buildings with heat from the discharge side of the steam turbine. Efficiencies for CHP installations vary based on site-specific parameters, but an intelligently designed CHP system typically operates with an overall efficiency of 65–85%.

Concentrating Solar Power Plants

A staggering 175,000 TW (terawatts or one trillion watts) of solar radiation strike the earth, and about half reflects into space or gets absorbed by clouds, leaving more than 89,000 TW to reach the earth's surface. When most people hear "solar power," they think of solar panels. The other form of solar power generation – concentrating solar power (CSP) – appears in this section because it also serves as a heat engine.

CSP technology uses mirrors to concentrate sunlight to heat a fluid to produce steam to drive a turbine like technologies using natural gas, coal, or nuclear fuel. Four approaches exist to configure the mirrors used for concentrating sunlight:

- Parabolic trough
- Linear Fresnel

- Solar dish system
- Solar tower plants

Parabolic trough systems, shown in Figure 3-17, are the most common. They have a parabolic arrangement of mirrors to concentrate sunlight onto a tube extending the length of the trough on the focal line of the focused sunlight. The tube contains a heat transfer fluid, such as a molten salt, to capture the thermal energy. The heat transfer fluid then boils water, produces steam, and powers the turbine.

Figure 3-17. Picture of Parabolic Trough CSP Mirrors.

Hydroelectric Power Plants

Hydroelectric power plants produce electricity like heat engine plants, but have much higher energy efficiency because they use the kinetic energy of moving water to turn turbine blades instead of using combustion gases or steam. Unlike fossil fuels, the plant loses no heat as exhaust. Typical efficiencies for hydroelectric plants average 90%[7] due to energy loss from friction and turbulence of water turning the turbine blades. Figure 3-18 shows a simplified drawing of hydroelectric power generation.

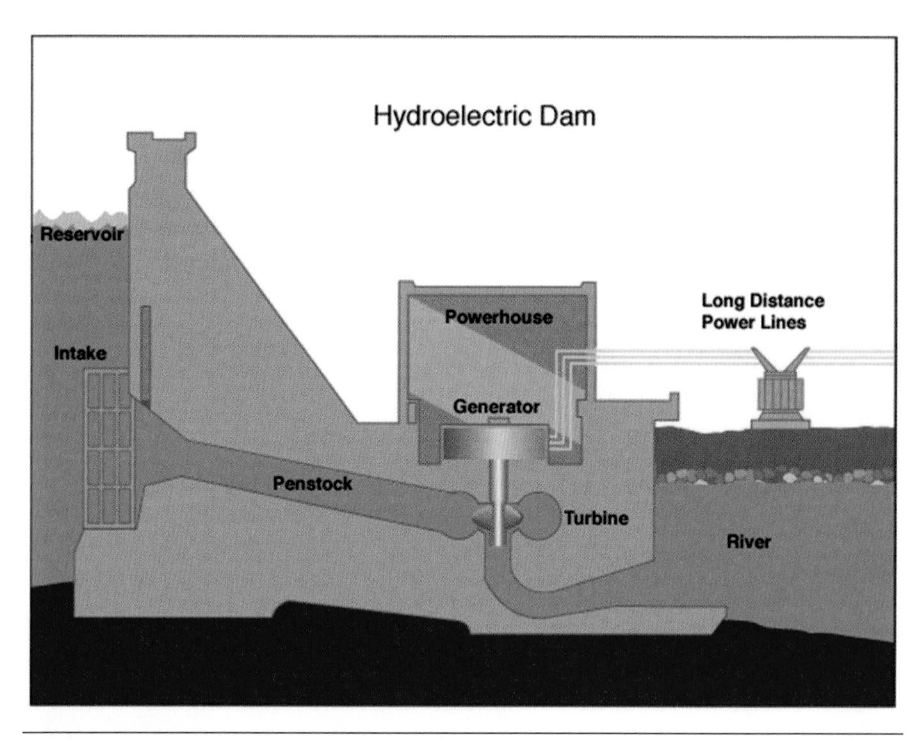

Figure 3-18. Hydroelectric Power Generation Drawing.

Plants use one of three approaches to generate electricity from moving water:

- Storage dam
- Run-of-river dam
- Pumped storage

We discuss the first two in this chapter and pumped storage in the chapter dealing with energy storage.

Storage Dam Hydrogeneration

Generating companies build dams across rivers at locations with a large elevation drop over a short distance, creating a reservoir. Gravity causes water to flow from the reservoir through a pipe called a "penstock" contained in the dam. The penstock directs the water past a turbine connected to an electromagnetic generator. Figure 3-19 shows the J. Percy Priest dam and generation facility near Nashville, Tennessee, which has a generation capacity of approximately 28 MW.

China produces the most storage dam hydroelectricity, followed by Brazil and Venezuela. Storage dam hydroelectricity is relatively cheap and does not pollute the air as much as other forms of generation. It does, however, involve building dams on rivers and streams, creating certain social and environmental problems.

Figure 3-19. J. Percy Priest Dam and Power Plant Building.

From a practical standpoint, the expansion of storage dam hydrogeneration has a limit when no more rivers or streams can support it.

Run-of-River Dam Hydrogeneration

As its name implies, this form of generation uses a river to generate electricity without creating a water reservoir like the storage dam. Some run-of-river plants, however, use small dams to ensure enough water availability to flow through the penstock. As with the storage dam, run of river generation uses a turbine driven by the flowing stream.

Wind Power Generation

Wind turbines extract kinetic energy from the wind and use it to turn electromagnetic generators. On a percentage basis, wind is the fastest growing source of electrical power generation in the world. China boasts the world's largest capacity for wind energy, totaling a little less than 300 GW in 2020. The U.S. comes in a distant second with less than half that much.[34] Interestingly, the oil- and gas-producing state of Texas has over 20% of its electricity capacity in wind power.

Wind Turbine Components

Wind turbines consist of four basic parts:

- Base
- Tower

- Nacelle
- Blades

The blades, typically two or three, spin a hub that transfers rotation to the nacelle which contains the electrical components, including the generator. The tower supports the blades, hub, and nacelle and houses conductors, and the base supports the entire structure. Figure 3-20 shows the parts of a wind turbine in more detail.

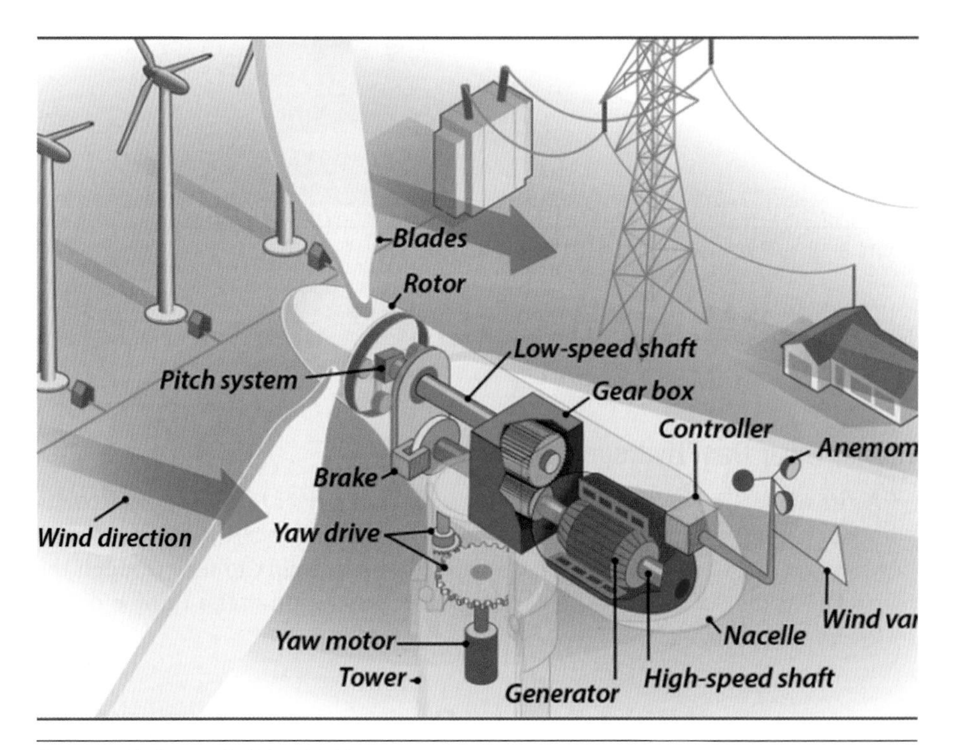

Figure 3-20. Parts of a Wind Turbine.

Wind Turbine Types

Wind turbines come in two basic types, horizontal and vertical, which correspond to the orientation of the axis around which they rotate. Horizontal is most common.

Wind Turbine Sizes

The size of wind turbines varies widely. The blade length is the biggest factor in determining generation capacity. Small wind turbines powering a single home may have an electricity-generating capacity of ten kilowatts (kW). The largest wind turbines in operation have electricity-generating capacities of ten megawatts, with larger turbines in development.

Wind Turbine Operations

Each turbine has a particular relationship between wind speed and power. For example, a 2 MW wind turbine may not start generating electricity until the wind speed exceeds seven miles per hour, called the "cut-in" speed. Then, as wind speed increases, output increases, perhaps achieving its 2 MW rating at around 25-30 mph. To prevent wind damage, a braking system will stop rotation at speeds around 55 mph, called the "cut-out speed."

Wind Farms

Wind "farms" cluster wind turbines to produce large amounts of electricity. A wind farm usually has many turbines scattered over a large area. At the time of this writing, the Horse Hollow Wind Energy Center in Texas, with 422 wind turbines spread over about 47,000 acres, is one of the largest wind farms in the U.S., with a combined generating capacity of about 735 MW (or 735,000 kilowatts).[35] Figure 3-21 is a picture of a wind farm in West Texas.

Figure 3-21. Wind Farm in West Texas. Note the oil pump in the foreground and the electrical power feeding it.

Wind turbines create air turbulence, and companies design them with spacing to prevent turbulence from affecting adjacent turbine performance – typically six-to-eight rotor diameters apart. Thus, if a wind turbine has a rotor diameter of sixty meters, the developer should place them around 400 meters apart.

Wind Farm Environmental Effects

Meteorological data in a recent study of the environmental effects of industrial size wind farms indicate they can significantly affect near-surface air temperatures due to the vertical mixing of air layers from turbulence caused by the rotors. The impacts of wind farms on local weather can be minimized by changing rotor design or placing wind farms in regions with high natural turbulence.[36]

Other Impacts of Wind Farms

Wind turbines have come under criticism for issues such as noise and visual impact. Their noise is constant (typically at the decibel level of a refrigerator) but becomes less of an issue as turbine heights increase. New wind turbines on the order of 100 to 150 meters tall also improve performance because wind speeds increase with elevation. Visual pollution remains a concern as wind turbines are visible from great distances and can affect the view for many miles.

Solar Photovoltaic Power Generation

Concentrating solar power (CSP) generation plants were included with heat engines because they produce steam to drive turbines. This section focuses on solar photovoltaic generation by what most people think of as *solar cells*. Solar photovoltaic (PV) generation comes in the form of large solar farms and smaller installations at homes and businesses.

Semiconductors

Photovoltaics directly convert light energy into electricity at the atomic level. In other words, the sun energy directly affects individual atoms to generate electricity. Photovoltaic materials, or cells, consist of semiconductors. The properties of semiconductors allow them to conduct electricity better than an insulator but not as well as a conductor – hence the name semiconductor.

PV Basics

When the sunlight strikes the semiconductor, the cell absorbs some of the sunlight causing electrons to break away from their parent atom and move freely to create an electric current. Figure 3-22 illustrates the operation of a basic photovoltaic cell, also called a solar cell.

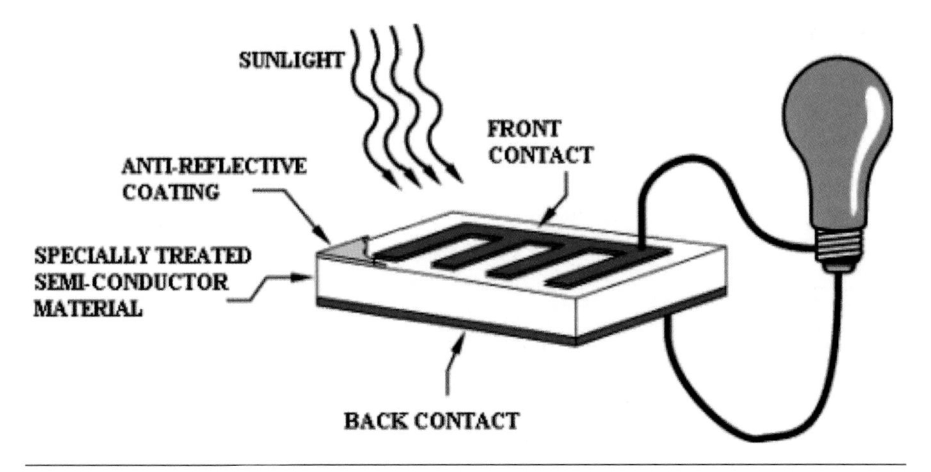

Figure 3-22. Operation of a PV Cell.

PV Extra Credit

This section is for extra credit – feel free to skip it. Typically, designers use silicon as the base semiconductor material in PV cells. The base material converts into two types of semiconductors: *p-type* and *n-type*.

The conversion happens because silicon has four valence electrons, that is, electrons that can form a chemical bond with other materials. Designers add boron or gallium, with only three valence electrons, to the silicon, and the materials combine chemically. Chemists say when the silicon with four valence electrons and the boron or gallium with three electrons bond together, they create a material with a *positive hole* or *p-type* semiconductor that wants to accept another electron.

Conversely, adding small amounts of elements with five valence electrons, such as arsenic or phosphorus, to the silicon creates an *n-type* semiconductor with an extra electron. When the solar cell absorbs sunlight, the two different types of semiconductors allow electrons to move through the *holes* and generate an electrical current.

Direct Current

Electrons flow in only one direction in solar cells, generating direct current (DC). Because the electric grid uses alternating current (AC), an *inverter* must change the DC generated by solar cells to AC. The inverter flips electric switches to alternate the incoming DC back and forth, creating AC.

Inverters contain inductors and capacitors to take the square DC wave and transform it into the smoothly alternating sine wave produced by electromagnetic generators.

Solar Farms

Solar farms consist of many solar panels connected to generate DC, convert it to AC, and feed it into the grid.

Figure 3-23. Solar Farm Near Nashville, TN.

Residential Solar

Homeowners are turning to rooftop solar panels for some of their own power. Solar panels are roughly five feet long and three feet wide (with small variations by manufacturer) and weigh about forty pounds. Figure 3-24 shows a rooftop solar installation.

Figure 3-24. Residential Rooftop Solar Panel Installation.

Environmental Issues

While solar panels do not cause atmospheric or other pollution during operation, some of the materials used to manufacture solar panels, such as boron and arsenic, are toxic. In addition, hazardous materials are used to clean and purify the semiconductor surface, and the manufacturing process can produce silicon dust. Those factors must be considered and managed properly.

Geothermal Power Plants

The phrase "geothermal energy" comes from the Greek words for earth (geo) and hot (thermos), referring to heat created by radioactive decay and frictional forces of the earth. Hot, or even molten, rocks encountering water can form reservoirs of hot water or steam. People have used that water (in the form of hot springs) for centuries for warmth, bathing, cooking and, perhaps, even healing.

The waters or steam from geothermal reservoirs can be extracted to power steam turbines to generate electricity in a similar manner to the heat engines already discussed in this chapter.

Four methods generate electricity from geothermal energy:

- Dry steam
- Flash plant
- Binary plant
- Enhanced geothermal systems

Dry Steam

This process pipes steam from below the earth's surface and uses it in a steam turbine to generate electricity. The user then condenses the steam to water and injects it back into the earth.

Flash Plant

This process takes hot water or steam from a reservoir and vaporizes (*flashes*) it from high pressure to low pressure, driving the steam turbine, and then returns the condensed water to the reservoir.

Binary Plant

In this process, water is not hot enough to produce steam, so the user adds a secondary fluid that boils at a lower temperature than water. Some examples include propane, isobutane, and an ammonia/water mixture. Because the secondary fluid does not come in direct contact with the steam or water, the user isolates corrosive brines or gases from the plant. A major advantage of this system is, even if the reservoir temperature is too low to flash water, it can still generate electricity with the secondary, or binary, fluid.

Enhanced Geothermal System

Enhanced Geothermal System (EGS) technology does not use hot water or steam near the surface of the earth but uses deep drilling to exploit hot temperatures. At depths of typically three-to-five kilometers, rocks may not be permeable or allow water flow, so EGS fractures the rock by injecting water and a small amount of chemicals. After heating the injected water, EGS generates electricity using either a flash- or binary-type technology.

Figure 3-25. Enhanced Geothermal Plant.

Needless to say, because of the drilling depths and required infrastructure, this type of plant has a higher capital cost than the other three.

Emerging Renewable Ocean Technology

Several emerging technologies harness energy from the ocean. The three most promising technologies are:

- Ocean thermal energy conversion (OTEC)
- Tidal energy
- Wave motion energy

Ocean Thermal Energy Conversion (OTEC)

This technology uses the temperature difference between the ocean's surface and deeper water. Deep water, typically defined as 1,000 meters, has a temperature around 5°C. At the surface, temperatures average 25°C. This 20°C temperature difference drives a turbine and generator. OTEC uses two approaches: open and closed.

The open system resembles a geothermal flash system. Injecting seawater from the surface into a vessel at low pressure causes it to vaporize. This vapor (low-temperature steam) goes to a turbine to power a generator. Cool seawater extracted from the deep condenses the steam before it gets returned to the ocean. Because vaporization leaves behind the salt, the condensation can also produce desalinated water.

The closed system looks like a geothermal binary system because it flashes a working fluid with a lower boiling point than water, such as ammonia, to drive the turbine. The working fluid condenses and goes through the cycle again. The 20°C temperature difference between surface water and deep water means the efficiency of this process hovers around 7%.

Tidal Energy

The gravitational force of the moon causes tides to cycle about every twelve hours. As tides change, water flows towards the shore and then away from the shore. This moving water turns a turbine attached to a generator. Tidal turbines work best in shallow water where tidal flow occurs. Because water is much denser than air, tidal water turbines capture more energy than wind turbines.

Another tidal system called a *tidal barrage* can capture tidal energy by using a dam-like structure to capture water during high tide. During low tide, it releases water through the penstock to turn the turbine to generate electricity.

Wave Energy

Wave motion also provides energy from the ocean. This text covers several ways to capture wave energy.

Floating Buoy: A floating buoy anchored to the bottom of the sea rises and falls with the waves. That up-and-down motion turns a shaft and moves a generator to make electricity.

Surface Floats: A device with multiple arms anchored to the ocean floor floats on the ocean surface. The arms flex with the wave motion and a hydraulic pump powers a generator to produce electricity.

Oscillating Water Columns (OWC): A partially submerged structure in the ocean allows incoming waves to enter at the bottom, and the rising water column pressurizes air in the structure's top. As the water recedes, the top part depressurizes and the pressure change pushes and pulls air through a turbine connected to a generator. A 500kW OWC has operated in Islay, Scotland, since 2000.

Internal Combustion Engines

Internal combustion engines (ICEs), like those powering cars, can also drive a generator. ICEs typically use diesel or natural gas as fuel, and the engine shaft connects to a generator. The ICE typically has a smaller engine size and a shorter startup time than other generation approaches and is, therefore, suitable for small power applications. It is also more portable than other forms of generation. One drawback is the ICE can use only a limited number of fuels.

ICEs do not normally support large-scale power generation but, rather, serve as backup, emergency power or in remote areas not easily connected to the grid. According to the EIA,[37] the ICE is gaining use for larger-scale power generation, especially where a substantial amount of the electricity generation comes from an intermittent source like wind or solar.

Specifically, when the wind does not blow or the sun does not shine, people must have either another source for electricity or suffer a power loss (blackout). In those instances, the user can rely on a quick start ICE that can also operate at partial capacity. In November 2018, the average capacity for an ICE generator was 4 MW compared to 56 MW for natural gas combustion turbines and 166 for NGCC. Some ICE units reach 20 MW and, if clustered together at the same facility, can deliver several hundred MW of power.

Fuel Cells

Fuel cells convert chemical energy from a fuel into electricity via a chemical process. Hydrogen serves as one of the most widely used fuel cell fuels but other fuels, including natural gas, methanol, formic acid, or borohydride can work, depending on the cell design. Fuel cells, like batteries, have three parts, including the positively charged cathode, the negatively charged anode, and an electrolyte. One of fuel cells' big advantages is producing electrical energy and water, both of which have value.

Fuel Cell Extra Credit

Feel free to skip this extra credit section. At the anode of a hydrogen-powered fuel cell, a molecule of hydrogen containing two atoms splits into two H^+ ions and two electrons. The electrons create electrical current, and the protons pass through the electrolyte to the cathode where they react with oxygen to produce water.

The electrolyte chemical allows the protons (but not electrons) to pass through. Several types of fuel cells exist, including polymer electrolyte membrane (PEM), alkaline, phosphoric acid, molten carbonate, and solid oxide fuel cells (SOFCs). One can generate electricity using methanol as a fuel (instead of hydrogen) in a direct methanol fuel cell (DMFC).

Fuel Cell Uses

Fuel cell power plants are typically small and used for backup power. One type of emerging vehicle also uses them – the fuel cell vehicle (FCV), which operates using hydrogen as the fuel. FCVs tend to use the PEM type of fuel cell.

Fuel Cell Challenges

Finally, nature does not provide hydrogen in pure gaseous form so it must be manufactured. Three common routes to produce hydrogen are natural gas reforming, coal or biomass gasification, and water electrolysis. Producing hydrogen consumes energy. Fuel cells also require a catalyst to make the chemical reaction occur, and the best catalyst is platinum, a substance almost as expensive as gold. Work continues, to make fuel cells more efficient and less expensive to produce.

Thermoelectric Generators

As the name suggests, thermoelectric refers to converting heat to electricity, a phenomenon first discovered in 1821 by Thomas Seebeck, who discovered that, when he heated one side of a piece of metal, electrons flowed from the hotter side to the cooler side, creating electrical energy.

As it turns out, good materials for thermoelectric generation, or TEG, are p-type and n-type semiconductors, discussed earlier in the section for solar energy. For the n-type semiconductor, electrons move from the hot to the cold side while, for a p-type semiconductor, positive charges move from the cold to the hot side. When joined at their ends, the buildup of charges in these two semiconductors results in current transmission directly proportional to the temperature difference.

The thermal efficiency for a TEG heat engine is quite low – typically less than 10% depending on the temperature difference. This low efficiency makes generating electricity very impractical. TEG can, however, convert waste heat to electricity, so a power plant with exhaust gases as wasted heat can use TEG to recover some of the waste heat and convert it to electricity.

Some companies are considering TEGs for vehicles. ICEs, with thermal efficiencies of around only 25%, waste a lot of heat. Recovering that heat and converting it into electricity improves efficiency, especially for hybrid vehicles powered by an ICE and electric motor.

Generation Capacity Factor

The United States EIA defines capacity factor as follows: the ratio of the electrical energy produced by a generating unit, for the period of time considered, to the electrical energy that could have been produced at continuous, full-power operation during the same period.[38] For example, a power plant having a nameplate capacity of 1,000 MW would generate 8,760,000 MWh in one year if it operated at full output all the time. If, during a year, that unit produces 7,556,000 MWh during a year, its operating factor for that year is 7,556,000/8,760,000 = 85%.

Operating factors can be deceiving as they compare unit nameplate capacity established by the unit manufacturer to actual production during a period of time. The operating factor for coal generation has steadily declined over the past ten years, not because they are capable of producing less electricity, but because they are only called upon as the generator of last resort, because they are fired by coal.

In general, generation types with high-capacity factors are considered more dependable than those with lower-capacity factors. Figure 3-26 shows actual (not nameplate) capacity factors by generation type.

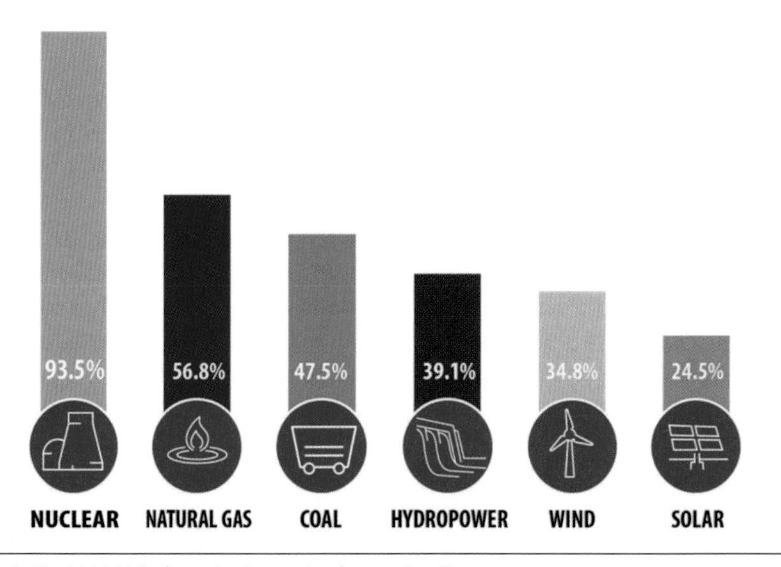

Figure 3-26. 2019 U.S. Capacity Factor by Generation Type.

Fossil Fuels and Nuclear Capacity Factors

Nuclear, natural gas, and coal capacity factors exceed those of wind and solar because their fuel sources do not depend on weather. Their capacity factors are not 100% due to:

- equipment failure;
- maintenance;
- lack of market; and
- low prices.

Fossil Fuel and Nuclear Actual vs. Potential Capacity Factors

The numbers on Figure 3-26 show the percentage of time each generation type *operated.* The numbers do not indicate the amount of time each generator was *available to operate,* a number normally higher than the capacity factor.

Coal Plants: Pulverized coal plants have available capacity factors of 85% and higher. Integrated gasification combined-cycle (IGCC) plants have available capacity factors somewhat lower (75-85%) because they have an extra turbine and an air separation unit that likely require more maintenance and downtime. Comparing the actual coal plant capacity factor in Figure 3-26 to the much higher potentially achievable factor (80%) demonstrates that, at many times, coal plants do not operate, for various cost, political, and environmental reasons.

Nuclear Plants: Nuclear plants have available capacity factors of about 90%, which compares favorably to their actual capacity factor.

Natural Gas Combined-Cycle Plants: The *available* capacity factor for NGCC units ranges from 80-90% but, because they operate primarily as "peaking" units (i.e., operating only during peak electrical demand), their *actual* capacity decreases due to lack of electricity need.

Hydro, Wind, and Solar Capacity Factors

The main reason for the difference between the available and actual capacity factors for hydro, wind, and solar is variation of energy input.

Hydroelectric

For hydroelectric, the average capacity factor for all U.S. plants varied from 35% to 43% from 2013 to 2018[39] due to variations in hydrologic conditions, regulatory restrictions, plant outages, and plant-to-plant variability. Drought conditions in the western U.S. caused the lowest value over that time.

Wind

Wind speed greatly affects the capacity factor for wind turbines and varies by geographical location, season, and even turbine height, as wind speed increases with altitude. Capacity factors for wind turbines vary from 25% to 50%, with higher values in windy areas like the Texas and Oklahoma panhandles. Winds typically increase offshore causing those wind turbines to have higher-capacity factors than onshore turbines.

Solar

Concentrating solar power (CSP) needs direct sunlight to generate electricity. As discussed earlier, the parabolic trough system uses curved mirrors in direct sunlight. The capacity factor for these systems can be improved by allowing the trough to rotate during the day to follow the sun moving east to west.

In addition, those systems can use thermal energy storage, or TES, to store thermal energy for use when the mirrors do not generate electricity. CSP systems have capacity factors ranging from 20% to 50%, with the high end representing systems with TES.

Solar PV can generate electricity (but not as much) during cloudy periods because it can still absorb the photons in the solar energy. Of course, sunlight varies during the day affecting the amount of electricity generated each hour. Capacity factors for solar PV range from 10% to 30% due to the large variable of latitude. Solar energy capacity factors increase closer to the equator.

Geothermal

Geothermal power plants have capacity factors from 70% to 80%, higher than other renewables because of the continuous flow rate of the water or steam. However, as the geothermal field grows older, the temperature and flow rate of the steam or water typically decline.

Other Technologies

For the different emerging renewable ocean technologies discussed, OTEC units tend to have high-capacity factors (75% to 95%) because the temperature difference between the surface of the ocean and deeper water does not vary much. Tidal energy has lower-capacity factors (20% to 40%) because tidal currents vary with time of day. Likewise, wave power capacity factors vary from 10% to 40% because wave energy varies with wind speed, wind duration, and fetch, the distance wind blows without encountering an obstruction.

Capacity Factor Challenges for Wind and Solar

Capacity factor is one of the key challenges for wind and solar. If they operate one-third and one-fourth of the time, as shown in Figure 3-26, companies must make huge investments in storage to provide electricity on calm days and at night.

Managing Emissions from Plants

Managing emissions creates a disadvantage for natural gas and coal plants due to increased capital and operating costs, although natural gas is much cleaner than coal.

Methane Emissions

A small percentage of methane either escapes accidently or is released intentionally during production and during transportation to the power plant. Methane, a greenhouse gas, has an estimated global warming potential twenty-eight to thirty-six times that of carbon dioxide (CO_2). Natural gas producers and transporters are working hard to limit and minimize both accidental and operationally-caused releases.

Managing Solids

Coal has several issues including coal ash, heavy metals, nitrogen oxides, and sulfur oxides. The coal combustion process leaves coal ash as a waste material that can contain heavy metals, such as arsenic, mercury, and cadmium. Companies capture and store some waste in landfills and ponds and use some to make concrete. If the ash is very small, it is captured and removed using an electrostatic precipitator (ESP).

Removing Nitrogen Oxides and Sulfur

Utilities must also remove nitrogen oxides from flue gas. At ground level, nitrogen oxide can react with sunlight and oxygen to form ground level ozone, one of the main components of smog. Both nitrogen and sulfur oxides can react with water in the atmosphere to form nitrous, nitric, sulfurous, and sulfuric acids – major components of acid rain. Coal power plants remove nitrogen oxides using *scrubbers*. Many processes can capture sulfur oxides. A wet sulfuric acid (WSA) process that produces sulfuric acid is one of the most popular processes.

Coal IGCC units and natural gas NGCC units using oxygen rather than air for combustion have an advantage when it comes to nitrogen oxide. Because nitrogen comes from the air and the plants do not use air for combustion, they produce no nitrogen oxides.

Carbon Capture and Sequestration

Carbon dioxide emissions are a growing concern, so companies capture and store (sequester) the produced CO_2. This process is, not surprisingly, called carbon capture and sequestration (CCS). CCS happens either post- or precombustion.

Postcombustion Capture: If the natural gas or coal power plant uses air for combustion, the CO_2 is extracted from the flue gas after combustion using a chemical process.

Precombustion Capture: IGCC plants gasify coal and other carbon-based fuels into synthesis gas (syngas) for combustion. The plants remove CO_2 and other pollutants from the syngas prior to generating power, resulting in lower emissions of sulfur dioxide, particulates, mercury and, in some cases, CO_2. With additional process equipment, a water-gas shift reaction can increase gasification efficiency and reduce carbon monoxide emissions by converting it to CO_2.

Sequestration: This process captures, compresses, and transports the CO_2 by pipeline to a geologic formation (like a depleted oil and gas reservoir) or other location to store it long term.

Summary

- Electricity does not occur naturally in commercial quantities and must be generated from other forms of energy.
- Currently, just over 60% of electricity generated in the world comes from converting chemical energy contained in fossil fuels (coal and natural gas) into electrical energy.
- Over 90% of the electricity used in the world today comes from electromagnetic generators.
- Electromagnetic generators consist of three main parts: rotor, stator, and exciter.
- Turbines convert kinetic energy in one of four flowing fluids – steam, gas, water, and wind – into rotational motion turning the generator shaft.
- "Heat engine" is a generic term for a device producing heat from a primary energy source, using part of the heat to create mechanical work, and discharging the remaining heat as a waste product.
- Combined-cycle plants use hot exhaust gas from the first turbine to produce steam and drive a second, team turbine.
- The capacity factor, expressed as a fraction or percent, is the ratio of a power plant generation over some time period compared to its rated capacity.
- In general, generation types with high-capacity factors are considered more reliable than those with lower-capacity factors.
- One of the key capacity challenges for wind and solar is their capacity factors.
- One disadvantage of natural gas and coal power plants is managing emissions, which adds capital and operating costs.

4

Electrical Transmission Systems

I believe that the science of chemistry alone almost proves the existence of an intelligent creator.

—Thomas A. Edison (1887–1931)

"Grandpa Tom, you told me electricity is made at generating facilities, but how does it get from there to my house?" asked Luke.

"See those poles along the street?" replied Grandpa Tom. "They are part of a local distribution system carrying electricity to our houses and other buildings from facilities called "substations," the grandfather explained. Then he added, "Substations receive electricity from transmission lines made up of big wires, called conductors, tall towers, and a lot of other equipment used to control everything."

"Do those wires bring the electricity directly to our city from where electricity is made?" Luke asked.

"Sometimes," Grandpa Tom replied, "but usually they connect to an entire network of transmission lines and substations, interconnected to ensure we can get the electricity when we want it."

"Wow, seems complicated," said Luke. "But I am glad they're there so I can use my computer to study and have my night light," he added.

Introduction

Most electricity is not generated where it is used, but at locations several to many miles away. This chapter discusses *transmission systems.* They transmit (move) electrical energy from its point-of-generation to distribution systems, as well as directly to large energy users like factories. Individual transmission lines do not always transmit electrical energy from generation to consumption points. Sometimes they interconnect with other transmission lines, forming a vast grid consisting of substations, towers, poles, wires, and a myriad of other components. The next chapter extends the topic of transmission to distribution, as it explains how electric energy

> **Energy vs. Power**
>
> Technically, transmission systems move electrical energy and not electrical power. Power is an expression of the rate (per unit time) at which electrical energy is transferred by an electrical circuit.
>
> Nontechnically, the two terms are often conflated and used interchangeably along with the term *electricity*.

is delivered from transmission systems to distribution grids and from there to consumers.

A transmission system analogy is the U.S. interstate highway system or German autobahn. Vehicles enter and leave highways at multiple points. As they travel along, vehicles leave one highway and enter another. The same thing happens with transmission systems. Distribution analogies include local streets and roads, municipal water systems and the natural gas local distribution grid.

Transmission systems move electrical energy between point A and point B as safely, reliably, and efficiently as possible, with due regard for the environment and the stakeholders along their routes. Most of the concepts, equipment, components, and challenges are the same or similar for transmission and distribution systems. The primary difference is voltage and size.

Electrical Transmission Systems

Electrical transmission systems consist of five parts:

- Transmission lines
- Substations
- Communication systems
- Control centers
- Protection and control systems, including "system control and data acquisition" (SCADA)

This chapter covers transmission lines, substations, and some parts of the control systems. Control centers, control systems, communication systems, and SCADA each have chapters dedicated to them. System protection is covered in various forms in this chapter, the distribution chapter, and the wide area controls chapter, owing to its importance and integrated nature.

Transmission Lines

Transmission lines start near generation facilities, typically at a generator step-up transformer, or at a high-voltage substation. The lines then extend to other high-voltage substations until nearing a distribution system (or end-use customer), at which time the voltage gets decreased at a step-down transformer. In other words, transmission lines transmit electricity between substations.

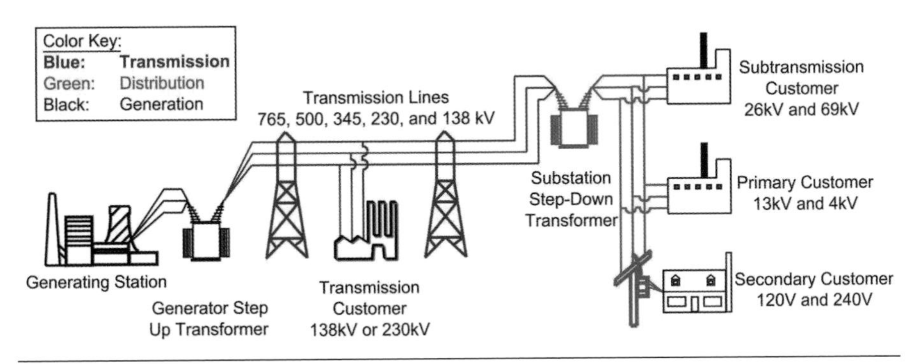

Figure 4-1. Transmission Line Schematic.

Transmission Voltage

Electricity can be generated across a variety of voltages ranging all the way from rooftop solar panels at 120V to utility-scale generation plants of up to around 35kV. Transmission voltages vary as well – from a little over 100kV up to as high as 1,100kV on the direct-current Changji-to-Guquan line, owned and operated by the State Grid Corp. of China.

The Power of High Voltage

Transmission lines transmit electrical energy at high voltages because the higher the voltage, the less power lost during transmission for the same amount of power transmitted.

Power into the System

The amount of power put into a transmission line at point A is called *apparent power* and depends on voltage and current:

$$\text{Power in} = \text{Voltage in} \times \text{Current in } (V \times I)$$

When voltage increases by a factor of two (doubles), the current can decrease by a factor of two (cut to half) and still move the same amount of power. Mathematicians say the amount of electrical energy carried by a circuit varies linearly with both voltage and current.

Power Lost during Transmission

Power lost as electricity moves from point A to point B relates to current and impedance:

$$\text{Power lost} = \text{Current squared} \times \text{Impedance } (I^2 \times R)$$

When the current reduces by a factor of two (cut to half), the power lost reduces by a factor of four (cut to one quarter). The relationship between power lost and current explains why transformers increase the voltage and reduce the current before electrical power enters the transmission line.

Mathematicians say the amount of electrical energy lost in a circuit varies exponentially with current and linearly with impedance. In other words, increasing the voltage by a factor of two reduces the current by that same factor. With the current reduced by a factor of two, the electrical energy lost during transmission is reduced by a factor of four.

Without doing the math, imagine how much more electrical energy would be lost if it were moved from the generation point to the customer at the standard household voltage of 120 volts versus transmission voltage of 765 kV, a factor of nearly 7,000 times.

Power out of the System

The electrical energy consumed by the load or loads connected to the circuit is called *real power, abbreviated P*, or *average power.* Impedance converts electrical energy into heat that dissipates into the surrounding environment. Because the electrical energy out is less than the electrical energy in, so are the voltage and current. Substations connecting one transmission line to another may contain transformers that step up the voltage back to the level needed by the connecting line.

Power Extra Credit

Impedance is a combination of resistance and reactance, where reactance is related to the amount of *reactive power* carried on the line. Reactive power (introduced in Chapter 1) is generated both on the transmission line itself (because the voltage and current alternate) and by reactive loads connected to the grid. Reactive power transmitting back towards the point of generation also experiences resistance manifested as additional heat loss.

Reactive Power

Reactive power itself is not lost, but its existence limits the amount of real power delivered from point A to point B. Figure 4-2 shows power, reduced by actual losses to conductor resistance (heat), and apparent power limited by reactive power.

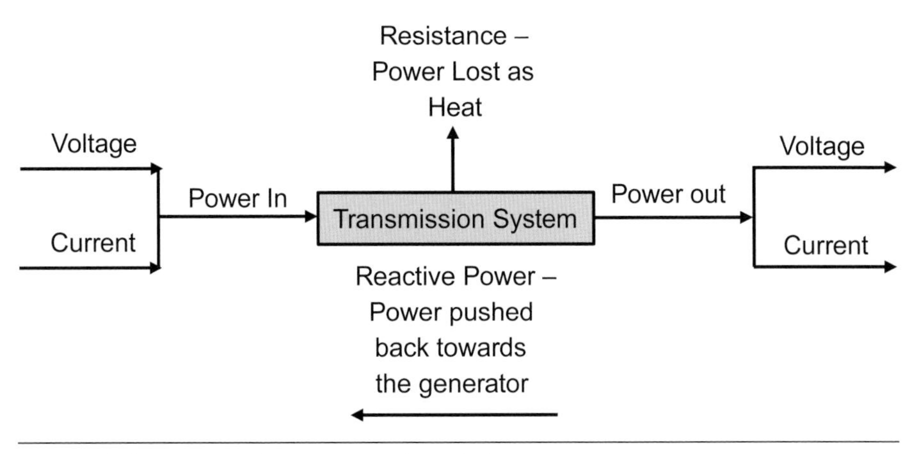

Figure 4-2. Transmission System Power Transmission.

Real Power

Calculating real power is not a simple equation. It involves what electrical engineers call the *power triangle*. Calculating real power requires understanding how much the voltage and current are out of phase with each other.

Power Factor

Power factors, which range from zero at the worst, to 1.0 at the best, are the ratio of real power delivered to the load divided by the apparent power of the circuit.

Volt-Ampere Reactive (VARs)

Reactive power is measured in terms of volt-ampere reactive, abbreviated VAR, and calculated the same as real power. VAR is a bit of Dr. Jekyll and Mr. Hyde; it resists transmission of real power, but is needed for grid stability.

Transmission grid operators work with generators and distribution companies receiving power to manage voltage, current, frequency and reactive power. More on VARs in Chapter 5.

Faults

The primary cause of transmission line outages are faults caused by equipment failures (transformer and rotating machine), human errors, and environmental conditions (wind and trees). When abnormal conditions occur and the voltage and current deviate outside normal operating ranges, system protection devices respond (open) and take the line out of service.

Line-to-Ground Fault

A line-to-ground fault happens when one or more of the phases contacts a low impedance path to ground. One example is a tree limb contacting the power line. Electricity is transmitted from the line through the tree to the ground, draining electricity from the line.

Line-to-Line Fault

As implied by the name, a line-to-line fault happens when two conductors from different phases touch each other. Both line-to-ground and line-to-line faults can be unsymmetrical or symmetrical.

Unsymmetrical Faults

These faults do not involve all three phases at the same time. Instead, one phase comes in contact with the ground, two phases come in contact with each other, or two phases come in contact with each other and the ground. Single line-to-ground and single line-to-line faults are the most common.

Symmetrical Faults

When all three phases transmit to ground or all three phases transmit to each other, the fault is classified as symmetrical.

Fault Protection and Location

Equipment installed along the transmission route and inside substations detects fault conditions and disconnects the faulted segment of the system from the remainder of the system. Some faults clear on their own. For example, a tree limb might fall on two phases causing a line-to-line fault, but then the wind blows the limb off and the fault clears. Others do not clear on their own and require repair by field personnel.

When uncleared faults occur, they must be located and repaired. Technicians may know what section of the line the fault is on, for example between substation X and substation Y, but they may not know the *exact* location and must locate the fault before they can repair it.

Transmission Line Components

The major components of transmission lines include :

- Towers, poles, and easements
- Conductors
- Insulators
- Shield and ground wires

Towers and Poles

Support structures carry the weight of the conductors, insulators, and other equipment fastened to them – called the dead load – and the live loads including wind and snow. Transmission designers in the U.S. use the National Electrical Safety Code (NESC) and codes developed by the American National Standards Institute (ANSI) and the Institute of Electrical and Electronic Engineers (IEEE) as guides for designing transmission lines. Other countries follow their own standards but, because the laws of physics apply universally, standards around the globe are similar, if not the same.

Height and Width: One of the guidance items is safe clearance (distance) between conductors. Another is distance between conductors and other objects, including the structures supporting the lines. Safe clearance standards directly affect support structure height and width. Generally, higher-voltage systems require greater distances than lower-voltage systems between conductors and between the conductors and structures on the ground.

> **Standards and Regulations**
>
> Industry develops standards, and regulators often turn standards into law by incorporating them into regulations.

Figure 4-3 shows a transmission line road crossing. Note the height of the supports and how the height decreases after the transmission line crosses the highway.

Figure 4-3. Transmission Line Crossing a Highway.

The next section discusses some of the most common support structures.

Lattice Towers: These structures, typically made of steel or aluminum and assembled on-site, are the largest tower type and carry the highest voltage. In remote or hard-to-access locations, helicopters carry tower sections to the site where workers bolt them together. Figure 4-4 shows several lattice towers carrying electricity generated at the plant in the background. Transmission towers normally support conductors in multiples of three (one for each phase). Additionally, one or more wires at the top of the structure provide lightning protection (shielding). Visible on the left tower in the foreground are six bundles of two conductors each supported by insulators arranged in a V configuration. Less visible in Figure 4-4 are two shield lines at the top. Bundling conductors (two or more conductors per phase) is common practice and increases the total amount of electricity transmitted.

Figure 4-4. Transmission Line with Lattice Towers. The generation plant is visible in the background.

Monopoles: Monopoles, as the name implies, are single poles. The chief advantage of monopoles is a smaller footprint than lattice towers. They may also cost less to install. Figure 4-5 shows a monopole transmission line under construction and a nearby lower-voltage line of a different, two-pole design.

Wood Poles: Wood poles sometimes support transmission lines and typically cost less to install but are limited in height, conductor size, and life span.

Figure 4-5. Monopole Line Under Construction. A two-pole design for a lower-voltage line appears next to the monopole line.

Support Considerations: Changing conductor direction puts more force on the towers and poles than just holding the conductors up and may require different pole designs and guy wires or other supports to ensure the support structure safely carries the weight and withstands the other forces.

Figure 4-6. Support Structure at Direction Change. Monopoles provide support where the direction changes, and lattice towers provide support on either side of the direction change.

Support structures come in different designs, and one veteran power transmission professional observed there are about as many different tower and pole designs and combinations as transmission system designers.

Easements (Rights-of-Way)

Transmission systems often exist on property owned by others, placing them in the *linear asset* category with pipelines, railroads, waterways, and other transportation modes. Linear assets come with an assortment of issues and challenges, as transmission companies seek to balance the need to provide transmission services with the needs of landowners.

> **Easement vs. ROW**
>
> A right-of-way (ROW) is a specific type of easement giving the holder the right to travel across the landowner's property.

Typically, transmission companies do not own the land the transmission lines traverse; rather, they purchase an easement from the landowner prior to constructing the line. Easements are legal rights granting the holder permission to use land for a specific purpose without transferring ownership.

Right-of-way (ROW) widths vary depending on the type and voltage of system on it. They are normally between 50 and 150 feet wide but can be wider when multiple lines are needed or for high-voltage lines. Transmission companies keep ROWs clear of trees, other vegetation, and structures that could pose a public safety risk or come in contact with the lines.

Figure 4-7. Transmission Line ROW.

Procuring Rights-of-Way.

No standard method of acquiring ROWs from landowners currently exists. Rather, transmission companies purchase them on a case-by-case basis using a variety of approaches. A number of state and local regulations apply to transmission lines, and transmission developers, at a minimum, follow prescribed regulations when negotiating with landowners.

Typical Provisions

Easements typically contain provisions including:

- ROW length and width
- Number of structures
- Structure height and design
- Wire height
- ROW clearing and construction practices
- Project schedule
- Postconstruction maintenance and ROW access
- Use of herbicide chemicals to control weeds and brush
- What the owner can do on the ROW

Some landowners ask for unique or special considerations. As an example, when rural electrification came to the U.S., a power company representative approached the grandfather of one of the authors about a ROW across his property. He reportedly told the power company representative he would agree to the ROW only if the power company agreed to connect his house to the new line.

Compensation

Saying transmission companies and landowners disagree regarding ROW values is an understatement. Typically, the utility determines the Fair Market Value (FMV) of the land and/or easement using appraisals. Ultimately, the parties work out the value in a negotiation process. In addition to compensation for the ROW, transmission companies often pay for crop damage or physical damage to property resulting from constructing and maintaining the transmission line. Various state and federal laws and regulations prescribe ROW purchase and management.

Eminent Domain

Sometimes, landowners do not want linear assets crossing their property even if they receive payment. They need electricity, but feel others benefit significantly more than they benefit from the transmission line. Sometimes landowners simply refuse to provide an easement at any price. When the parties cannot agree on the ROW's value, governments give the power company the right of eminent domain.

Many years ago, the judicial system developed the concept of eminent domain, which gives governments (or their agents) the right to take private property for public use. However, the Fifth Amendment to the U.S. Constitution, applied to the states through the Fourteenth Amendment, requires "just compensation" for a taking of private property for public use. Thus, eminent domain involves filing a lawsuit against the landowner to get a court-approved ROW. During the lawsuit, the transmission company provides the court evidence the public interest outweighs the landowner's rights. Cases decided in favor of the transmission company also establish the ROW's value.

Conductors

Most people call them *wires*. Transmission line designers choose the conductor materials that move the most electrical power at the lowest overall system cost, while meeting safety, reliability, and environmental requirements. Most often, they use aluminum (often supplemented with steel for strength) for transmission lines.

Voltage Rating

Safe voltage for aerial (versus underground) lines depends on physical separation (spacing) between conductors, structures, other components, and equipment, and between the system and the public.

Ampacity Rating

The amount of current a conductor moves (ampacity) depends on the resistance of its material and its cross-sectional area. Ampacity rating is the maximum current the conductor can transmit without exceeding conductor temperature ratings. Temperature and other environmental factors increase or decrease conductor ampacity rating and, consequently, the line rating.

Temperature Rating

As discussed elsewhere in this book, resistance opposes electricity transmission and generates heat. Too much heat melts conductors or makes them brittle. Industry standards specify maximum temperature ratings designers use when selecting conductor materials and sizes.

Conductor Materials

Copper is an ideal conductor but has a low strength-to-weight ratio, meaning poles must be placed closer together than if a stronger material (such as steel) were used. Aluminum has relatively good conducting properties and lighter weight, making it the most common choice for transmission line conductors. For added conductor strength, a steel core is typically used in conjunction with aluminum.

Conductor Types

Conductors are either solid or, more commonly, stranded, meaning several smaller solid conductors are combined, forming a bundle. The bundle is twisted during manufacturing to keep the strands together.

Hybrid Configurations

Transmission companies now use a combination of materials to improve the strength-to-conductivity ratio. One example consists of a multi-stranded, carbon-fiber composite core providing greater flexibility and sag/tension performance, and an expected service life of more than forty years.

Maximum Power

The maximum power any conductor can carry is its ampacity rating times its voltage – an important consideration when planning transmission systems.

Conductor Sag

As a conductor's temperature changes, its length expands or contracts. A conductor operating at maximum ampacity on a hot afternoon sags more between poles than the same conductor with no current on a cold summer night – another important consideration for transmission planners.

Insulators

Transmission lines operate at high voltages, but the structures supporting the conductors are at the same electrical potential as the ground (zero). Difference in electrical potential (voltage) transmits electricity, causing it to seek routes for leaving the conductors and arcing to the support structures. Insulators constructed of highly resistive materials support the conductors and prevent electrical transmission to the structures. In addition to preventing transmission to the structures, insulators also maintain physical distance between the conductors and structures, preventing an electrical arc between the two.

Electrical Arcs

Lighting provides an example of electrical arcs. The same thing can happen if a tree limb or other object gets too close to an electrical conductor.

Insulator Materials

The atomic structure of insulating materials means electrons remain with the individual atoms rather than swarming like the outer electrons of conductors. Electrons interacting with each other cause electrical transmission. When the electrons do not interact, they cannot transfer electrical energy.

For many years, porcelain served as the preferred insulating material for transmission and distribution lines. Owing to recent advances in material science, polymers have replaced porcelain as the material of choice. Silicone rubber is currently the most widely used insulator material for new high-voltage transmission lines, but many porcelain insulators remain on existing transmission lines. Figure 4-8 shows porcelain insulators, and Figure 4-9 shows polymer insulators.

Figure 4-8. Porcelain Insulators Arranged in a String.

Figure 4-9. Polymer Insulators, Also Called Composite Insulators.

The insulators shown in Figures 4-8 and 4-9 look simple but are actually highly engineered with multiple connected parts.

Modular Insulators

Because voltage drives current, higher-voltage lines require more resistance and, therefore, more insulators between the conductors and towers than lower-voltage power lines. Multiple disc insulators connected to each other in series provide higher-insulation levels. Figure 4-8 shows two modular insulators with each constructed from a "string" of identical disc-shaped insulators attached with metal *clevis pins* or ball-and-socket links.

Modular insulators use more (or fewer) disc-shaped insulators or composite discs depending on the line's voltage. Figure 4-8 shows eight individual discs, meaning the system likely operates at 115kV. The composite insulators in Figure 4-9 are longer, meaning the line voltage is considerably higher.

Insulator Orientation

Transmission insulators normally orient as strain insulators or suspension insulators. The line in Figure 4-8 turns a corner, so the conductors connected to the two insulators pull in opposite directions and must handle the force – hence the label *strain insulators.* The two conductors terminating at each strain insulator connect to each other with a conductor loop.

Suspension insulators (Figure 4-9) suspend conductors and need not handle lateral forces. As with towers and poles, insulators come in a variety of configurations and get mounted in various arrangements to meet the project's needs and the designer's preference.

Shield and Ground Wires

Figure 4-4 showed two lines at the top of the lattice towers. These wires are called shield wires, and they protect the system from lightning strikes. They are attached to support structures via wires running down the poles to the ground. Any voltage surge from lightning travels through them to the ground, thereby preventing damage to the lines, structures, or equipment. One or more shield wires are typically installed above power lines so any lightning is more likely to strike them than the conductors.

Grounding

The earth has essentially an infinite ability to dissipate electrical energy. Electrical engineers say the earth provides a "return path" for the circuit. Grounding, connecting a wire to a rod pounded into the earth, is a safety precaution present in nearly all electrical systems. If there is a problem with the circuit, any excess electricity is dissipated by the grounding system.

Underground Transmission Lines

Sometimes transmission lines are installed underground and require extra care and planning because they do not use air as an insulator and, unlike above-ground lines, can suffer excavation damage.

Underground lines do not require insulators but require *insulation.* Common practice involves installing insulated cables through conduits or directly in the ground. Splicing cables together requires extra skill and training.

Figure 4-10 shows two monopoles on one side of an elevated freeway and two poles on the other side. Power travels to the monopoles on aerial conductors and then goes through insulated, buried cables under the freeway to the other side of the road where it is transmitted back onto aerial conductors.

Figure 4-10. Buried Freeway Crossing.

Transmission Substations

Substations sit at each end of transmission lines and at intersections and interchange points along the transmission path. Substations transform voltages; stop, start, and direct current; monitor and manage process variables; and, generally, enable safe and efficient operations. This text covers the major equipment and components in substations and discusses substation types. Figure 4-11 is a picture of a typical substation.

Figure 4-11. Typical Substation.

Substation Equipment and Components

Substations contain a variety of equipment and components including:

- Lightning arresters
- Disconnect switches
- Circuit breakers
- Reclosers
- Insulators and bushings
- Bus bars
- Transformers
- Inverters and converters
- Instrument transformer
- Intelligent electronic devices (IEDs)
- Communications systems
- Control house
- Ground and shielding system
- Protection and control systems
- Station batteries
- Extra credit – managing reactive power
    - Capacitors
    - Reactors
    - VAR compensators

Lightning Arresters

Substation components (and nearly everything else electrical) want nice, steady power and do not want sudden surges (like those caused by lightning strikes). Consequently, the first component electrical energy encounters on its journey through substations is usually lighting arresters (one for each phase). One end of the arrester connects to the line and the other connects to a conductor running to ground. The material between the two connections is essentially a resistor with enough resistance to prevent normal voltage from driving current through it.

Lightning arresters, despite their name, do not arrest lighting. Rather, they divert it to the ground where it dissipates. When voltage surges happen, the higher voltage across the resistor overcomes the resistance, and electricity transmits through the arrester's body to the grounding system, which safely dissipates it. When voltage drops back to normal, the resistance prevents any more electrical transmission.

Figure 4-12 shows three phases of incoming power that progressively go past lightning arresters, disconnect switches, and a circuit breaker. Also shown is the control house.

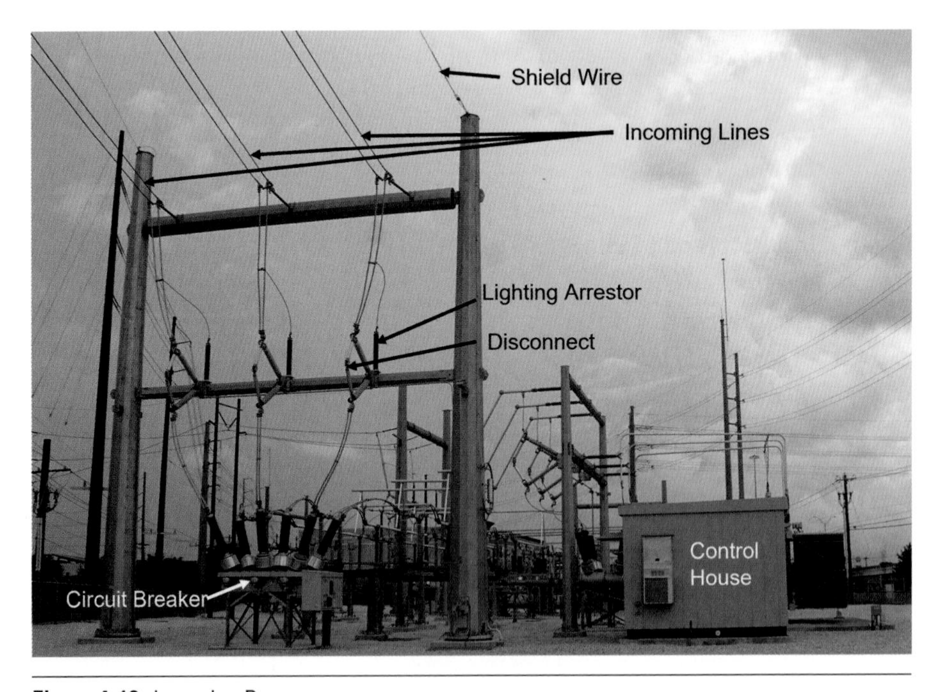

Figure 4-12. Incoming Power.

Disconnect Switches

Disconnect switches (often just called *disconnects*) are typically located immediately after lightning arresters on the station input side and immediately before lighting arresters on the station outgoing side, allowing isolation of the substation from connecting lines for operating and safety purposes. Disconnect switches are sometimes located before and after major pieces of station equipment to ensure they remain de-energized during equipment maintenance activities. Disconnects appear in Figure 4-12.

Disconnects come in a variety of designs and provide a similar function to light switches, with one important difference: they are normally not designed to operate *under load*. In normal operations, incoming and outgoing current is shut down prior to opening the disconnect switch. Some can be opened and closed only manually, while others have actuators, allowing remote operation. The open disconnect prevents electrical transmission to the station or piece of equipment. The most common disconnect is an air-gap switch, meaning the air around it provides the insulation from any arc that might form when it opens or closes.

Under Load

Under load is the same as energized, that is electricity is being transmitted. De-energized means electricity is not being transmitted.

Circuit Breakers

Another type of switch, called a circuit breaker, opens or closes under load just like a light switch. Because circuit breakers normally operate under load, opening and closing them creates an electrical arc. This arc could damage nearby equipment and conductors, so the area containing the arc should be protected (insulated) from surrounding equipment and conductors. The most common insulator – air – simply isolates the spark from surrounding materials just like with an air gap switch. As voltage and accompanying arcs increase, air no longer provides sufficient insulation and component designers turn to three other materials for insulation:

- Vacuum
- Gases
- Oil

Figure 4-12 shows a gas circuit breaker – so called because a gas serves as the insulating media.

Circuit breakers are intentionally operated (opened or closed) for operating purposes. They also serve as a protective device, automatically opening to protect the circuit in the event of electrical faults.

Reclosers

Sometimes short circuits occur for a short time but quickly resolve themselves. When that happens, a recloser – a type of circuit breaker – comes into play by opening for a short time and automatically reclosing to return electrical energy to the circuit (hence the name "recloser"). Reclosers can open and close several times during a short period of time, attempting to continue service. After opening and closing several times, they typically remain open, requiring closure from the control room or by an on-site technician. Electric utilities commonly use reclosers in the distribution system more than the transmission system.

Insulators

Figure 4-13 shows two types of insulators. The angled ones in the upper left of the figure are strain insulators, similar to those shown in Figure 4-8. The other type of insulator is called a bushing. Bushings are located on pieces of equipment.

Bushings

Bushings connect to a piece of equipment on one end and to incoming or outgoing conductors on the other end, providing a path for electricity to travel from the incoming or outgoing conductors to surrounding equipment without arcing. The main difference between insulators and bushings is that current is transmitted through conductors contained inside the bushing, whereas insulators support only an external conductor.

Figure 4-13 shows one set of six bushings directing electricity into and out of a circuit breaker.

Figure 4-13. Bushings, Transformer, and Busbars.

Busbars

Figure 4-13 shows busbars (or bus bars), which are conductors (most commonly aluminum) produced in a variety of shapes. Busbars are typically mounted well above the ground and connect to insulators mounted on supports. They serve as a handy place to connect incoming and outgoing lines and various pieces of equipment.

Transformers

Figure 4-13 also shows a transformer. As discussed in Chapter 1, transformers contain a primary core and a secondary core. Around each core are multiple windings of a conductor – one for the primary and one for the secondary core. The ratio of the number of times the conductor winds around the primary core to the number of times another conductor winds around the secondary core determines how much the transformer increases or decreases voltage. If the secondary core has twice the number of windings as the primary core, it discharges twice the voltage supplied to the primary, as shown in the following equation:

$$\text{Voltage out} = \text{Voltage in} \times \text{Primary Windings}/\text{Secondary Windings}$$

Transformers come in many different designs and sizes, and electrical designers select the best type for the equipment configuration. Of all pieces of equipment in a substation, transformers are generally the most expensive single component. Not all substations contain transformers.

Taps

Some transformers come with multiple secondary-core *taps*, or connection points. Changing the tap used for the outgoing conductor connection changes the number of windings the current travels through before leaving the transformer. Electrical engineers select the connection tap based on the voltage entering the transformer and the desired voltage from the transformer.

Cooling

The transformer process involves converting electrical power to magnetic fields and back to electrical power. That process generates heat and, thus, transformers need cooling systems. Cooling systems come in two primary forms: air-cooled and oil-cooled. Figure 4-13 showed an air-cooled transformer in a substation. The cooling fins and fans are visible on the left side of the transformer. Figure 4-14 shows an oil-cooled transformer. Note the oil tank on the top and the circuit breaker to the left of the transformer.

Figure 4-14. Oil-Cooled Transformer.

Inverters and Converters

> **Semiconductors**
>
> The electric industry has made great strides using semiconductor materials as inverters and converters, replacing older and less efficient equipment and techniques.

Most power is generated as AC, but the rising influence of photovoltaic (PV) solar generation, producing DC, and the use of high-voltage DC transmission lines makes including inverters and converters in this text essential – although very few transmission substations contain either at this writing.

Inverters: Inverters of varying design convert DC to AC. Fundamentally, inverters receive DC and, through a series of switches, convert DC to *square AC waves.* A series of devices then converts these square waves into smoothly oscillating waves matching the amplitude and frequency of AC.

Converters: These devices convert AC to DC. As with inverters, converters come in various designs. Most electronics use DC power and contain small converters. Because DC does not oscillate like AC, DC transmission systems do not generate reactive power, making them more efficient than AC systems. This increased transmission efficiency is driving development of long-distance DC lines, which, along with the rise in PV-generated DC, is driving converter (and inverter) technology and design.

Instrument Transformers

Instrument transformers are installed at various points in substations, often in conjunction with circuit breakers, to provide voltage and current measurements used to protect, control, monitor, and meter transmission lines and substation components. The most common types of instrument transformers are *current transformers* (CTs) and *voltage* or *potential transformers* (PTs).

Intelligent Electronic Devices (IEDs)

The National Institute of Standards and Technology (NIST) defines IED as "any device incorporating one or more processors with the capability to receive or send data/control from or to an external source (e.g., electronic multifunction meters, digital relays, controllers).[40] More about IEDs in the chapter dealing with wide area control.

Communication Systems

Communication systems carry data and commands between devices located within the local substation and between substations and other locations including the central control room. More about communications in the chapter dealing with wide area control.

Control Houses

Some of the instruments, equipment, smart devices, and communications devices at substations require protection from the elements and reside inside a control house. The control house also contains protection and control equipment (including SCADA and remote terminal units – RTUs) through which technicians interact with equipment and devices via a human-machine interface (HMI). More about controls and HMIs in the chapter dealing with wide area controls. Figure 4-12 shows a substation control building.

Station Ground and Shield Systems

All conductors inside the substation are insulated from the equipment, components, and supports, and all equipment, components, and support structures are "grounded." The term "grounded" means they have zero voltage potential to earth and zero voltage potential to each other. Improper grounding can result in unintended voltage potential between objects, allowing electrical arcs.

Grounding involves connecting all equipment, components, support structures, and even the fence, to form a "grid." The grid firmly connects to the earth in several locations, minimizing resistance to transmission from the grid to the ground. While this may sound simple, grid and grounding require thoughtful design and careful construction.

Protective Devices

System equipment is designed for specified voltage and current ranges, and deviating from those ranges can damage the equipment. Accordingly, various protection devices are installed in substations and along the line. These devices detect electrical anomalies caused by faults, including overcurrent, undercurrent, overloads, and reverse currents. When they detect a problem, they automatically disconnect equipment from the rest of the grid or disconnect entire sections of the line. Relays are one type of protection system; they use a small amount of power to switch a large amount of power.

Electromechanical Relays: Electromechanical relays – essentially switches – were first used as signal amplifiers in long-distance telegraph operations. Later they formed the basis of electrical protection systems. They take the signal from the incoming circuit and transmit it to a second circuit (hence the name "relay").

Electromechanical relays contain electronic parts including: (i) an electromagnet to control opening and closing; (ii) an armature (the moving part), which opens and closes; (iii) a spring to force the relay back to its original position after each revolution; and (iv) a set of electrical contacts to transfer the power. There are several different types of electromechanical relays, and which type is used depends on the specific mechanical device it controls.

Static (Solid State) Relays: Static (or solid state) relays, first introduced in the 1960s, were the next generation after the electromechanical type. The term *static* indicates the relay has no moving mechanical parts. Static relays have longer life spans than electromechanical relays, respond faster, and make less noise when they operate.

Static relays perform the same protective functions as electromechanical relays but incorporate analog electronic devices (transistors, capacitors, and small microprocessors) instead of magnetic coils and mechanical components. Instead of comparing operating torque (force) like an electromechanical relay, static circuits consist of comparators and level detectors, which perform measurements. Static relays monitor the incoming voltage and current waveforms using analog circuits and compare the analog values to its settings. When the measured quantity reaches a certain defined value, the relay triggers and trips (opens) a circuit breaker.

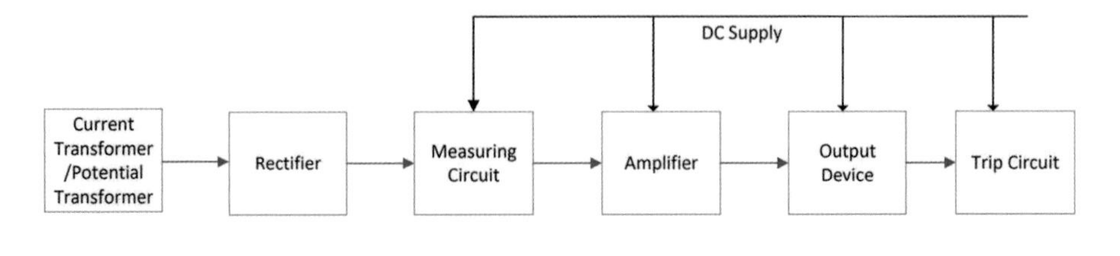

Figure 4-15. Static Relay Block Diagram.

Even though more advanced than electromechanical relays, just like electromechanical relays, static relays perform only a single protection function, meaning complex protection schemes require multiple static relays. The benefits of static relays over electromechanical relays are flexibility in settings and saving space.

Microprocessor-Based Relays: Digital relays (also called microprocessor-based relays) are rapidly replacing the two older relay types. They use microprocessors and compute exact values, meaning they are more accurate and reliable than the older style relays. They also allow remote monitoring and remote programming. Digital relays detect electrical faults using software-based protection algorithms and often include multiple protection functions in one unit, meaning they can monitor and report on multiple characteristics.

Digital relays introduce measured quantity analog to digital conversion (A/D conversion) and use a microprocessor to implement them algorithmically. Advantages include:

- High level of functionality integration
- Additional monitoring functions
- Functional flexibility
- Wide ranges of working temperatures
- Complex function and increased accuracy
- Self-checking and self-adaptability
- Ability to communicate with other digital equipment (peer-to-peer)
- Longer life

Uninterruptible Power Supply

Many substation components require power, so substations include batteries and chargers, and sometimes even backup generation, for when the station is disconnected from the grid.

Extra Credit – Managing Reactive Power

Reactive power is not *lost* (although it does produce heat from resistance as it moves) but is also not *real power* available for consumption. One way to think of reactive power is *stored* power. The more power stored in the system, the less power available for use.

Reactive power is important for AC system operation and must be properly managed. Some substations contain equipment to manage reactive power, including capacitor banks, reactors, and VAR compensators. Reactive power comes in two essentially opposite forms: inductive and capacitive. Inductive power and reactive power can be thought of as cancelling out each other.

Capacitor Load Banks

Groups of large capacitors absorb and store inductive reactive power as an electric field and put it back into the system as needed. Capacitor load banks can remain connected to the grid or switched on and off depending on the need to manage reactive power.

Reactors

High-voltage inductors are often called *reactors* and are used to manage capacitive reactive power and, like capacitor banks, can be switched on and off.

Static VAR Compensators (SVCs)

As previously described, reactive power is measured in VARs. SVCs are an arrangement of capacitors, inductors, and switches that actively manage reactive power.

Substation Types and Functions

Substations contain some or all of the components discussed earlier, depending on their function.

Step-Up Substation

AC power plants typically generate voltages below transmission voltages, so the voltage is *stepped up* (increased) from generation voltage to transmission voltage. These substations typically reside within the geographic boundary of the plant and serve as the beginning of many transmission systems. They contain large, expensive transformers often referred to as GSUs or Generator Step-Up units.

Switching Substation

These substations perform only switching (directing electricity) and do not have transformers. They connect and disconnect transmission systems to and from each other.

Some switching substations add voltage regulation. For connecting systems operating at different voltages, transformers increase or decrease voltages as needed for each system. In the case of connecting systems operating at the same voltage, station transformers increase voltage to make up for voltage lost through impedance.

Inverter and Converter Substations

PV solar farms feeding AC transmission systems convert DC current generated by PV cells into AC power using inverters. Once inverted, transformers step-up the voltage to transmission levels for AC circuits.

High-voltage direct current (HVDC) transmission lines require converting AC to DC and increasing voltage for transmission – converters change the current and transformers increase the voltage.

Distribution Substations

These substations contain similar components to step-up substations, but they step-down the voltage to distribution levels and are typically located:

- at the *ends* of transmission systems;
- along a transmission system's routes at connection points with other transmission systems; or
- along the route at connection points with distribution systems.

Transmission System Planning and Design

Generally, transmission system lines are built or modified for the following reasons:

- Capacity – new or increased demand (e.g., a city experiencing rapid growth).
- Economic – increased supply or market geographic options.
- Customer-driven – large customer service (oil refinery, large industrial customers).
- System improvements – overall system reliability, stability, redundancy, or performance.
- Generation interconnection – new supply (such as a new wind or solar farm) or energy storage facilities.

Transmission planners and designers consider these factors and search for the safest, most reliable, environmentally responsible, and efficient system designs to link power demand and supply. Understanding future demand and supply requires planners and designers to work closely with economic development, local and regional governments, and other entities, to understand and meet those needs.

Because projects require regulatory approval, transmission planners work with regulators, especially if the project exceeds certain dollar limits or requires environmental permits and/or new easements. This approval process can take a great deal of time, is not guaranteed, and often involves review and input from other utilities, along with public and stakeholder comment.

Line Design

Line designers start with beginning and ending locations and delivery/receipt points along the way. They consider voltage, current, and a variety of other factors as they lay out the route and choose the support structures, conductors, and other components. In the words of one veteran designer, laying out the route and selecting the components has its challenges, but the two largest challenges and the ones taking the longest are permitting and acquiring ROWs. Transmission engineers seek safe, reliable, resilient, secure, and cost-effective designs, and work with a myriad of stakeholders towards achieving that goal.

Acquiring Rights-of-Way

Transmission lines can extend hundreds of miles and involve thousands of landowners with individual needs and wants. Some work diligently with the transmission companies for an acceptable arrangement. Some just do not want a transmission line built on their property (often referred to as the "Not in My Backyard" philosophy or NIMBY). They see that many others benefit while they bear the lion's share of the burden.

This text acknowledges the challenge of acquiring ROWs but offers no elegant solutions other than starting early, conducting thorough analysis and justification, meeting with regulators well in advance, and communicating with transparency. The authors acknowledge and admit they like clean and abundant power but would prefer not having a transmission line or substation in their backyards.

Permitting

Line construction requires many permits. In the United States, permitting can stretch from the Federal Energy Regulatory Commission (FERC) for an overall permit to the county commissioners who approve road crossings.

The National Environmental Protection Act (NEPA) may require an Environmental Impact Statement (EIS) or Environmental Assessment (EA). Transmission planners wade through a labyrinth of federal, state, and local laws, rules, regulations, procedures, and public comment periods, attempting to balance the viewpoints and needs of numerous stakeholders, all with their own point of view.

Substation Design

Substation designers start with the substation's function and location. They consider many other factors and select the components and best control scheme for the needed functions at the specified voltage and current. Designers consider how to arrange components on the substation property and create detailed drawings reflecting that arrangement. They also consider how the components connect and provide those details on plan drawings. Like transmission engineers, substation engineers work to provide safe, reliable, resilient, secure and cost-effective designs to satisfy stakeholder needs.

Protection and control designers, SCADA designers, and communication system engineers work in concert with substation designers to ensure each component properly connects to the control system and that the data output by instruments accurately appear on scada screens many miles away in central control rooms.

Transmission System Operations and Maintenance

After the lines and substations are finished, they become the responsibility of the operations and maintenance departments.

Operations

Generating companies supply electricity into the grid. Transmission companies and independent system operators (ISOs) interface with power users, including large industrial and commercial entities and local distribution companies. They share operational responsibility for the grid and divide scheduling, monitoring, directing, and controlling requirements. More on operations in the chapter dealing with ISOs, who, like conductors leading an orchestra, provide overall direction for grid operations.

Transmission control typically takes place in highly secure 24/7 manned control centers, where scada, energy management systems, and other critical operational and communication systems ensure safe, reliable, efficient, and resilient transmission systems.

Transmission field operations personnel, working along the line and at substations, provide control centers with additional operating knowledge and perform local operating tasks.

Maintenance

With the growth of technology, the definition of "maintenance" changed over the years from "fix it when it breaks" to "monitor component health and take actions to prevent failures before they happen." Transmission maintenance teams monitor asset condition, calibrate instruments, and repair/replace components (hopefully) before they fail. Crews also actively control vegetation in the ROW to ensure access and prevent flora from growing near lines.

All this otherwise routine activity turns into emergency response when storms ravage the grid and restoring power becomes an "all hands on deck" situation. Electric utilities enter into regional mutual aid agreements with surrounding utilities. In the words of one transmission maintenance technician, "There is no above average; continuous up time is average, and everything less counts as a failure."

Key Performance Indicators

Most utilities track their performance using various *key performance indicators* (KPIs), including common reliability metrics like transmission *line fault index* (the annual average transmission line faults – or failures – per 100 miles of line), *system average interruption duration index* (the length of a typical outage on a utility's system, abbreviated as SAIDI) and *system average interruption frequency index* (average number of times – or frequency – a customer's electric service gets interrupted during a fiscal year, abbreviated as SAIFI).

SAIDI and SAIFI are the most common system reliability metrics to benchmark and compare a utility's performance with other utilities. Things like severe weather, wildlife, and equipment failure affect both indexes.

Summary

- Transmission lines transmit electricity between substations.
- Transmission uses high voltages because the higher the voltage, the less power lost during transmission for the same amount of power transmitted.
- The primary causes of transmission line outages are faults caused by equipment failures, human errors, and environmental conditions (wind, animals, and trees).
- Transmission systems often exist on property owned by others, and transmission companies balance the need to provide transmission services with the needs of landowners.
- Transmission designers start with beginning and ending locations and delivery/receipt points along the way. They consider voltage, current, and a variety of other factors to choose a route and support structures, conductors, and other components.
- Substations transform voltage; stop, start and direct electricity; monitor and manage process variables; and protect the system from faults.
- System equipment is designed for specified voltage and current ranges, and deviating from those ranges can damage the equipment. Accordingly, various protection devices are installed in substations and along the line.
- Substation designers start with the substation's function and location, consider numerous other factors, and select the components and control scheme for the needed functions.

5

Electrical Distribution Grids

Science can amuse and fascinate us all, but it is engineering that changes the world.

—Isaac Asimov (1920–1992)

"Grandpa Tom, I know those poles along the street are electric poles and the wires on top of them carry the electricity, but what is that can-like thing hanging on the pole at the street and what does it do, and what is that gadget next to it with wires connected to both ends?" Luke asked as they drove to the zoo.

"That can-like thing is a transformer," Grandpa Tom replied. "If you look closely at the gadget, you can see one of the wires carried by the cross member on the pole connects to one end of the gadget and a wire runs from the other end of the gadget to the transformer. That gadget is called a disconnect – a technician opens it to disconnect the transformer from the line if they need to work on it," Grandpa Tom explained. "The transformer receives high-voltage electricity from the wire on the line and reduces it to lower levels," he added. "The voltage of the electricity going into the transformer is called primary distribution voltage, and the lower voltage coming out is called secondary distribution voltage. The secondary voltage is safe for us to use in the house."

"Ok, but where does the electricity come from," asked Luke.

"Electric companies build large towers to support wires to transmit electricity from the place it's generated – for example, a coal-fired power plant, wind farm, or nuclear generating facility – to the places it's needed, like cities and towns," Grandpa Tom explained. "Transmission lines provide electricity to substations. The wires, or ones like them, connect to the substation and extend all around town delivering electricity so we can stay warm in the winter and cool in the summer, cook our food, watch TV, use our computers, and do a lot of other things we enjoy."

"Gosh, getting electricity from where they make it to our neighborhood sure seems complicated," stated Luke.

"It sure is," replied Grandpa Tom. "We don't usually think about any of that until something goes wrong and the lights go out!"

Power Supplied to Distribution Grids

Electrical distribution systems (normally arranged in one or more grid patterns) receive electricity from a variety of generation facilities and safely, reliably, and efficiently transmit it to customers at the required voltage, current, and frequency – with due concern for the environment and a myriad of stakeholders. Entities called distribution companies, electric companies, or utilities typically own and operate distribution systems, although in some locations local or state governments own and operate them.

Customers expect all the electricity they want, at the correct voltage, every second of every hour of every day. Over the past one hundred and fifty years or so, electrical distribution systems evolved and continue evolving. A brief explanation of that evolution follows.

Local Generation

At the beginning of the twentieth century, more than 4,000 individual electric utilities operated in the U.S. They existed largely in isolation from each other, generating and distributing their own electricity, and did not interconnect.[41] Many of these local generating plants produced steam and heated nearby offices, apartment buildings, hotels, and stores with the exhaust heat.

Utility-Scale Generation

As global electrical power demand grew, especially after World War II, these utilities constructed larger power plants (called *utility-scale* plants) and closed smaller, less efficient plants. According to the U.S. Energy Information Administration (EIA), as of December 31, 2020 there were 23,417 electric generating units, located at about 11,070 utility-scale electric power plants, existed in the U.S. Utility-scale power plants have a total nameplate electricity generation capacity of at least 1 megawatt (MW)[42] and exclude the myriad of local and small generators like residential rooftop solar panels.

Transmission Lines

Transmission lines that transmited electricity from utility-scale sources, sometimes many miles, gradually replaced local generation. These transmission lines interconnected with each other allowing utilities to share the economies of scale enjoyed by the larger plants. Interconnections also provide back-up sources of supply – if one generation plant goes off-line, either other connected plants ramp up to cover the load or another plant is brought online, depending on the ability of the online plants to cover the lost generator. Large plant economies of scale combined with interconnected transmission lines meant some generation plants were built as joint ventures between various utilities. The trend of replacing several smaller plants with one larger plant extended to the rest of the world.

Cogeneration

Crude oil prices entered the 1970s below $20 per barrel and exited the decade at over $100 per barrel – increasing by a factor of five. The price of oil took the cost of all other energy forms with it. Vastly higher oil prices led to all sorts of energy conservation and production ideas. One of the ideas was *cogeneration,* where industrial plants generate electricity using heat from the manufacturing process that would otherwise go to waste.

In cogeneration, the plant uses most of the electricity it generates, but some remains available to export to the electric grid. The idea of a small amount of electricity entering the grid from sources other than dedicated generation plants was a new one in a heavily regulated industry. Many countries developed new rules and regulations to deal with the electricity sources distributed along the grid. Historians point out this "new idea" essentially flipped on its head the earlier, local generation practice of generators heating nearby buildings with their exhaust heat.

Distributed Generation

Over the years, the idea of distributed generation and small generators evolved, driven by economics and the desire to move from fossil fuels to renewable resources. Figure 5-1 shows a residential rooftop solar system. In this case, the homeowner purchases all electricity from the local utility and sells electricity generated by the solar panels to the utility at the highest *avoided cost* of generation.

> **Avoided Cost**
>
> The cost an electric utility pays to generate or purchase electricity – with the exception of electricity purchased from the renewable resource generators.

Figure 5-1. Residential Rooftop Solar System. This system is rated at 3.8 kW.

Microgrids

Distributed generation gave rise to the concept of *microgrids* – self-sufficient energy systems serving a discrete geographic footprint, for example, college campuses, hospital complexes, business centers, or neighborhoods. Microgrids receive their electric power from one or more distributed generators (solar panels, wind turbines, combined heat & power, generators); they commonly contain energy storage and connect to larger grids for reliability reasons.

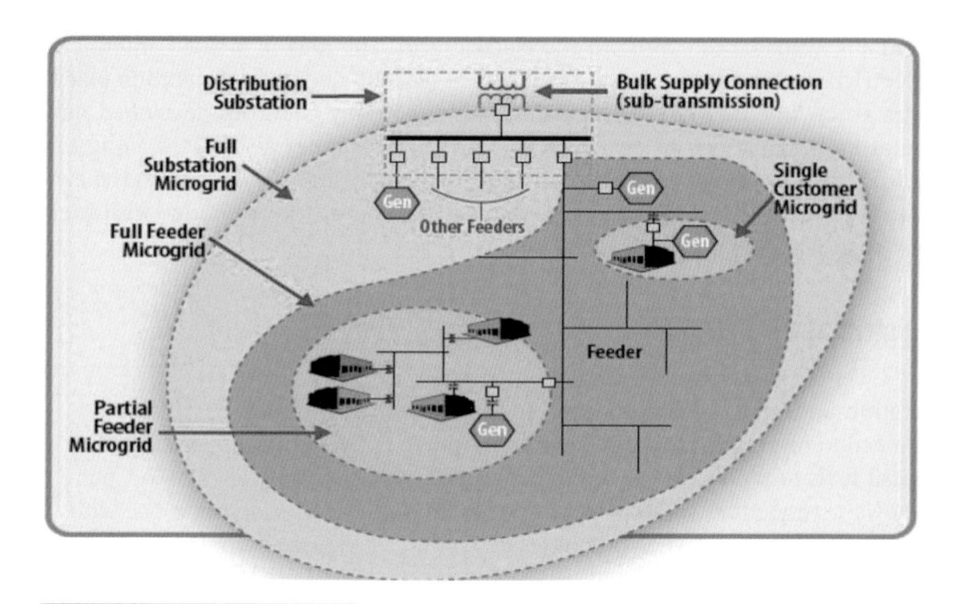

Figure 5-2. Various Types of Microgrids.

The ability to disconnect microgrids from other power sources means microgrid owners can decide how much they want to invest in storage to improve reliability. As an example, one microgrid owner might decide to invest in a significant amount of storage. The microgrid could remain connected to the main grid during times of normal operations, ensuring the storage is always fully charged. When the main grid loses power or during times of peak demand, the microgrid can disconnect and use the storage to maintain reliability or save money. Microgrid distribution systems operate much like larger distribution systems but on a smaller scale. At its smallest, a microgrid could be one house with solar generation and a battery bank.

Substations

Distribution systems begin at substations where they receive electricity from transmission lines (and other sources) and reduce (step-down) voltage from transmission levels to distribution levels. These substations are appropriately called *step-down substations* and often connect to more than one transmission system, increasing supply flexibility and reliability. (For a review of substations, their functions, and components, refer back to Chapter 4). Figure 5-3 shows a step-down substation.

Figure 5-3. Step-down Substation that Reduces Higher Voltage to Lower Voltage.

After leaving the step-down substation, electricity is normally transmitted through one or more additional substations on its way to the customer. Those substations:

- isolate faulted components or sections of line;
- disconnect components or sections of lines from the rest of the system for maintenance or repair;
- change (transform) voltage levels;
- provide *reactive power* to correct power factors;
- control voltage;
- provide grid operators data (voltage, current, power).

In addition to receiving electricity from transmission lines, distribution systems receive electricity at a variety of voltages from small-distributed generation facilities of various types (like rooftop solar installations). Electricity generated by small generators may enter the distribution grid directly rather than through substations. Figure 5-4 shows the distribution system connection for the rooftop solar system shown in Figure 5-1.

Figure 5-4. Distributed Connection to Rooftop Solar. The leftmost box is an inverter and converts solar DC to AC. The meter next to the inverter measures electricity from the solar panels into the distribution system, and the meter on the right measures electricity from the distribution system to the house.

Incoming Power

Electricity from transmission lines enters step-down substations as three phases of alternating current (AC), each carried on one or more conductors. As explained in Chapter 3, the combined voltage across the three phases equals zero (assuming equal loads). Electrical engineers say the three phases complete the circuit for each other, meaning they do not need a dedicated return path back to the source of generation.

Distribution Voltage

Annoyingly, the terms transmission voltage, primary distribution voltage, secondary distribution voltage, high-voltage, and low-voltage have different definitions depending on who uses them. This text treats *transmission voltage* as the voltage between the step-up transformer at the generation facility and the step-down transformer at the distribution substation, *primary distribution* voltage as the voltage between that distribution substation and final transformer, and *secondary voltage* as the voltage from the final transformer to the end-use customer.

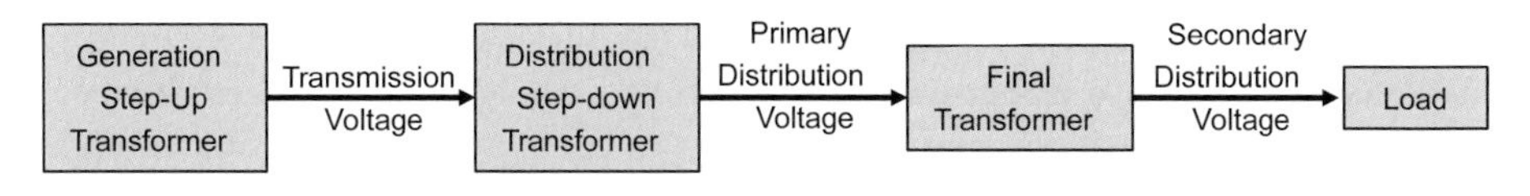

Figure 5-5. Transmission, Primary Distribution, and Secondary Distribution Voltages.

Electricity follows the same laws of physics around the world, and engineers all over the world apply those laws in similar fashion except in two areas: frequency and distribution voltage.

Early generators ran at different speeds, resulting in different frequencies. Around 1893, when interconnecting those systems became critical and required standardized frequencies, the American Westinghouse Electric Company and Germany's AEG set their generation equipment to function at 60Hz and 50Hz, respectively, creating what would become the world's two standard AC frequencies (commonly referred to as the North American and European Systems, respectively). The European and North American systems operate at similar generation, transmission, and primary distribution voltages but deliver electricity to the ultimate customer at different voltages. Customer voltages vary between countries; the most common are 240 volts and 120 volts.

The internet contains many zealots arguing one system is "better" than the other. In reality, both systems have pros and cons, and both have their own nuances.

Primary Distribution Voltages

Primary distribution voltages vary between and within countries based on the particular distribution system, and the voltage ranges do not matter as long as the same system uses the same voltages. Distribution engineers use more precise terms but, in general, primary distribution voltages vary from about 4,000 to about 40,000 volts.

Primary distribution lines often deliver electricity directly to commercial or industrial customers through substations owned by the distribution company or the customer, for large loads, or through pad-mounted transformers, for medium-size loads. These loads normally use three-phase systems and operate across a range of standardized voltages. Higher voltages allow lower currents and smaller conductors – a fact distribution designers consider in their designs.

> **Voltage and Current**
>
> Power depends on voltage and current – 120 volts at 10 amps delivers the same 1,200 watts as 220 volts at 5 amps.

Secondary Distribution Voltages

Residential secondary distribution voltages are standardized at 220V, 230V, and 240V (European System) or 120V, 208V, and 240V (North American System) and are typically single-phase. Commercial loads use these same voltages when connected to single-phase systems. Connecting to three phases allows other voltages. Again, designers consider connected loads as they select transformers and design systems.

Distribution Transformers

Transmission lines deliver three-phase, high-voltage AC to step-down substations. Those substations transform the high voltage to residential, commercial, and industrial customer voltage levels using either dual-winding (traditional) transformers or single-winding (auto) transformers.

Dual-Winding Transformers

AC on the primary side winding of dual-winding transformers produces a changing magnetic field that interacts with the secondary windings, producing AC in them. The voltage of the electricity produced in the secondary side winding depends on the ratio of the number of turns on each winding. In other words, transformers transfer (move) energy and transform (change) voltage. Because the windings do not electrically connect to each other, they are called *magnetically coupled* and *electrically isolated*. Dual-winding transformers are the most widely used type of transformer for electrical energy. Transformers were discussed in Chapter 1 where Figure 1-16 shows a dual-winding transformer.

Autotransformers

> "Auto" (Greek for "self") refers to the single coil acting alone, not to any kind of automatic mechanism.

Autotransformers have only one winding wrapped around one core and work on the same principle as dual-winding transformers. The primary side winding extends along the entire length of the core, and the secondary side connects (taps) the primary winding at different numbers of winding turns from the end of the core. Tap location determines the voltage output from that tap.

Most autotransformers have at least three taps, each with a different voltage output. Because autotransformers use less material, they are smaller, lighter, and cheaper than dual-winding transformers. Unlike dual-core transformers, the primary and secondary sides electrically connect, meaning they do not provide electrical isolation between them.

Autotransformers are most often used for making small voltage changes. Because autotransformers can have multiple taps, they are often installed on long lines as voltage regulators and restore voltage reduced by impedance. Restoring

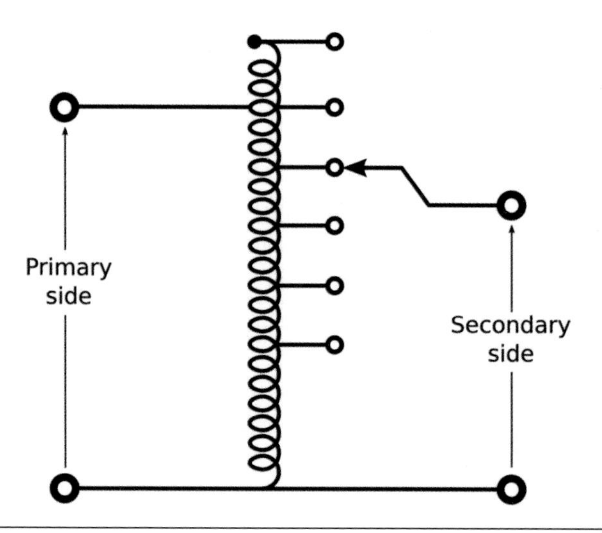

Figure 5-6. Schematic of an Autotransformer.

voltage means customers at the far end of the line receive the same average voltage as those closer to the source.

Transformer Connection Patterns

Three-phase voltage transformation uses either one three-phase transformer – a single device with all windings constructed on a single iron core contained inside an enclosure, as shown in Figure 5-7, or three, single-phase transformers connected externally, one to each of the three phases. One three-phase unit is normally cheaper than three single-phase units, but using three independent single-phase units means that, if one fails, it can be replaced. Some critical applications keep one spare to reduce downtime in the event of a failure.

The three primary conductors and three secondary conductors of the three phases are not connected to each other inside the transformer. Instead, the three ends of the primary conductors connect to each other, as do the three ends of the secondary windings, in one of two patterns, delta or wye (also called star).

Figure 5-7. Three-Phase, Pad-Mounted Transformer. This single unit transforms all three phases.

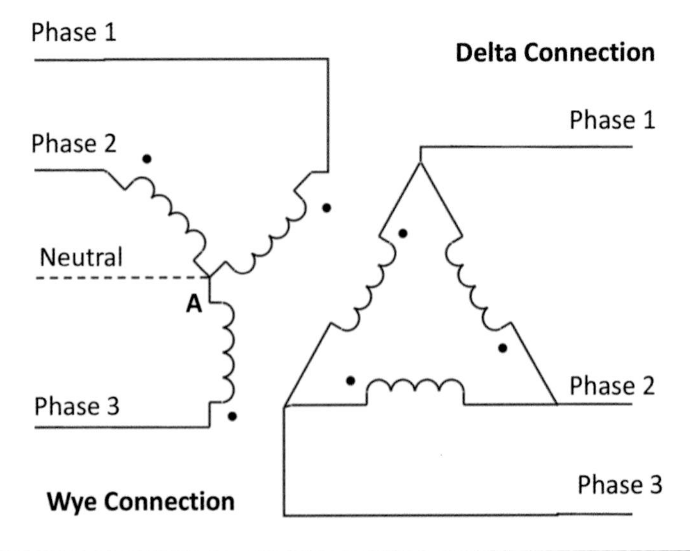

Figure 5-8. Wye and Delta Connections.

Wye Connection

The wye connection in Figure 5-8 has four conductors, three phases connected together at point A, and a neutral wire also connected to point A. As long as all three phases are equally loaded and 120 degrees out of phase with each other, the voltage at point A is zero – no current is transmitted on the neutral wire, and it only serves as a precaution if the phases become unequally loaded. Single-phase loads, however, require a neutral line.

Delta Connection

The delta connection in Figure 5-8 has three conductors connected sequentially. If all three phases are equally loaded and 120 degrees out of phase with each other, no neutral wire is needed, as the three phases complete the circuit for each other.

Transformer Configurations

There are four primary transformer configurations:

- Delta-delta: Common for transmission lines and distribution circuits with three-phase loads, but can also be used for a limited number of single-phase loads.
- Delta-wye: The most common distribution transformer. Three phases enter the transformer on the primary side and connect in the delta formation inside the transformer. The secondary winding adds the neutral wire. The three secondary phases can supply three-phase loads, and single-phase loads are served by one of the three phases plus the neutral wire. The three phases plus the neutral wire exit the secondary winding on the left.
- Wye-wye: Common for distribution loads served by three-phase four-wire, three-wire single-phase, and two-wire single-phase systems. (These systems are discussed later in this chapter).
- Wye-delta: The least common configuration. The primary side neutral wire compensates for unbalanced phases on the primary side.

The nomenclature places the primary connection type first and the secondary connection type second.

Center-Tapped Transformers

Center-tapped transformers provide two voltage levels for single-phase loads and a hybrid, 208V, for three-phase systems.

Single-Phase, Center-Tapped: As shown in Figure 5-9, this single-phase transformer has one phase plus a neutral wire on the primary side and three conductors on the secondary side. Center-tapped transformers can supply both 110V for lighting and small electrical appliances and 220V for larger loads, such as clothes dryers or heating and air-conditioning units, by combining the two 110V legs from the transformer.

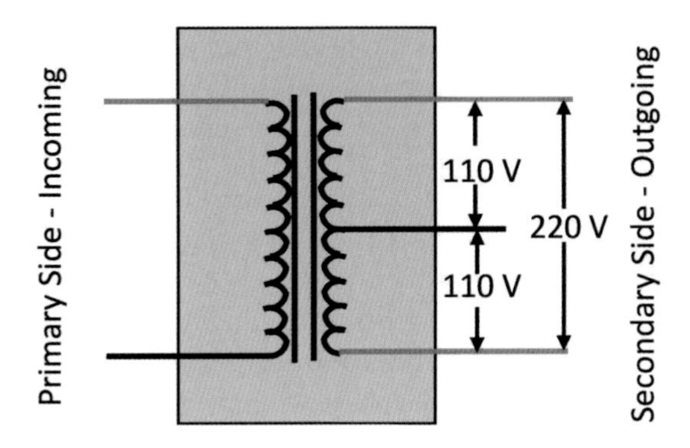

Figure 5-9. Center-Tapped Transformer.

Three-Phase, Center-Tapped: Delta secondaries can also be center-tapped, but, in this case, the tap is connected where the three phases connect to each other. This means the center is at zero to each of the phases. Connecting the load between the neutral and one 120V phase, yields 120V of single-phase. Connecting the load between two 120V phases yields 208V.

Center-Tap Extra Credit: In the case of single-phase, the center tap connects to the exact middle point of the secondary winding. Because AC alternates from positive to negative at 60 Hz, the voltage along the coil alternates between positive 170V and negative 170V, passing through 0V twice each cycle. Another way to think about this is that the two ends of the winding are 180 degrees out of phase with each other, so the voltage difference between them is 240V and the center tap is always 0V relative to the ends.

Because the three phases are 120 degrees (and not 180 degrees) out of phase from each other, the algebraic summation between any two phases of center-tapped delta transformer secondaries is 208V and not 240V, as in the case of center-tapped wye secondaries.

Distribution Grids

From step-down substations, conductors (commonly called *feeder lines* or simply *feeders*) transmit electricity at primary distribution voltages to transformers located near each load. From the final transformer, secondary feeders transmit electricity to the load connection point. Even though many primary distribution lines and service connections are below ground – particularly in cities and towns – the connection point is still commonly called the *service drop*, harkening back to the days before buried feeders. Along the route, an assortment of switches, circuit breakers, regulators, meters, and other devices enable safe and reliable electrical service.

Feeder Patterns

Nomenclature associated with arranging and connecting feeders varies by country, company, background, experience, and personal preference. Therefore, discussing the two pattern concepts – radial and loop (network) – is better than getting caught up in the exact names of the circuit types.

Radial

The first feeder pattern – radial – is the simplest. Feeds leave substations, radiating out like tree branches. The system's simplicity requires the least equipment and devices, minimizing equipment and labor costs thereby yielding capital costs lower than the more complex loop (network) systems. Because they connect to only one supply source, however, their simplicity also limits reliability.

Conventional Radial Systems

These systems receive electricity from only one source. Loss of that source results in lost customer service. Additionally, radial system maintenance usually requires shutting down the system completely, cutting customer service. Utilities use radial systems when the need for low initial cost, simplicity, and space economy outweighs reliability needs. Equipment used for radial systems typically consists of one substation with a primary switch, a properly sized transformer for the load, and a low-voltage switch.

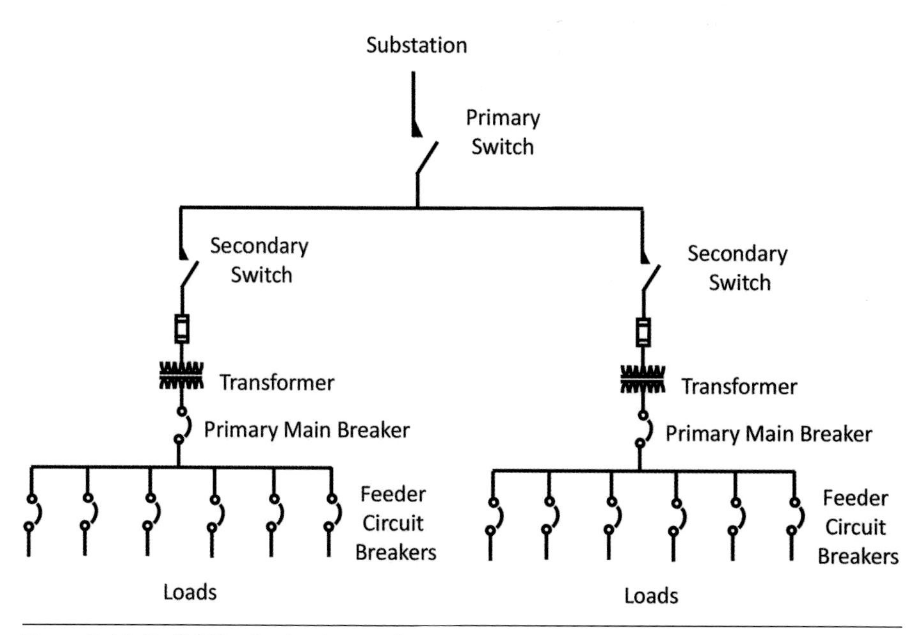

Figure 5-10. Radial Distribution System Schematic.

Primary Selective Radial Systems

Primary selective radial systems are simple radial systems connected to two power sources. The additional power connection provides similar economic advantages to the simple radial system but with greater reliability. In the event one power source fails, the other takes over (thus, "selective"). A brief outage takes place between losing the primary source and switching to the alternate source – unless the sources run parallel to prevent the outage. Losing the secondary transformer or distribution equipment still results in lost service, as does performing maintenance.

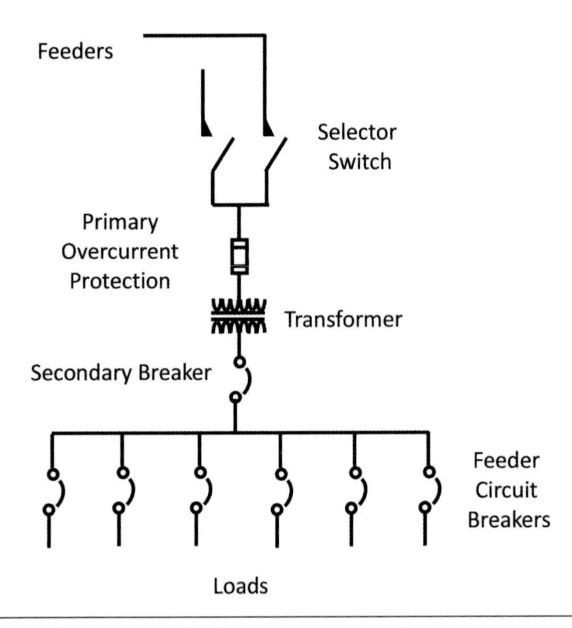

Figure 5-11. Primary Selective Distribution System Schematic.

Loop

Loop (also *network* or *grid*) patterns take the selective idea one step further – they have multiple power connections. If one connection fails, one of the others immediately takes over, usually automatically and with no loss of service. Loop systems are basically circles with connections. Unlike radial systems, power is transmitted from either direction.

Networks also add switches to isolate any section experiencing a fault from the rest of the grid. Isolating the faulted section and diverting power feeds around that section limits service interruptions. Some distribution professionals differentiate between *simple loops* and *networks,* saying simple loops are only one loop and networks have a series of interconnected loops. In either case, the goal is the same – continuity of service.

Figure 5-12 shows two primary selective distribution systems converted to a loop arrangement. Either transformer can feed either or both radial distribution systems connected to the loop. If a fault occurs in either of the two radial systems connected to the loop, that system can be isolated, allowing continued service to the other and limiting the number of loads losing service.

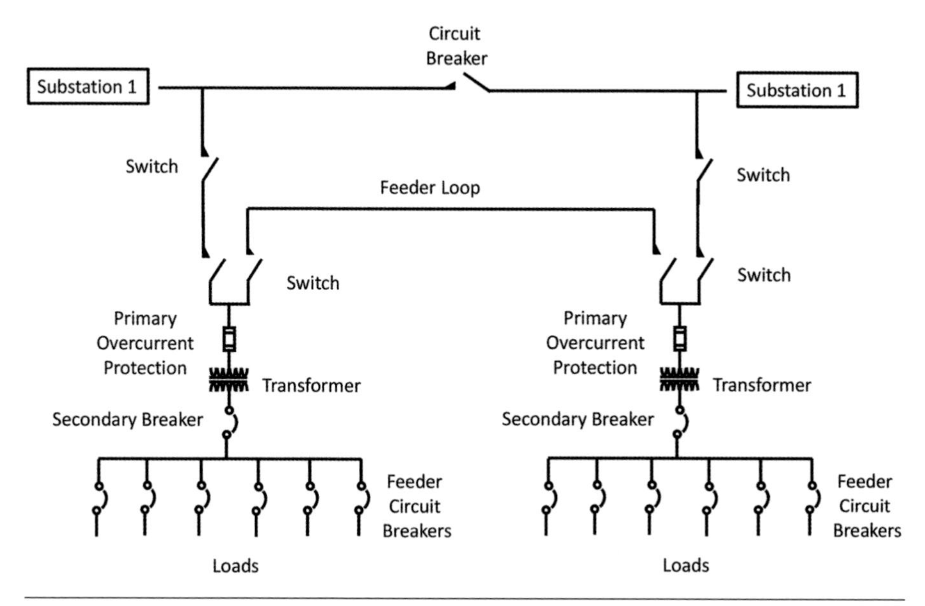

Figure 5-12. Primary Loop Distribution System Schematic.

One word of caution: loops and networks normally provide increased flexibility and reliability but, if demand suddenly exceeds supply, some connected systems must be immediately dropped to maintain grid stability and prevent it from crashing.

Composite

Distribution system designers use a combination of radial and loop (network) arrangements throughout their distribution area, as they balance cost and reliability.

Feeder Arrangements

Depending on the supply and load, distribution systems use a variety of pole, insulator, and conductor arrangements. The most common conductor arrangements are:

- Four-wire, three-phase
- Three-wire, three-phase

- Three-wire, single-phase
- Two-wire, single-phase

Distribution systems begin at substations and transmit three phases of electricity to large loads and branch off as single-phase systems for smaller loads.

Three-Phase, Four-Wire Systems

These systems are served by delta-wye transformers – delta on the incoming side and wye on the outgoing side. Figure 5-8, a schematic of a three-phase transformer, shows four conductors exiting the secondary side of the transformer on the right – three for the three phases and one neutral line. Three-phase, four-wire systems commonly supply a mix of three-phase and single-phase loads. Figure 5-13 shows a three-phase, four-wire system. Single-phase loads connect the distribution transformer to one of the three phases and to the neutral line. The total load on each phase is monitored, and single-phase taps are moved from one phase to another, keeping loads on all phases balanced.

Figure 5-13. Three-Phase Four-Wire System Leaving a Substation. The neutral line is the lowest on the pole, with three conductors, one for each phase above it.

Three-Phase, Three-Wire Systems

Delta-delta transformers serve these systems. As the name states, they have only three conductors – one for each phase. Distribution systems use three-phase, three-wire feeders mainly for three-phase loads, but they can also supply single-phase loads, in which case a delta-wye distribution transformer creates the neutral line.

Single-Phase Distribution

Homes and small businesses do not require as much power as commercial and industrial establishments and commonly receive electricity from a single phase. Single-phase systems typically have either two or three wires.

Single-Phase, Two-Wire Systems

Two-wire systems commonly run short distances, serving small loads, although they can run many miles in rural areas with widely spread loads.

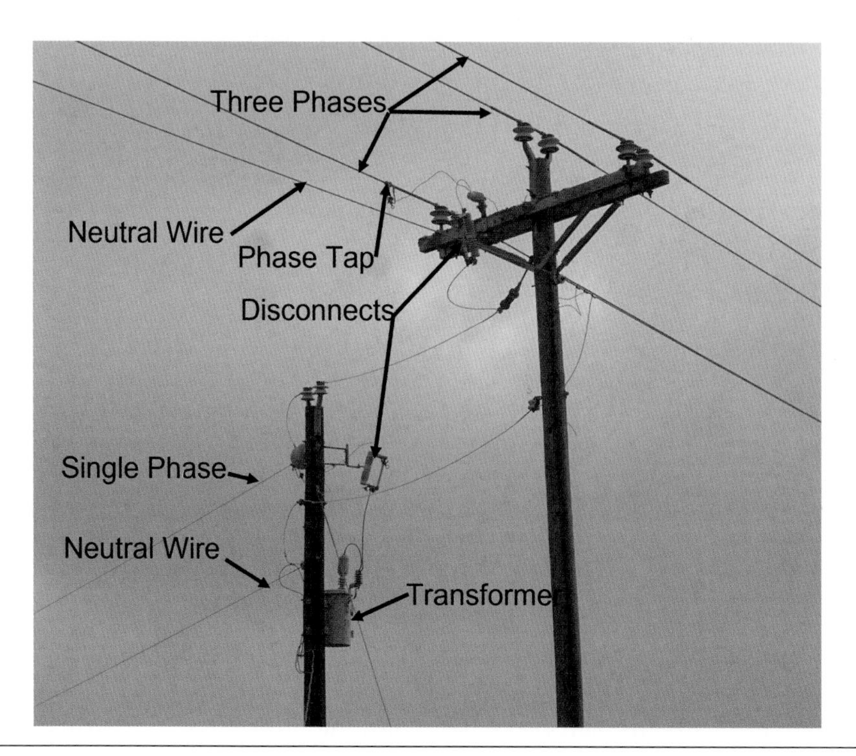

Figure 5-14. Rural Single-Phase, Two-Wire Distribution Line. Connected to a Three Phase Distribution Line.

Single-Phase, Three-Wire System

Whether two- or three-wire, single-phase systems are used depends on total load size and economics. Single-phase, three-wire systems are more common as secondary, rather than primary, distribution. In the North American System, they start at center-tapped transformers delivering 120V and 240V to residential loads, including electric dryers, furnaces, and air-conditioning units.

Figure 5-15. Three-Wire, Single-Phase Secondary Distribution. The three wires enter the riser at the center bottom of the picture.

Grids using the European System do not typically use center-tapped transformers because they already operate at the higher (230–240) voltage. The higher voltages used in European System grids also mean they use fewer, but larger, transformers than grids using the North American System. Fewer transformers mean longer secondary feeders. Because they operate at higher voltage, European System grids typically use smaller-diameter feeders than the North American System.

Distribution Components

Various components connect to distribution systems along their routes. This section discusses grid protection devices, including circuit breakers, automatic distribution circuit reclosers, sectionalizers, and disconnects, as well as voltage regulating transformers, capacitors, and meters.

Circuit Breakers

As distribution lines exit the substation, circuit breakers protect the substation and feeders by separating the substation from the line in the event of a fault (abnormal current). Circuit breakers have fixed and moving contacts (electrodes) in a closed chamber containing a fluid (either liquid or gas) to smother any arc between the contacts. In normal conditions, the contacts stay closed. When a fault occurs, the contacts open manually or by remote control (when needed). When the grid experiences a fault, the breaker trip coils energize and pull apart the contacts to open (break) the circuit.

Because circuit breakers are designed to operate under load, grid operators and field technicians use them to de-energize substations prior to opening substation incoming and outgoing disconnects.

Automatic Distribution Circuit Reclosers

Normally just called *reclosers,* this device, like a circuit breaker, detects faults and opens to interrupt the fault current. Unlike circuit breakers, reclosers automatically reclose (hence the name recloser) if the fault clears quickly – for example, when a tree limb momentarily touches a power line and then falls to the ground.

Sectionalizer

Sectionalizers work in concert with reclosers and keep track of how many times reclosers open and close. When the number of cycles reaches a preset value, the sectionalizer opens and remains open until reset (manually or remotely). Sectionalizers cannot operate under load and need an upstream circuit breaker or recloser that can operate under load. Because the sectionalizer opens before the recloser closes, it does not have to operate under load. Preset cycle values vary by application and are often only one cycle for underground lines because repetitive faults might damage the underground conductor's protective coating.

Reclosing and sectionalizing are functions that, in the past, were performed by two separate devices. Technology improved, and reclosers can now be programmed to also act as sectionalizers and remain open after a preset number of cycles. Figure 5-16 shows a protective device on a single-phase line.

Figure 5-16. Protective Device for Isolating a Line Section.

Disconnects

Some disconnects are designed to operate under load. Rather than just depending on an open circuit breaker, technicians (or grid operators) open disconnects to provide clear separation of the facility from live current. Figure 5-17 shows a three-phase, gang-operated, load-break disconnect switch used to isolate the three phases on the right of the switch from the three phases on the left of the switch. The term "gang" means all three phases are disconnected simultaneously, and "load break" means it operates under load.

The conductors on each side of the pole's crossmember terminate at strain insulators isolating one side from the other. From the terminations, wires run to each side of the disconnect. The rotating pole connects to a handle at ground level and to the slide bar on the crossmember. Three rotating insulators connect to the slide bar via a linkage. When the technician on the ground turns the handle connected to the rotating pole, the slide bar moves to the left and the linkage connected to the slide bar rotates, disconnecting its blade from the cradle connected to the stationary insulator.

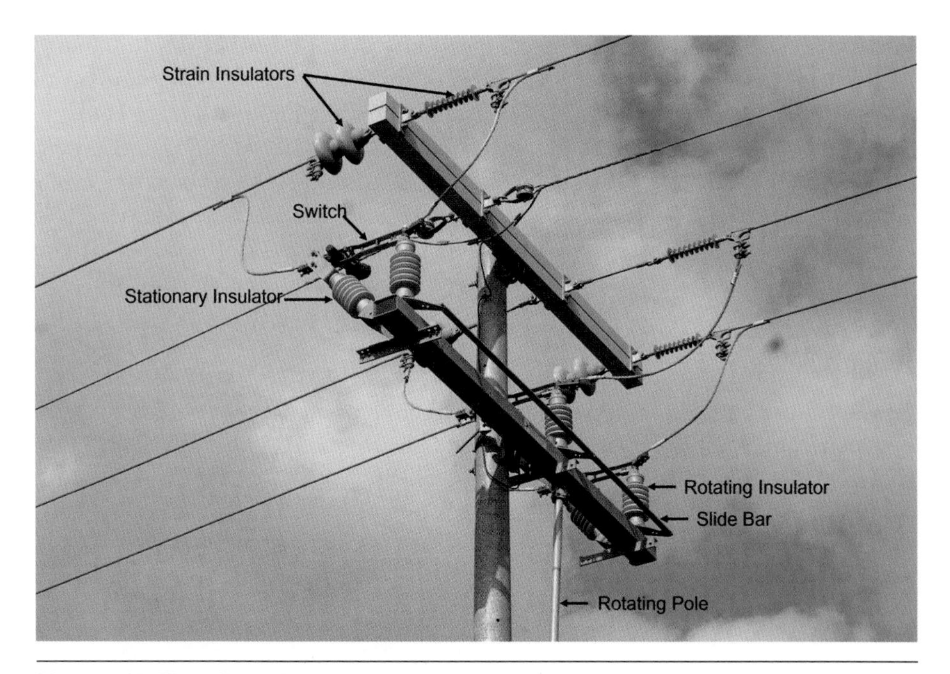

Figure 5-17. Three-Phase Disconnect

Voltage-Regulating Transformers

Line impedance reduces voltage on longer lines, creating a need for voltage-regulating transformers (regulators) to increase the voltage back to required ranges.

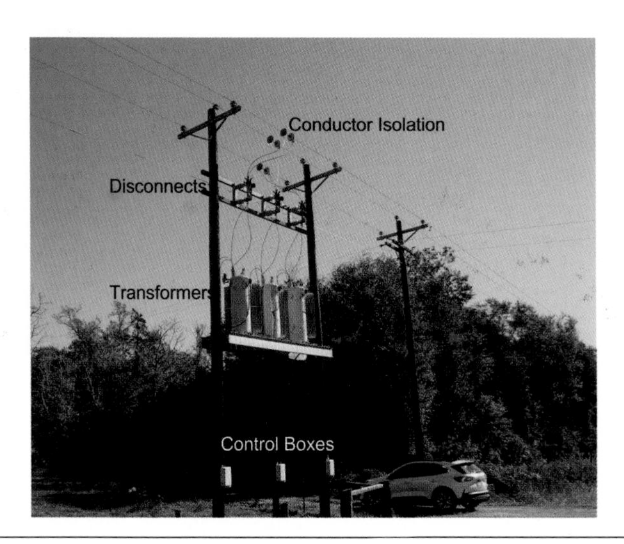

Figure 5-18. Three-Phase Transformer Bank

Tap Changers

Tap changers vary the internal transformer taps to alternate the primary-to-secondary-voltage ratios.

Electricity is transmitted from the line to the transformer primary winding and from the transformer secondary winding into the next section of line. If it is an autotransformer, its secondary winding has taps at various voltages to determine the ratio of primary-side to secondary-side voltage. Control boxes located at road level allow control of the ratios.

Capacitors

Capacitors manage reactive power (VARs). Figure 5-19 shows capacitors on a distribution pole. The vacuum disconnects allow connecting or disconnecting these capacitors from the grid as needed.

Figure 5-19. Capacitors for VAR Management

Meters

The amount of electric power used (in kilowatt-hours – kWh) depends on voltage, current, and time. Electric meters monitor these three factors using various techniques and calculate kWh delivered, for billing purposes. The following sections discuss various meter types.

Mechanical Meters: Utilities used these meters for many years but are replacing them with more modern technologies. They contain two conductor coils that create magnetic fields. *Current* going across the conductor affects one coil while the *voltage* going across the conductor affects the other coil. Together, the magnetic fields turn a thin aluminum disc at a controlled rate (aluminum, though not magnetic, moves in this case through a principle known as an *eddy current*). The disc turns a series of gears moving the five dials recording electricity usage in kWh.

Eddy Current

All conductors have resistance, and when AC is transmitted through the conductor, eddy current induced in the conductor opposes transmission. The increased impedance results from the AC and its resistance.

Digital Meters: The first digital meters were merely mechanical meters using electronic (rather than mechanical) displays. Modern digital meters measure volts and amps directly and calculate power from those readings. High- and medium-voltage metering involves reducing voltage and current using potential (voltage) transformers (PTs) and current transformers (CTs) connected to the line. The meter then measures the lower (safer) levels of volts and amps and uses them to calculate power. The resultant calculation is increased by the transformer ratio, arriving at kWh delivered. Figure 5-20 shows an industrial service drop with PT and CT meters.

Figure 5-20. Industrial Service Drop with PT and CT Metering.

Smart Meters: So-called *smart* meters remotely provide detailed and accurate data on electric usage in real time (or pre-established intervals). The U.S. Department of Energy estimates almost 100 million American homes use smart meters as of 2019.[43] Figure 5-21 shows point-in-time usage data from a cell phone app connected to a smart meter.

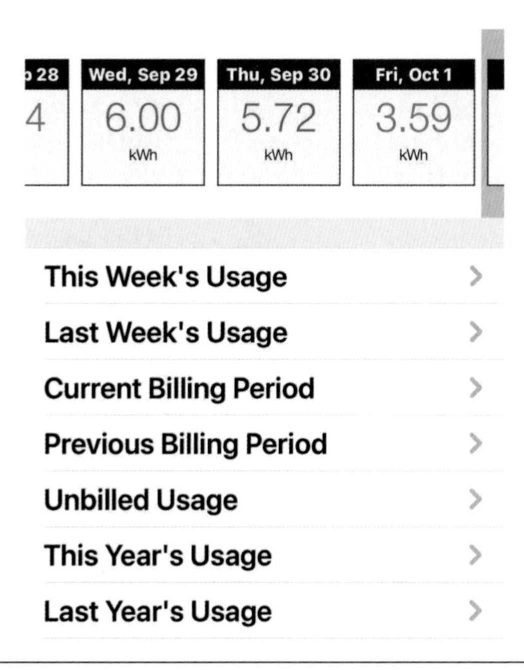

Figure 5-21. "Smart Meter" Usage Example.

Real-Time Metering

Obtaining real-time electricity consumption data using smart meters allows immediate measurements and eliminates monthly estimates. Smart meters are read remotely, eliminating physical travel to locations for the sole purpose of visually reading the meter. Smart meters also allow connecting and disconnecting service remotely from a company's energy control center without needing to send a field crew to initiate or disconnect service. Smart meters also provide information to the control center to warn the utility of impending equipment failures and to allow the utility to pinpoint outages, making it easier to dispatch repair trucks when and where needed.

Some electric utilities use smart meters to institute *time-of-use* pricing (charging less during times of low demand/low wholesale prices – like overnight – and charging more during times of high demand/high wholesale prices). Utilities may also avoid constructing new power plants if they can monitor and control energy usage in real time.

Despite the benefits of smart meters, some people have concerns over the personal data smart meters collect, considering them a violation of customer privacy. Landlords, for example, might monitor their tenants' usage patterns, inferring from those patterns if a tenant has broken a lease provision.

Cable, Communication, and Other Connections

Distribution poles provide convenient locations for cable, communication, and other equipment, and distribution companies often lease space on their poles to local telephone, cable, fiber, and cellular providers, enhancing revenues.

Figure 5-22. Distribution Pole Supporting Other Cables.

Secondary Distribution

Conductors from the final transformer to the customer (as used in this text) are commonly called *secondary feeders*. They follow the same concepts and patterns as primary feeders but operate at lower voltages, are generally smaller in size, and extend from a transformer to a service drop.

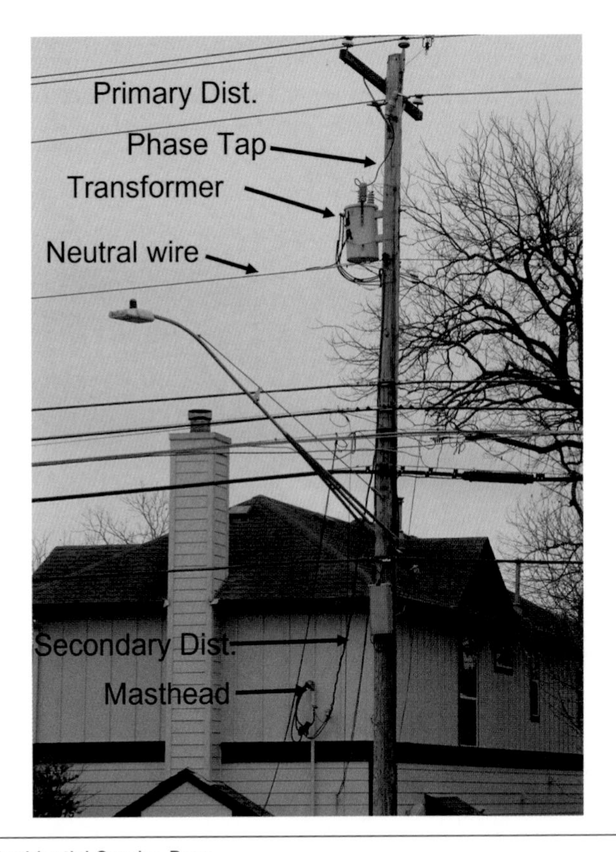

Figure 5-23. Residential Service Drop

Overhead service connections are called *weatherheads, mastheads, or service heads.*

Customers

Nearly all electrical customers belong to one of four categories:

- Residential
- Commercial
- Industrial
- Transportation

Residential

The residential sector includes single-family homes and multi-family housing and accounts for more than one-third of electricity usage in the U.S. The primary uses of electricity in the residential sector are heating and cooling, lighting, water heating, and appliances/electronics. Summer electricity demand in the residential sector tends to be highest on hot afternoons due to increased air-conditioning use, followed by evenings, when lights are turned on. Winter residential peaks also depend on regional temperatures.

Commercial

The commercial sector includes government facilities, service-providing facilities and equipment, and other public and private facilities. This sector also accounts for more than one-third of U.S. electricity consumption. The primary uses of electricity in the commercial sector are lighting, heating, ventilation, and air-conditioning. Electricity demand in the commercial sector tends to peak during operating business hours and decreases substantially on nights and weekends.

Industrial

Industrial facilities and equipment use electricity for processing, producing, or assembling goods, including such diverse industries as manufacturing, mining, agriculture, and construction. Overall, this sector uses less than one-third of the nation's electricity. More than half of the electricity used in manufacturing goes to powering various motors. Other large uses include heating, cooling, and electrochemical processes. Electricity use in the industrial sector tends to not fluctuate through the day or year as much as in the residential and commercial sectors, particularly at manufacturing facilities that operate around-the-clock.

Transportation

The transportation sector consumes most of its energy by directly burning fossil fuels such as gasoline, diesel, and jet fuel. The current electric vehicle market is small, meaning transportation activities account for less than 1% of total U.S. electricity use. This sector is forecast to grow at a faster rate than the other three sectors due to the emergence of electric vehicles (EVs).

Distribution Design

During the design process, design engineers match new and existing customer demand (residential, commercial, industrial, and transportation) to new and existing supply with due concern for reliability, safety, efficiency, and the environment.

Over many years, the safety codes and regulations on which engineers rely have evolved. Distribution design is a lot like transmission design, with three important differences. The first is obvious – voltage. The second is less obvious – the number of connections. Distribution has many more supply points and demand loads than does transmission. Finally, new transmission lines are long, linear assets that involve securing rights of way from many landowners. Transmission lines also involve more permitting and a higher level of environmental study.

Distribution Design Process

The distribution design process involves three phases described below:

Planning

The planning phase is the first step in designing distribution facilities and includes the following steps:

- Developing future load models to plan feeders and equipment
- Modeling electrical load to serve and impact on the distribution system
- Identifying source and operational/switching requirements to serve the load

Preliminary Design

After the planning phase, the design process enters the preliminary design phase, which includes:

- Identifying detailed design requirements from planner or customer
- Visiting field location to verify existing equipment and determine route
- Developing a preliminary design including equipment locations, line routes, and one-line diagram
- Providing a rough estimate of cost and schedule

Final Design

After performing a preliminary design, the process enters its last phase, final design, which includes:

- performing load calculations like voltage drop/flicker, line clearances, cable pulling and fault current;
- developing a final detailed design from verified equipment location, line routes, and calculations;
- developing equipment details, pole memos, trench details and supporting documentation; and
- providing a bill of materials and detailed cost estimate to the customer.

National Codes and Standards

Each country has its own codes, standards, and regulations. In the case of the United States, the National Electrical Safety Code (NESC) is used primarily for distribution design, while the National Electrical Code (NEC) is used on the customer side of the meter.

NESC

The NESC provides standards for the safe installation, operation, and maintenance of transmission, substation, distribution, and communication facilities. Included in this standard are safe work practices to protect workers and the public. Although the NESC is a voluntary standard, many localities adopt it as law in their jurisdictions. Additionally, cities and towns often develop local codes.[44] As they make design decisions, designers carefully check them against the applicable national codes.

NEC

The NEC covers customer installations. Generally, building wiring systems include interior and exterior wiring from the service point or power source to the outlet(s).

Local Utility Codes

Cities and towns often have local utility codes defining the type of service (underground or overhead), utility operating voltages, service point location requirements, electric facility ownership, equipment maintenance responsibilities, conductor colors, requirements for facilities crossing property lines, and a myriad of other details.

Construction Document Preparation

The final phase translates all the work of the previous stage to detailed drawings and specifications in sufficient detail for contractors to develop bids. After selecting the contractor, the same documents apply to constructing, overseeing, and managing the project, including inspection and quality control.

Customer Process

When they want to obtain service, customers typically submit a service request to the distribution company. Service requests include a request for the building permit and plot plan, elevation drawings, and surveys identifying existing electric facilities, gas lines and meters, and other pre-existing utilities.

Once customers receive approval for electric service, they determine the meter location and other logistics. Typically, the utility will schedule the service after the following:

1. Application for electric service
2. Application approval
3. Electric permit and other required permits
4. Work is completed per the utility's requirements
5. Inspections
6. An account established with the business office

Field Operations

Electric utility field operations involve construction, maintenance, vegetation management, meter installs, and restoration after a service outage. Figure 5-24 shows a utility contractor replacing a pole after a drunk driver damaged the original pole.

Figure 5-24. Contractor Replacing a Distribution Pole.

The picture on the left shows the pole before the crew began work. The gash created by a car hitting the pole appears on the bottom of the pole. The picture on the right shows the replacement pole with a new transformer. In this case, the crew left in place the damaged pole and the masthead because the lines did not have sufficient slack from the bottom of the riser to the connection points for a seamless connection.

Utilities spend tens of millions of dollars each year on operating expenses and capital expenditures. They use tools like geographic information system (GIS) technology as they track equipment, identify usage patterns, make connections, and track relationships between materials, workers, and fleets. Distribution grids cover wide areas, so field crews use various communication methods (radio, cellular telephone) as they coordinate work. They must also conduct such mundane and detailed activities as managing spare parts, routing trucks, and moving equipment to work sites.

Key Performance Indicators

Utilities typically establish key performance indicators (KPIs) for work management, construction effectiveness, and operating/maintaining equipment safely, while adhering to policies, procedures, guidelines, and external regulatory requirements. Utilities establish quality control measures and KPIs tracking effectiveness and correcting gaps or compliance issues. A common issue for modern electric utilities is recruiting, developing, and retaining high-performing workers.

Safety

One of the most important aspects of field work is safety. Protecting field workers requires developing protocols to ensure control room operators de-energize lines before field crews work on them. When field staff must work on energized lines, they use extra safety precautions, including *arc flash-rated* clothing, insulated gloves, and insulated blankets laid over live lines. Arc flash severity depends on three factors:

- Proximity to the hazard
- Temperature
- Time for circuit to break

Arc Flash

A flashover of electric current that leaves its intended path and travels through the air from one conductor to another or to ground, often with a violent result. If a person is near the arc flash, it can result in serious injury or death.

Utilities protect workers in several ways. Some methods protect qualified employees working on electrical equipment, while other methods focus on employees working nearby. Protective methods include:

- De-energizing the circuit
- Using safe work practices
- Installing extra insulation
- Guarding circuits
- Installing barricades
- Using ground fault circuit interrupters (GFCIs)
- Installing grounding (secondary protection)

When crews work on live circuits, safety-related work practices include:

- Personal protective equipment (PPE)
- Insulated tools
- Written safety program
- Job briefings

Summary

- Electrical distribution systems receive electricity from a variety of generation facilities and transmit it to end-use customers at the required voltage, current, and frequency.
- Distribution systems begin at substations, where they receive electricity from transmission lines (and other sources) and reduce (step-down) voltage from transmission levels to distribution levels.
- From step-down substations, conductors, commonly called *feeder lines* (or simply *feeders*) transmit electricity at primary distribution voltages to transformers located near each load.
- From the final transformer, secondary feeders transmit electricity to the customer connection point.
- Nearly all electrical customers belong to one of four categories: residential, commercial, industrial, and transportation.
- One of the most important aspects of electrical distribution is safety.

6

Wide Area Controls and Cyber Security

Simple good, complex bad.

—Samuel Thomas (1953 –)

"Grandpa Tom, what is that thing we drive by on the way to the ranch – the one with the big fence and lots of poles, wires, and other equipment?" asked Luke.

"That is an electrical substation," Luke's grandfather replied.

"But what is it used for?" Luke asked.

"The equipment inside that fence controls the transmission of electricity and makes sure we have electricity around the clock," replied the grandfather.

"But where are the people making it work?" asked Luke.

"Equipment inside that substation, equipment inside other substations, and equipment connected along the electric lines are all monitored remotely," explained Grandpa Tom. "Monitoring means collecting what engineers call 'process data' from the equipment, so monitoring happens in the substation and along the lines. Control means opening, closing, starting, and stopping equipment directing electricity where and when it is supposed to go," added the grandfather. "Some of the control happens inside the station automatically, some happens automatically from a central control room, and humans oversee everything," added the grandfather.

"That must take a lot of equipment and really smart people!" exclaimed Luke. "Can they operate all the equipment in our town from just one control room?" he asked.

"Even bigger than that," Grandpa Tom said. "The state of Texas has one operator, called the Electrical Reliability Council of Texas (ERCOT), that operates the electrical grid for most of the state and is called an Independent System Operator or ISO," added the grandfather.

"What about the rest of the world, does it have ISOs also?" wondered Luke.

"Yes, lots of them," Grandpa Tom replied.

"Wow," exclaimed Luke.

The Grid

The electrical grid can be thought of as a vast machine composed of generation facilities, transmission lines, and distribution systems. As with any machine, understanding and controlling the grid are critical to ensure generation facilities (from large nuclear plants to rooftop solar panels), conductors (operating at a variety of voltages stretching across large expanses), assorted instruments and pieces of equipment (located at multiple substations, along lines, and in control centers), and, increasingly, storage facilities all operate safely, reliably, and efficiently. Figure 6-1 is a picture of the Electric Reliability Council of Texas, Inc. (ERCOT) control room.

Figure 6-1. ERCOT Control Room.

This chapter discusses the instruments, equipment, software, and smart devices used for grid monitoring and control and the related topic of cybersecurity. The following chapter covers the people and processes making it all happen.

SCADA

The person who came up with "Supervisory Control and Data Acquisition" got it reversed. Data is acquired first, and then process control occurs. But it is too late to change to "DAASC," so SCADA it is.

The U.S. National Institute of Standards and Technology (NIST) defines SCADA as:

> A generic name for a computerized system...capable of gathering and processing data and applying operational controls over long distances. Typical uses include power transmission and distribution and pipeline systems.

SCADA was designed for the unique communication challenges (e.g., delays, data integrity) posed by the various media that must be used, such as phone lines, microwave, and satellite.[45]

According to this definition, SCADA is a *computerized system* that gathers, processes, and monitors information, allowing operations control. Not included in this definition, but critical to the system, are the *instruments, equipment, communication systems,* and *smart devices* – also called intelligent electronic devices (IEDs) – generating data and performing controls. Figure 6-2 is a SCADA schematic.

Abbreviation or Word?

SCADA systems are ubiquitous, and, like *scuba* (self-contained underwater breathing apparatus), the abbreviation SCADA may become a word – scada – by the next edition of this text, but for now it is SCADA.

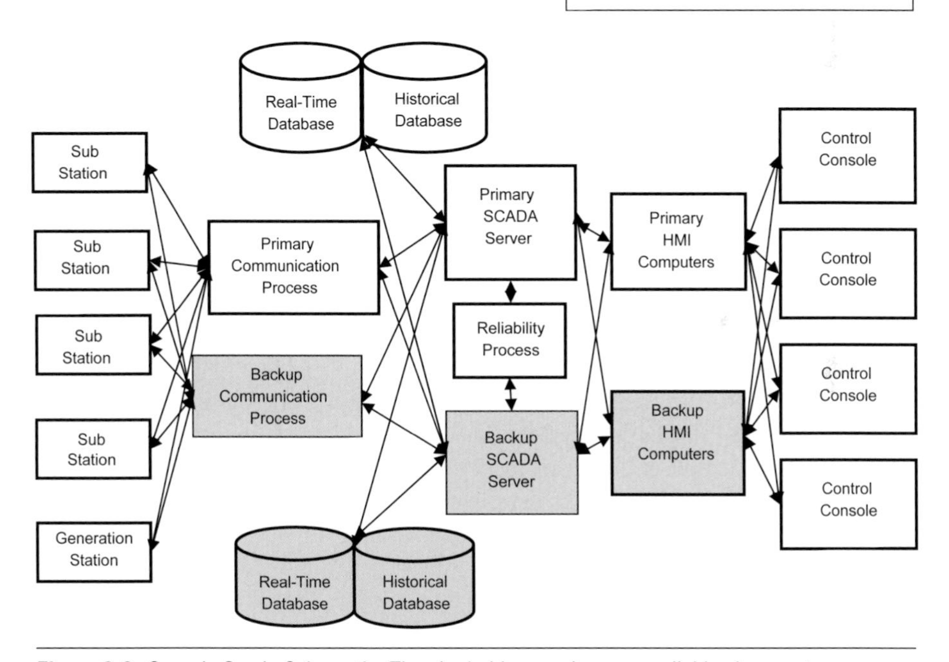

Figure 6-2. Generic Scada Schematic. The shaded boxes show a parallel backup system.

Reading the figure from left to right, the instruments, equipment, and IEDs are largely located at substations. The lower left box in Figure 6-2 adds a generation facility. Increasingly, SCADA reaches directly into the generating facility, where it interfaces with the process control system operating the plant. This direct connection to the generators

IED

In this text, the abbreviation IED refers to integrated microprocessor-based controllers of power system equipment, such as circuit breakers, transformers, and capacitor banks.

provides data to an energy management system (EMS), allowing operational optimization across the grid.

IEDs collect data from instruments and equipment and perform preliminary analysis. Based on this preliminary analysis, smart devices may operate station equipment by sending commands. From the station, one or more communication systems send the station data to a central smart device, labeled in the figure as a SCADA server. In small systems, the central smart device might be a personal computer.

> **Smart Device**
>
> This text uses the term broadly to mean devices that receive, manipulate, and store data and share that data with other devices and users, regardless of the size of that device.

The arrows in Figure 6-2 represent communication systems transmitting process data (volts, current, frequency, equipment status) collected by IEDs to the primary SCADA server and transmitting commands back to field-based equipment. Communication systems, like other parts of SCADA, have backups and redundancy.

The central smart device has a real-time database containing a current value table holding the most recently received data values from the field devices and receiving data updates as frequently as the communication system allows. When a new value comes in, the previous value gets archived in a historical database. Just like the smart device at the substation, the central smart device analyzes the data and, depending on its programming, may send commands back to the substation smart device for execution at the substation.

The smart device arranges information and displays it on a series of displays on a "control console" staffed by a system operator. Based on their analysis, operators send commands to the central smart device, which sends those commands through the communication process to the station smart device, which, in turn, passes the command to the appropriate piece of equipment for execution. Increasingly, as technology improves, control functions move to computers, and the operator roles shift to higher-level decision-making, planning ahead, and responding to emergencies.

Substations

Substations, and the equipment, instruments, and components comprising them, were discussed in Chapters 4 and 5. Much of the equipment located at the substation is automated, providing monitoring and control functions to either technicians at the station or system operators in the control room. An on-site technician must physically open, close, start, or stop unautomated equipment (for example, some switches). Station instruments read process data, including voltage, amperage, frequency, and device status, as well as a myriad of other data. Most station instruments can be read remotely, and the data collecting process is called *data acquisition.*

Local SCADA

Local SCADA is often included as a convenience at a substation for operations or maintenance people, providing them an easy way to interact with the equipment. Many times, local SCADA vendors are different from those supplying the centralized SCADA. Most local SCADA systems have one or perhaps two displays and a keyboard. Displays allow the viewing of equipment status, and keyboards allow data input and the issuing of commands. Figure 6-3 shows a block diagram of local station SCADA.

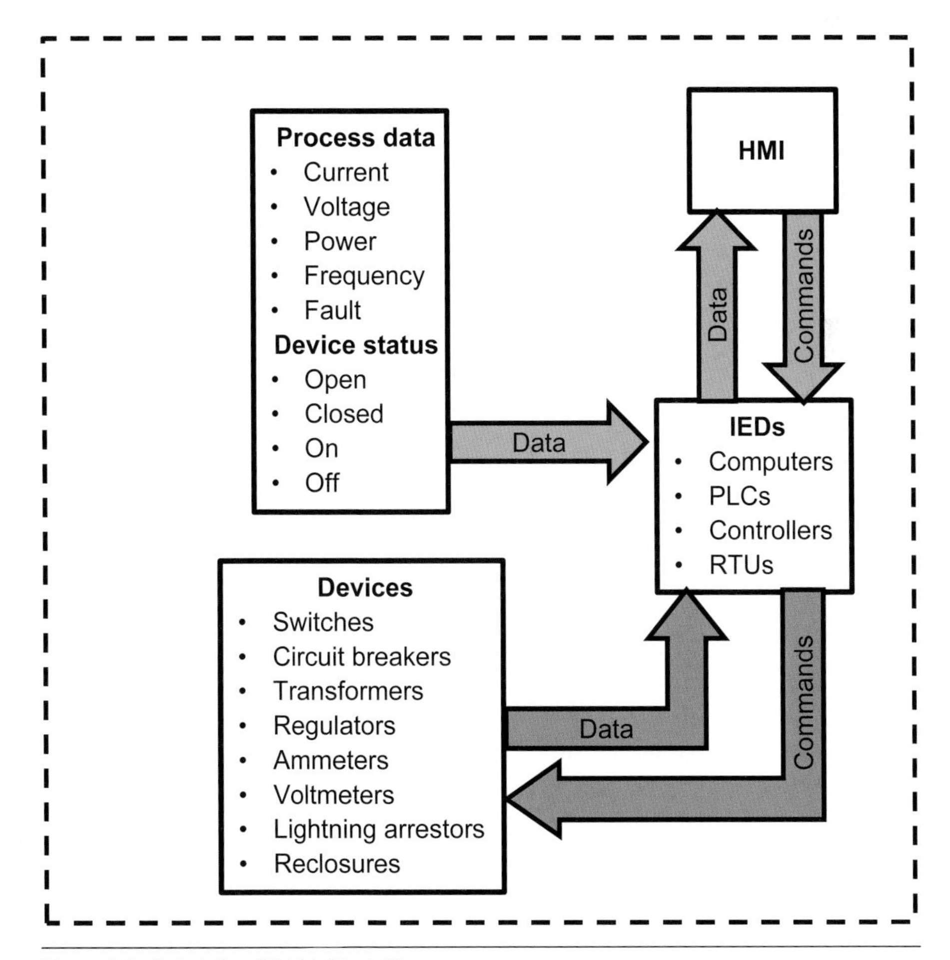

Figure 6-3. Substation SCADA Block Diagram.

Local substation SCADA provides the front and back ends of the overall SCADA system, as it collects process data and executes commands.

Communications

Network Protocol

A networking protocol is a set of rules for formatting and processing data – like a common language for computers. The computers within a network may use different software and hardware but protocols enable them to communicate.

Standardized protocols are like a common language computers can use, like two people who speak different languages but can communicate using a shared third language.

Transmission and distribution substations are located miles apart and, in the case of transmission, can be hundreds or even thousands of miles from the central control room, making robust and secure communications a priority. From the broad range of options – leased lines, microwave, radio, dedicated phone lines, the internet, local area and wide area networks – SCADA engineers choose what works for the location and control system in question. Most communication systems combine technologies. Various smart devices and communication systems use different protocols. SCADA designers must ensure all devices can communicate with each other. As mentioned earlier, redundant communication systems provide increased reliability.

Centralized SCADA

At their heart, SCADA systems are large databases collecting, organizing, storing, and analyzing data, displaying it, and then processing commands. Many different industries use the same SCADA systems. They all use the same database architecture, communication protocols, display software, and command processes and are configured to fit the applications of the industry.

Databases store:

- Values – volts, amperes, frequency
- Status – open, closed, on, off
- Commands issued by smart devices or humans – open, close, turn on, turn off
- Events – breaker opened or ground fault occurred
- Data quality – the trustworthiness of the data (e.g., “stale” because the remote terminal unit (RTU) supplying the data was off-line for some time)

All the data is time-stamped, meaning the system recorded the time it happened. Figure 6-4 adds the central SCADA to Figure 6-3.

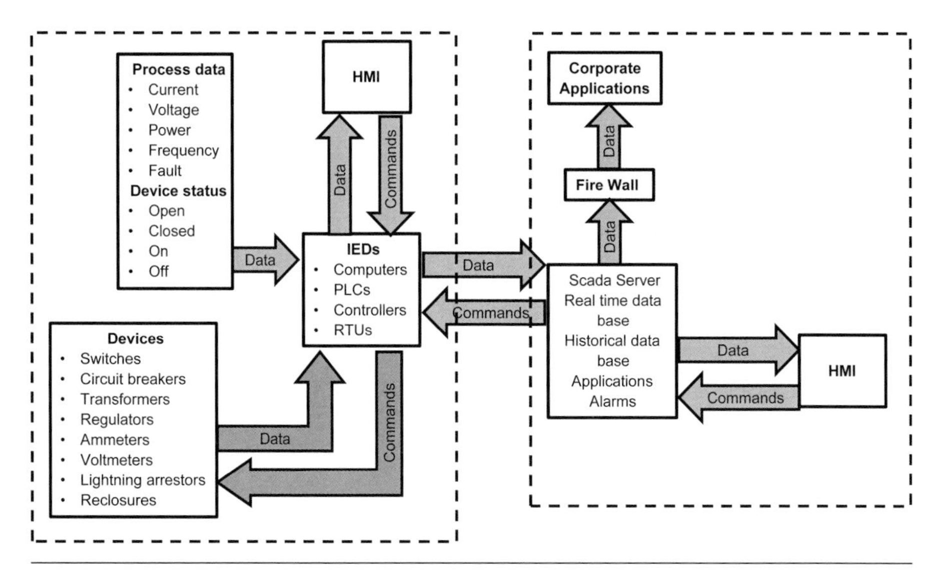

Figure 6-4. SCADA Block Diagram.

Real-Time Database

The database arranges voltages, currents, frequencies, and other process data to allow access when needed. Graphics packages present this information on displays from which operators visualize operations. The system compares operating values to preconfigured limits and issues warnings when values exceed these limits on either the high or low side.

Historical Database

Real-time data gets transferred to a historical database on a set time frame, providing powerful tools to optimize the power systems for maximum efficiency. Trending tools arrange data in various ways. Graphical interfaces simplify analysis. Historians store data from past events, facilitating comparisons among various operating scenarios.

Analyzing trends and historical data also helps maintain desired power factors, voltage levels, and other system parameters at desired levels. These tools become more critical as grid complexity increases, driven by increased numbers of small generators and storage facilities.

Points

The notion of "points" covers a lot of information. The origin and destination of data going to and coming from a device are *points*. For example, the incoming voltage at the Montopolis substation (in Austin, Texas) is one point. Each point links to a unique identifier. Process data – in this case, voltage – goes from the origin point (the incoming station voltmeter) to a destination point (the place in the computer that captures and stores it in the real-time database).

The point of all this discussion of points (pun intended) is to emphasize the care SCADA engineers must exercise to keep all the data interacting correctly. Electrical control systems have hundreds to thousands of points. A points database or system administrator keeps track of the points. The points database contains a list of all devices, commonly shown on a process and instrumentation diagram (P&ID) for easier reference. During SCADA installation, technicians carefully verify all points are listed and connected properly at each end.

Following original construction, SCADA technicians update the points database when they change the system. If they add a new, remotely controlled connection point, they must correctly add it to the points database and test the devices to ensure proper connections. Because of the potential consequences of errors in the points database, access is strictly restricted to authorized personnel.

Update Frequency

The SCADA system receives data from thousands of devices that travels along satellite channels, fiber optic strands, microwave systems, or even phone lines. Some values change rapidly and frequently. Other data, like circuit breaker status, does not change frequently during normal operations. Data requiring frequent updates usually has multiple, parallel communication circuits so the system can poll data as fast as possible. The SCADA system manages all the data to ensure its availability as needed and that it does not become garbled in the process.

Gathering and communicating all this data in an orderly fashion requires an established protocol dictating how often (every five seconds, for example) and when (7:00 a.m. each day, for example) to gather data. The frequency and manner of data gathering fall into one of three categories:

- Polling (also called scanning)
- Reporting by exception
- Fixed interval reporting

During a poll or scan, the central system requests from a device a complete value/status report (often called an integrity scan) or a partial report, based on point type or priority. Scan tables establish how frequently the system polls devices. Some data is reported quite often, other data only once per day.

In a "change report," a device reports only a change in value or status since the prior poll. For example, if a pump turns off, it alerts the control system, but sends no status update if it keeps running. In "report by exception," the device initiates communications, but the SCADA system does not poll it. Such an event also sends an alert to the operators. Some method of periodic integrity checking usually takes place for "report by exception" to ensure that data changes are captured.

Fixed-interval reporting is self-evident. A device reports its data on some regular basis (some say fixed-interval reporting is simply another example of "report by exception" – if the set time has not elapsed, the device does not report).

Applications

While many industries use the same generic SCADA, including the same database, communication protocols, display software, and command processes, they do not all use the same software applications. In an electrical power system, a command might open a breaker. In a pipeline system, the same command might start a compressor. SCADA engineers build applications for each industry and customize or configure them to the specific operation.

Configuration Tool

Any database must be configured for the specific application. Like a spreadsheet, the real-time database starts as a blank page with rows and columns awaiting data. The users configure it (the spreadsheet or the SCADA database) by writing instructions (or commands in the case of the SCADA database and equations in the case of the spreadsheet), using functions and macros. The functions and macros for writing instructions and commands make up the configuration tool. The easier and more intuitive the tools, the better users like the spreadsheet or SCADA database.

Developing, updating, and maintaining SCADA systems calls for, at minimum, a graphics editing program to build the screen graphics and a database configuration tool. Like a spreadsheet, the easier and more intuitive the tools provided by the vendor are, the better the developer and, probably, the SCADA user like it.

Alarms

Alarms play an important role in operating the grid because they alert operators to problems or potential problems. Some of these problems require immediate attention. Alarm management teams carefully consider, in advance, the urgency and criticality of each alarm and develop alarm ranking systems, ensuring the most critical and urgent alarms appear at the top of the display, with less critical and less urgent alarms listed lower on the display.

Application Interface

Some applications are outside the SCADA system but use data stored in the real-time database. This requires an interface or "hook" to allow data to move from the real-time database to the application through a fire wall to protect the security of the data and system.

These hooks are analogous to the interface allowing customers to download their personal banking data from a bank's server to the money management software on their computers. The customer does not see the hook but cares a great deal about whether the interface works well.

Like banking data, the SCADA interface must be secure. Any time an application accesses the real-time database, the interface must protect the real-time database from outside intrusion.

Reports and Logs: Reports and logs relate closely to data sharing. SCADA systems generate various reports automatically and distribute them electronically (or in paper form) within the company and to customers. Within the company, various groups can request generation of *ad hoc* reports. The company's data sharing protocols greatly impact the decisions around report generation and distribution.

Logs are a bit different. Logs capture events and commands and are an important part of the historic database for troubleshooting, optimization, and training. The SCADA system routinely generates them and (normally) distributes them to only those with a "need to know." However, when an incident occurs, logs are of great interest to all types of stakeholders within the company and outside (e.g., regulators, consultants, vendors).

Cybersecurity[46]

SCADA systems make extensive use of computer and networking technologies borrowed from the information technology (IT) world and vulnerable to malicious cyberattacks and malware. The critical and differentiating factor in addressing electrical SCADA system cybersecurity is the potential consequences of a system compromise. If an attacker compromises an insurance company computer system, the company and its customers may suffer financial consequences. If an attacker compromises a SCADA system controlling electric power systems, businesses and households can lose electric power and everything that depends on electric power: light, heat, water, refrigeration, communications (to name a few). In other words, people's health and even lives could be put in jeopardy, so electrical power systems must have strong cybersecurity protections.

SCADA systems have:

- centralized assets, such as servers and workstations where operational personnel monitor and perform supervisory control of field-based equipment;
- field-based equipment located at various facilities throughout the system; and
- communications equipment connecting the central facility to the field-based equipment.

SCADA systems usually have a high level of redundancy in their design, to provide 100% availability. The company may also have an alternate (or backup) site with a set of duplicate assets to serve as the backup control center if the primary control center becomes unavailable or unusable for any reason. And, of course, various field sites may contain anything from a small RTU up to a stand-alone control system and a communications system to the central facility.

Backup Requirement

NERC (in Reliability Standard EOP-008-2) requires a fully functional backup control center (BUCC) in case the primary control center is lost. These BUCCs cannot have only "basic" duplicate assets. They must be able to completely control the entire system.

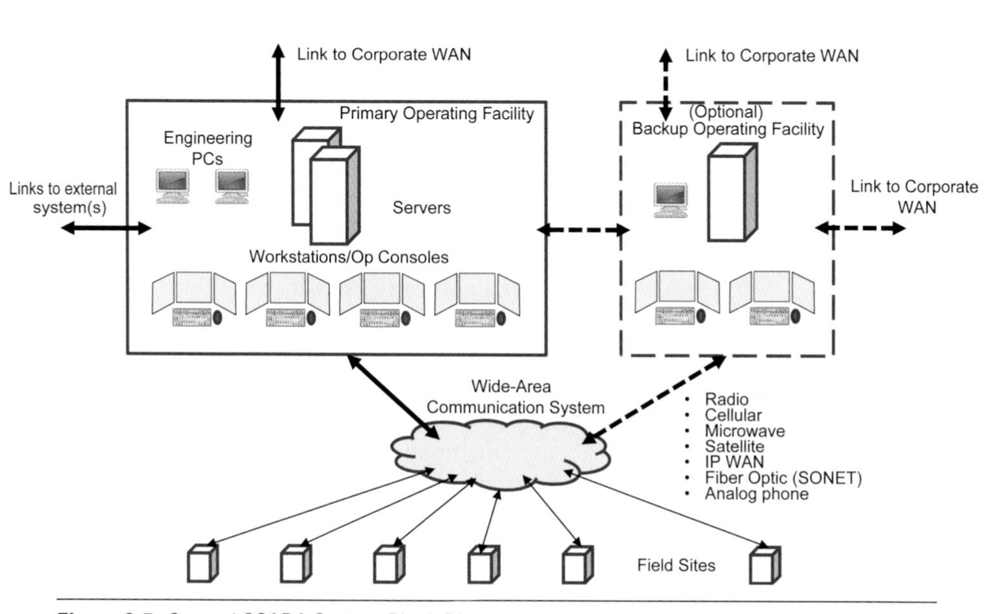

Figure 6-5. General SCADA System Block Diagram

From a cybersecurity perspective, the primary question is: "How could an adversary gain access to the SCADA system to attempt a compromise?" The second question is: "What is the worst-case consequence if an attack were successful?"

Attack Pathways

The first question deals with *attack pathways* – the means for gaining access to attack the SCADA system. Even if a system is full of exploitable vulnerabilities and weaknesses, an attacker cannot take advantage of and exploit them unless a viable attack pathway exists. Cybersecurity professionals generally recognize six attack pathways:

1. Wired network connectivity to the system
2. Wireless network connectivity to the system
3. Physical access to the system
4. Use of portable computer-readable storage media on the system
5. The use/connection of portable computer-based devices to the system
6. The supply chain for the system

The first two items refer to Internet Protocol (IP)-based local or wide-area wired and wireless networks that do not usually include low-speed, serial, or asynchronous (or bit-oriented) RTU polling circuits employing a fixed-command-set SCADA/RTU protocol. This is because, at worst, an attacker can hijack such a circuit and send false information to the SCADA system (or commands to the field-based equipment) but cannot compromise the SCADA system (i.e., deliver malware, modify program code, or alter system functions). This does not mean those circuits do not need protection, just that they are not considered attack pathways to compromise the SCADA system itself.

Likewise, there are special-purpose industrial local area networks (LANs or a "fieldbus"), not based on Ethernet or Transmission Control Protocol/Internet Protocol (TCP/IP) technologies, that are not usually considered an attack pathway, essentially for the same reasons.

If a company protects attack pathways, an attacker will most likely never succeed in compromising the SCADA system. The question, then, is how to best protect attack pathways.

The *Sunburst* supply chain attack in 2020 demonstrated that even well-protected systems can have an "Achilles heel" (the supply chain). That attack was not prevented but, fortunately, was eventually detected, pointing out the need for protective measures *plus* detective measures in case protections fail.

Sunburst Attack

A massive, fifth-generation cyber attack waged against U.S. government agencies and technology companies leading to the compromise of systems in over forty government agencies.

Worst-Case Scenarios

Many people fail to fully explore worst-case scenarios because they interpret the term *compromise* as "shut-down" or "break."

The proper interpretation of the term "compromise" is to envision an event in which an attacker takes control of an operator's human-machine interface (HMI) and issues commands and changes the settings of the field-based equipment. In an electric power transmission system, such an event could mean tripping or closing circuit breakers, thereby damaging bulk power system equipment. Properly assessing the worst-case consequences provides a legitimate basis to justify the manpower and funding to implement a strong cybersecurity program.

Cybersecurity Countermeasures

Few computer systems and communication networks have integral mechanisms for cybersecurity. Cybersecurity is usually layered on by applying *countermeasures*, which fall into three broad categories:

- Physical
- Technical
- Administrative

An equipment enclosure with a keylock is a *physical* countermeasure. Installing a firewall between a corporate wide area network (WAN) and SCADA system's LAN is a *technical* countermeasure. A policy prohibiting the use of any Universal Serial Bus (USB) flash drive on the SCADA system is an *administrative* countermeasure. Countermeasures can be protective, detective, or corrective, or a mixture of these functions. ("Corrective" measures help remediate the impact of a cyberattack or aid in recovering from a cyberattack.)

The best cybersecurity usually comes from a combination of appropriate countermeasures applied to the attack pathways defined above. Of course, eliminating an attack pathway is best, if possible, but not always easy to accomplish. Administrative countermeasures apply to personnel (including contractors and vendor personnel) in the form of policies, procedures, and training.

Why SCADA Systems are Vulnerable

Up to the 1990s, very few SCADA systems connected to corporate business/IT systems, and connections to regional/regulatory entities were via leased telephone circuits using some SCADA protocol or proprietary applications. Today, connections to the corporate WAN using Internet Protocol (IP) networking, and even using the internet to communicate with those same regional/regulatory entities, is common.

Many servers and workstations run Microsoft Windows[47] operating system variants, making common the use of USB flash drives and communication via Ethernet and TCP/IP. Laptop and desktop PCs also have USB connections and include integrated "Bluetooth" and "Wi-Fi" capabilities. In other words, systems generally, and unfortunately, contain one or more attack pathways.

Fortunately, vendors of SCADA technology, as well as corporate IT organizations, are becoming more aware of the need for cybersecurity and somewhat more capable of applying countermeasures. Cybersecurity is, however, more important than ever as hackers gain experience and skills.

Applying Countermeasures

After eliminating as many attack pathways as possible, the next step is adding protective and detective countermeasures for remaining attack pathways. For wired or wireless network pathways extending beyond the SCADA system itself, companies generally install firewalls.

Firewalls

> **Malware**
> Software specifically designed to disrupt, damage, or gain unauthorized access to a computer system.
>
> **Signature**
> A specific pattern that allows cybersecurity technologies to recognize malicious threats (such as a byte sequence) in network traffic or known malicious instruction sequences used by malware families.

Today's "Next Generation" firewalls have extensive abilities to detect and block malicious and suspicious traffic. Some even detect and block malware propagation (and have malware "signatures" needing periodic updates). If integrated with a security event information management (SEIM) system, these firewalls can not only block attacks but can provide to designated personnel near real-time notifications and alerts of nefarious activities. For maximum SCADA security, every network connection to external entities (including connections to a corporate WAN) should have such a firewall.

Any *permanent* links to external entities, especially if using IP for connectivity, should have the additional security of a virtual private network (VPN) between the external entity and the SCADA system with the firewall's VPN-gateway functionality enabled. This arrangement also requires a gateway at the other end of the communication channel. It also applies to connections between a primary and backup operating site.

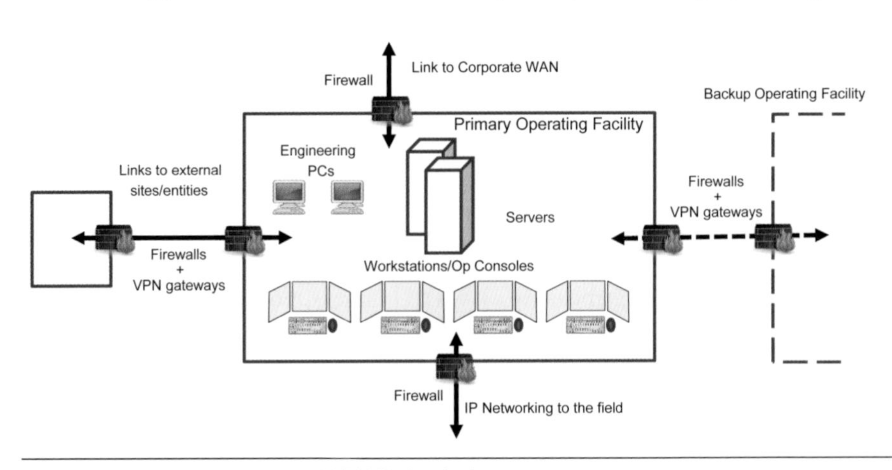

Figure 6-6. Applying Firewall and VPN Technologies.

Wi-Fi Connections

In general terms, cybersecurity professionals discourage wireless (Wi-Fi) technology with a SCADA system or within the overall facility and LAN. Wi-Fi networking is one of the hardest pathways to defend, but the easiest to eliminate. Unless a SCADA system requires a Wi-Fi connection, the company should prohibit its use. When Wi-Fi is allowed, devices and access points (wireless routers) should use the latest level of wireless security technology: currently WPA2, using AES encryption (unless products with the promised WPA3 standard have become available). Even if using the latest and most secure wireless security, a firewall should reside between the wired LAN and the wireless access point(s).

Internet Connections

The three primary ways attackers gain access to computer systems are:

- "Phishing" emails containing a link to a malicious website
- Employees browsing to a malicious website
- Malicious email attachments containing malware

Improper responses to these three attack methods give a hacker a path into a computer, which he can then use to attack other computers and systems.

SCADA systems should have *no* pathway to the internet allowing either email or web browsing from any computer on the SCADA system or on a network shared by the SCADA system (which differs significantly from having a VPN communication session traversing the Internet). SCADA systems should also be isolated from the corporate WAN, at least with a high-capability firewall, because that corporate WAN almost certainly provides internet access to corporate users. If personnel in the control center need web or email access, that access should come via a separate computer not connected to the SCADA system's LAN.

Physical Security

Physical access pathways are generally protected by one or several of four types of physical security countermeasures:

- Locked rooms with restricted keycard access
- Locked cabinets with a key sign-in/out procedure
- Video surveillance
- Security guards/officers

Companies should limit access to computer/server rooms and telecom/network cabinets to only those who need access, and they should establish and enforce policies and procedures for escorted access. Lack of adequate physical security compromises cybersecurity protections.

Physical security is more difficult at remote field sites than at central sites because they are usually not staffed and typically have only minimal physical security – often a chain-link fence and a locked gate.

Figure 6-7. Control Building at a Remote Field Site.

SCADA systems using IP networking at field sites create a potential attack pathway. If an adversary gains physical access to a field site, that adversary could establish an IP network connection and create an attack pathway to the SCADA system through that connection. Consequently, both physical security and a strong firewall that can detect and block attacks on that IP network pathway are critical at unstaffed field sites.

Preventing Portable Media Attack

Portable media such as USB flash drives are ubiquitous and used extensively for data transfer and software update, but a large percentage of malware gets into computer systems via these drives. As discussed earlier in this chapter, common practice includes limiting USB connections as much as possible to eliminate them as attack pathways. Portable media attack pathways, however, cannot be eliminated because some patches and software updates, signature updates, addition of new applications to the SCADA system, and installation of cybersecurity technical countermeasures require portable media.

Two common approaches exist to protect against attacks via portable media. First, many organizations prohibit the use of any non-approved (unknown, uncontrolled) portable media on the SCADA system or any associated sub-system.

Second, they use color-codes and labels to clearly identify permitted media, as well as a scanning system (a "kiosk" to transfer files – after antivirus or AV scanning – from untrusted media onto known, approved, trusted media), then used by the SCADA system and associated sub-systems.

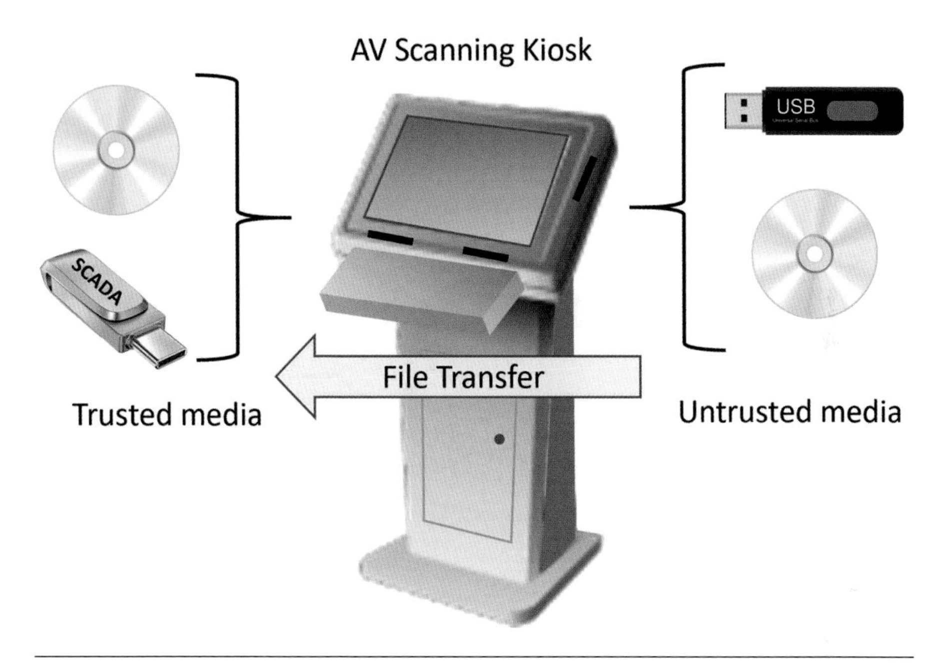

Figure 6-8. Portable Media Scanning Station.

> **Heuristic Scanning**
> A form of scanning not dependent on a malware signature to perform detection.

These scanning stations usually employ several AV scanning engines with at least one capable of performing *heuristic scanning*. The scanning station may also scan files for anomalous content (e.g., JavaScript in a Microsoft Word® document) and block specific file types (like a Visual Basic script, PowerShell script, or executable file). Scanning stations require periodic updates to their malware signature database to protect against newly identified malware.

Preventing Portable Computer Attacks

Portable computing devices (e.g., laptop PCs or tablets) used for testing, support, or diagnostic functions are sometimes connected to the SCADA system. These connections include LAN connections, connections to a "console" port, or connections to a diagnostic/maintenance interface. These connected devices can contain malware and use the connectivity to infect/attack the SCADA system.

Vendors sometimes bring laptop PCs containing test/diagnostic software and request a connection to a device, network component, or computer on the SCADA system. These connections, if permitted, create the possibility of an inadvertent attack. To prevent these attacks, some organizations require delivery of any outside laptop or tablet in advance and put it through an AV scan just like portable media. Hard drives are removed from the computers and plugged into a USB adaptor module that can scan them using the same scanning kiosk as used for portable media to verify the hard drive contains no malware.

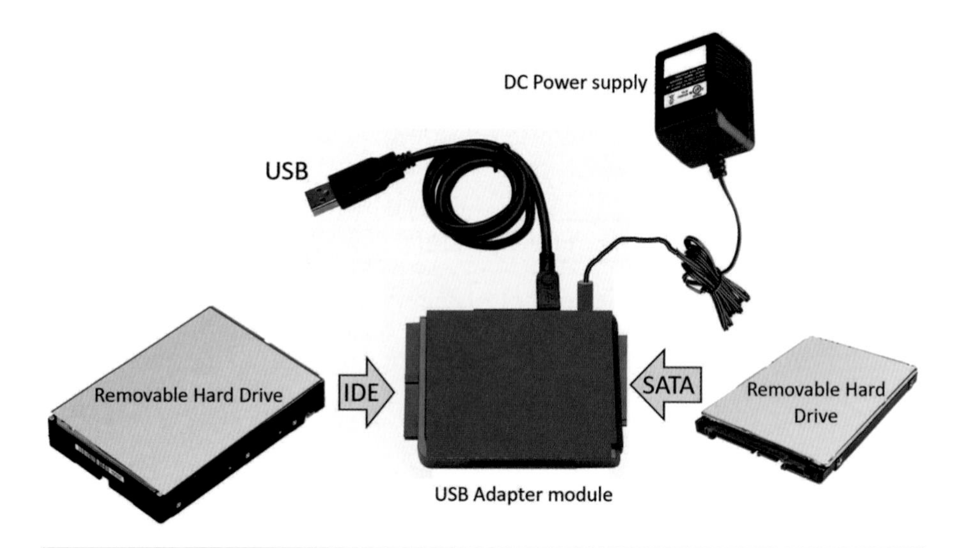

Figure 6-9. Hard Drive USB Adapters.

> **Root Kit**
> A set of software tools enabling an unauthorized user to gain control of a computer system without being detected.

One good feature of this scanning approach is that the presence of a *root kit* or hidden malware in an alternate data stream (ADS) or in disk slack space (which can pose a problem for AV software on the laptop PC) will not hinder the scan.

Protecting Against Supply Chain Attacks

Protecting against supply chain attacks is one of the most difficult challenges – as demonstrated by the Sunburst attack previously mentioned. The SCADA supply chain includes receiving and applying patches, signature updates, and software updates to various system components. Supply chains also include purchased equipment and software that become part of, or are connected to, the SCADA system, giving vendor personnel physical access and, possibly, remote access to a

SCADA system. (Allowing vendors to use their own computer media and portable devices as part of this access was discussed earlier in this chapter.)

Most organizations address the supply chain issue with policies and procedures such as escorting and observing vendor personnel and performing incoming testing and inspection of equipment (and, of course, AV scanning of computers and computer media). These serve as examples of "detection" and, to a degree, "prevention" countermeasures on the supply chain attack pathway.

Mitigating Countermeasures

In addition to eliminating attack pathways and preventing attacks, companies often employ mitigation measures to lessen the impact or lower the likelihood of an attack.

One common mitigation activity is "hardening" the servers and workstations comprising the SCADA system. Systems hardening consists of a collection of tools, techniques, and best practices to reduce vulnerability and security risk by eliminating potential attack vectors and reducing the attack surface by removing superfluous programs, account functions, applications, ports, permissions, access, etc., giving attackers and malware fewer opportunities to gain access to a system.

Removing or Disabling Applications

One aspect of hardening is removing or disabling applications, services, and utilities not required for the essential SCADA functions. For example, Windows comes with X-Box services, FAX services and telephony services (among others) not needed for SCADA functions. Removing or disabling these applications and services means a hacker cannot use them as attack pathways. Hardening also includes eliminating unnecessary user accounts and support accounts and strictly controlling administrative accounts.

Eliminating or Disabling Remote Access

Another hardening strategy is eliminating or disabling remote access services and blocking TCP/UDP ports not needed for the SCADA system functions.

Remote desktop support in Windows is a dangerous and exploitable service that should be removed unless absolutely needed.

TCP/UDP

TCP provides applications a way to deliver (and receive) an ordered and error-checked stream of information over a network.

Software applications use the User Datagram Protocol (UDP) to deliver a faster stream of information by doing away with error-checking. When configuring some network hardware or software, a technician should know the difference.

Attackers look for open ports and known-vulnerable services on computers if they get onto a network. Therefore, removing or blocking them eliminates some exploitable vulnerabilities.

Policies and Procedures

Hardening also includes developing and implementing policies and procedures for timely patching and updating of the SCADA system. Security patches (and sometimes updates) should be routinely and systematically applied to eliminate known and exploitable vulnerabilities, making the system less vulnerable to attack. Patches and updates, however, should be thoroughly understood before being applied because installing the patch or update can potentially result in system damage.

Patches should be thoroughly tested for compatibility with the SCADA system and applied to the backup system (or a test system) before the primary system. If a patch applies to parts of the system removed during hardening, it should not be applied. Applying patches and updates on a test system is recommended, and some organizations make virtual machine (VM) copies of their major SCADA system components and use those for testing prior to installing any patches and updates on the operating SCADA system.

Hardening a system reduces its *attack surface*, meaning it eliminates potentially vulnerable software. If, for example, an attacker penetrated the SCADA defenses and tried to exploit a known weakness in the Windows media player on one of the SCADA system computers, but that application had been disabled or removed, the attack would fail. These hardening suggestions just scratch the surface – many online sources discuss hardening recommendations in more detail.

Implementing Endpoint Protection Software

Whitelisting

A cybersecurity strategy allowing only an approved list of applications, programs, websites, IP addresses, email addresses, or IP domains to run in a protected computer or network. Users can access applications or take actions only with explicit approval by the administrator. Anything outside of the list is denied access.

Another mitigating strategy is implementing endpoint protection software on the SCADA servers and computers. This is a form of host-based intrusion detection system (HIDS) that acts like a firewall and AV scanner and that may, if it supports whitelisting, even prevent unauthorized software (such as newly installed malware) from executing. With whitelisting, even if an attacker manages to install malware on a computer, the whitelisting can prevent it from executing, which prevents it from performing its malicious functions. Windows supports a capability called “AppLocker,” a form of whitelisting.

Implementing Full Disk Encryption

Systems containing sensitive information can implement full-disk encryption (called an encrypted file system or EFS), making any extracted information useless to the attacker. SCADA systems, however, generally do not contain information of a sensitive nature, making EFS of questionable value. Windows supports a capability called "BitLocker" – a form of EFS.

System Recovery and Restoration

A final mitigation measure is establishing thoroughly documented, vetted, well-tested system recovery and restoration procedures. Support personnel should receive specified training (and periodic retraining) in performing system backups and restorations and maintaining and updating the necessary backups to perform restoration and recovery (e.g., making new backups after applying patches or installing updates).

Detective Countermeasures

The threat of attack never ends, and attackers' methods and techniques keep evolving, which means protective measures may fail. Attackers may also bypass defenses. For example, an employee might bring in to the office infected media (e.g., a USB drive) and insert it into a networked computer, allowing malware to propagate through the system. The possibility of a successful attack requires detection measures to provide alerts in the event of a defense breach.

Network-Based Intrusion Detection System

One common detection tool is a network-based intrusion detection system (NIDS). The NIDS collects message traffic passing through the network and examines that traffic for known or suspected malicious content, usually by setting up a Switched Port Analyzer (SPAN) on every network switch. The SPAN sends all message traffic to another switch that aggregates the messages and passes the aggregated message stream to a *sensor* – a very high-speed specialized computer that examines and analyzes the traffic for known or suspicious patterns.

When the automated system detects suspicious message patterns, it sends a message to a dedicated NIDS console so cybersecurity staff can review the alert and perform additional analysis to assess the threat. The message patterns used by the NIDS are called *signatures*. Evolving hacker approaches requires periodic update of the NIDS signature.

Security Information and Event Management

Another monitoring approach is called Security Information and Event Management (SIEM) technology. SIEM involves software products and services combining security information management (SIM) and security event management (SEM) to provide real-time analysis of security alerts generated by applications and network hardware. These systems can also log security data and generate reports for compliance purposes.

Combined NIDS and SIEM

Detection can be improved by combining a NIDS with a SIEM and having every network-connected SCADA system asset (e.g., servers and PCs) forward its logs to the SIEM. In addition, all other network components (such as the firewalls and switches) should send syslog messages to the SEIM. The SIEM is configured with rules, somewhat like a firewall, directing it to process and correlate the (Sys)log output looking for suspicious, anomalous, and malicious activity. With a SIEM, a separate NIDS console is usually not needed because the SIEM provides alarms and alerts. For this approach to work well, the SCADA system must have a network-based time source keeping all equipment time synchronized. Figure 6-10 shows the layout of a combined NIDS and SIEM system.

Syslog

An acronym for "System Logging Protocol" which is used to send system log or event messages to a specific server, called a syslog server. It primarily collects various device logs from several different machines in a central location for monitoring and review.

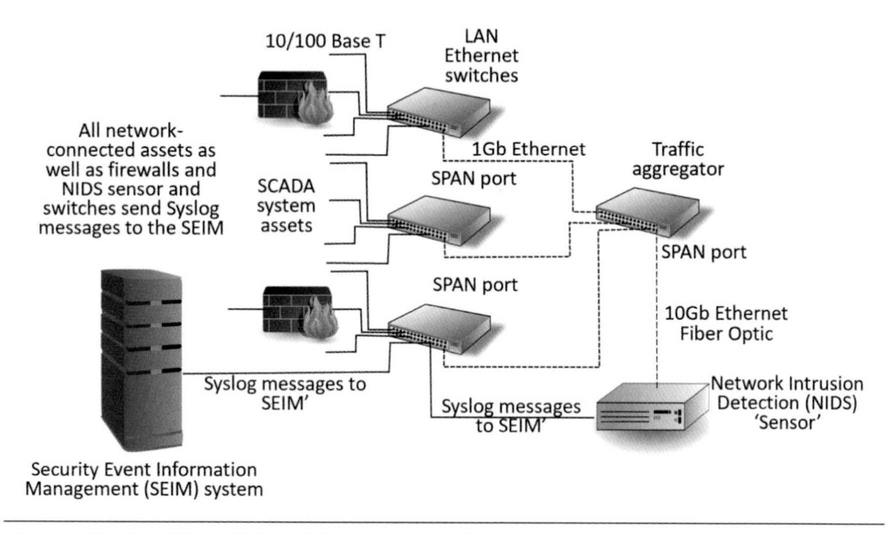

Figure 6-10. Simplified NIDS and SEIM Configuration.

Field Site Communications

> **Serial Communication**
> A communication method using one or two transmission lines to send and receive data continuously sent and received one bit at a time.

Communication between the SCADA server and field-based equipment can use a wide range of communication technologies – wired, wireless, leased, or self-constructed. Currently, connection with field-based equipment is either an IP-based WAN connection using an IP-based industrial protocol (e.g., Inter-Control Center Protocol (ICCP) or Utility Communications Architecture Version 2.0 – UCA2) or a low-bandwidth *serial* connection employing some form of SCADA, RTU, or Programmable Logic Controller (PLC) asynchronous (or bit-oriented) communication protocol such as DNP3.0, Modbus-RTU, or a hybrid combination of the two. The first option offers an attack pathway into the SCADA system, whereas the second does not.

Access to a communication channel also provides a pathway for attack. Such an attack happened to a water/wastewater SCADA application in Australia using radio communications.[48] In theory, such an attack could also happen on communications using wired or fiber-optic communications.

Firewalls

This text previously discussed placing a secure firewall between the SCADA system and the IP network extending to field-based equipment. That approach, however, protects only the SCADA system. Entities may also protect field-based equipment by placing firewalls at field sites. Those firewalls could also implement a VPN with the central firewall to encrypt message traffic. Specialized firewalls exist to understand industrial protocols and support highly detailed, protocol-specific rules.

Encryption

Serial communications (wired and wireless) require encryption to prevent unauthorized access. The current best solution is "bump-in-the-wire" link encryption devices (at both ends) that scramble message traffic. Figure 6-11 shows use of firewalls in the case of IP communication, and encryption in the case of serial communications. Both approaches provide a high level of assurance an attacker who gained access to a communication circuit/channel would not be able to send false messages or commands to the field-based equipment or the SCADA system, thus eliminating that attack pathway.

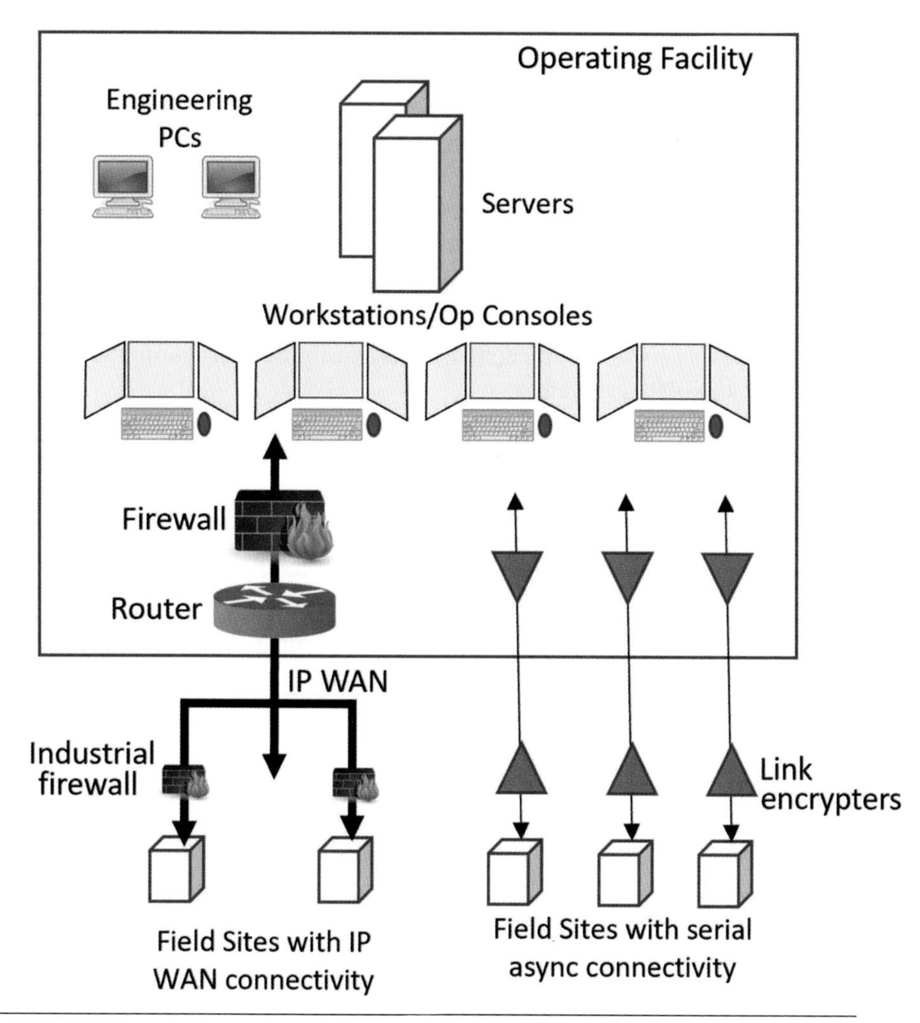

Figure 6-11. Employing Firewalls and Link Encryption at Field Sites.

Conclusion

SCADA systems are known targets of cyberattacks by a range of threat actors, including foreign governments and terrorist organizations. Such attacks have already occurred, including on an oil pipeline[49] that was using a popular SCADA system also used by the electric power industry.

When establishing cybersecurity for a SCADA system by applying a suitable combination of physical, administrative, and technical countermeasures, the primary goal should be eliminating (to the fullest extent possible) viable attack pathways, protecting the remaining pathways, and establishing detection measures on the remaining pathways.

Summary

- Many industries use Supervisory Control and Data Acquisition (SCADA) systems to monitor and control processes and equipment.
- SCADA systems gather data from various devices throughout a system into a centralized location for monitoring and control by specially trained system operators.
- SCADA systems use various types of communication methods to transmit data from field locations to a central control room.
- Many types of network protocols help devices using different computer languages communicate with each other.
- Cybersecurity is an important part of designing a SCADA system, to prevent hackers from accessing and controlling equipment.
- Applying countermeasures (physical, technical, and administrative) is an integral part of cybersecurity.
- Removing or limiting hackers' access to attack pathways is one of the most important and effective countermeasures.
- Utilities also use firewalls, encryption, access-control lists, and whitelisting as countermeasures.
- Utilities must also use physical access controls (e.g., card keys) to prevent unauthorized access to networked computer devices.
- Portable media and transient devices (e.g., USB drives, laptops) can provide an attack pathway and, therefore, must be controlled and limited.
- The supply chain also provides an attack pathway needing countermeasures.
- Utilities can use "hardening" tactics to protect networked devices.

7

Control Rooms

It is not enough to do your best; you must know what to do, and then do your best.

—W. Edwards Deming (1900 –1993)

"Grandpa Tom," said Luke, "On our way to the ranch the other day, you told me about the people making sure electricity was always on. You said they worked in a big room, and you used three letters. I think those letters were UFO. Can you tell me more about what those people do?" asked Luke, as they passed the same substation on their way home.

"Of course," replied Grandpa Tom. "You are close on the letters – they are ISO, an abbreviation for independent system operator – the company that manages the electric grid," the grandfather explained. "Control rooms look a lot like the NASA control room we toured at the Johnson Space Center during spring break. They have huge video walls and lots of computers and displays on control consoles," Grandpa Tom added. The grandfather went on to explain that grid operators monitor and control the grid.

"What is the difference between monitoring and controlling?" Luke questioned.

"Good question," Grandpa Tom chuckled. "Monitoring involves *watching* grid voltage and frequency and keeping an eye out for problems; controlling involves *making changes* to equipment so the grid operates better or when something unexpected happens. The wide area control system – software, computers, and equipment – perform most of the routine tasks, and humans handle the more complex tasks," Grandpa Tom added. "The people in the control room can even tell exactly how much electricity we used at the ranch this weekend because we have a 'smart' meter there," the grandfather said.

"Cool, I think I want to work for the electric company when I grow up!" exclaimed Luke.

The Electric Grid

This text takes a broad view of an electric *grid* as including all generation, transmission, distribution, and end-use customers. Grid operators monitor and control electric grids from control rooms. Figure 7-1 depicts at a high level how electric energy historically moved from generation sources through transmission and distribution lines to end-use customers (load).

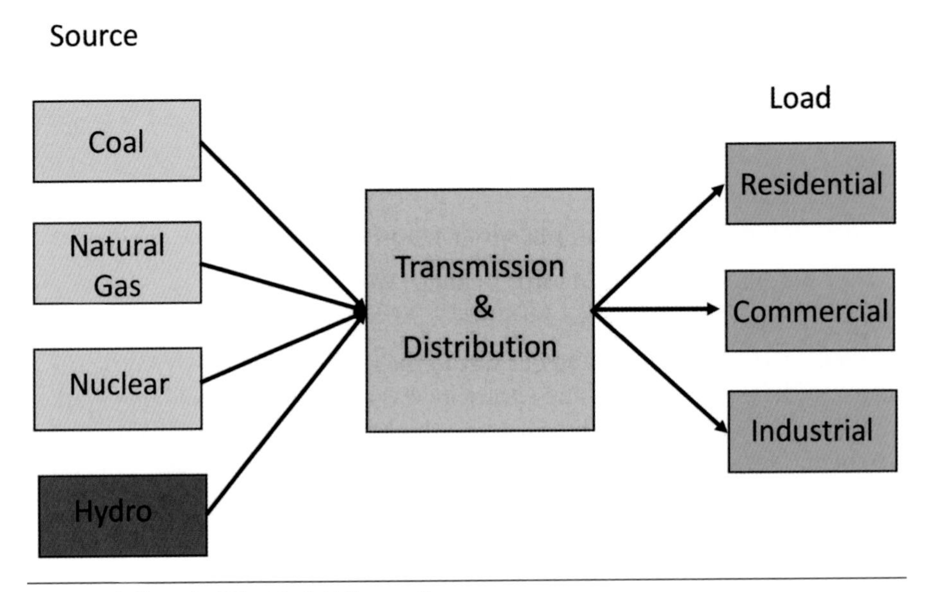

Figure 7-1. Historical Electric Grid Energy Flow.

Operators at many control rooms work together to facilitate generating, transmitting, and distributing electricity to end-use customers. The number of control rooms involved depends on how many generation resources, transmission lines, and distribution systems operate at any point in time. Like a symphony, all the generators, transmission lines, distribution lines, and loads act in concert. One sour note can ruin the performance. Over time, the grid has evolved and become more complex. Increasing grid complexity combined with low-capacity-factor wind and solar generation resources increases the difficulty of monitoring and controlling the grid. Ensuring grid reliability and resiliency in the face of increased complexity is a key challenge facing the electric industry (discussed in more detail in Chapter 12).

Vertically Integrated Operations Approach

When one entity owns and operates the entire electrical power value chain – generation, transmission, and distribution – economists call it *vertical integration.* Vertical integration served as the electrical power business model in the early days of the industry. Utilities typically had control rooms with three *desks* – generation, transmission, and distribution. Operators at each desk coordinated activities affecting their respective assets.

Gradually, individual utilities interconnected, making them physically and commercially more complex. Physical complexity relates to size and number of connection points, and commercial complexity relates to number of asset owners connected to and owning parts of the grid, as well as the number of transactions across the grid.

Operating Entities

Understanding the challenges of operating the overall electric grid requires some background knowledge about the entities involved. Figure 7-1 (above), loosely depicts the interconnected nature of the overall electric system (generation, transmission, distribution and end-use customers).

To maintain overall grid stability, generation, transmission, and distribution must maintain voltage and frequency within safe operating margins. If system voltage or frequency exceeds or drops below established safety thresholds, protection systems (e.g., circuit breakers and switches) begin to automatically *shed load* to bring the variables back into safe operating ranges.

Shedding Load

The terms "load shed" or "shedding load" are industry terms for disconnecting customers from the grid. Most lay people refer to this as a "blackout" or "brownout," depending on severity.

Historically, individual utility companies developed their own "footprint" of operations based on a grant of monopoly authority over a geographic area (through a *certificate of convenience and necessity*). A utility with a CCN *balanced* the generation connected to its transmission/distribution system with the load (customers) connected to its system, giving rise to the term *balancing area.*

Certificate of Convenience and Necessity (CCN)

A CCN is a certificate issued by a government agency granting a company authority to operate a public service (especially a utility or transportation company).

Figure 7-2 shows several separate utilities in proximity to each other.

To increase reliability, these separate utilities with their separate electric systems connected to their neighboring utilities to support each other when needed. For example, if a generator in Utility 1's area went off-line unexpectedly, generators in Utility 2's area could produce more electricity and export it to Utility 1 to keep the lights on. Figure 7-3 depicts the interconnection of adjacent utilities to improve reliability.

Typically, at the end of a time period (e.g., one month), the utilities that imported and exported electricity to/from each other reconciled the electricity transferred between their areas, and whichever utility imported more electricity from another would "settle up" the account (i.e., pay for the net amount of electricity it imported) at an agreed-upon price. That process became known in the industry as the *settlement process.*

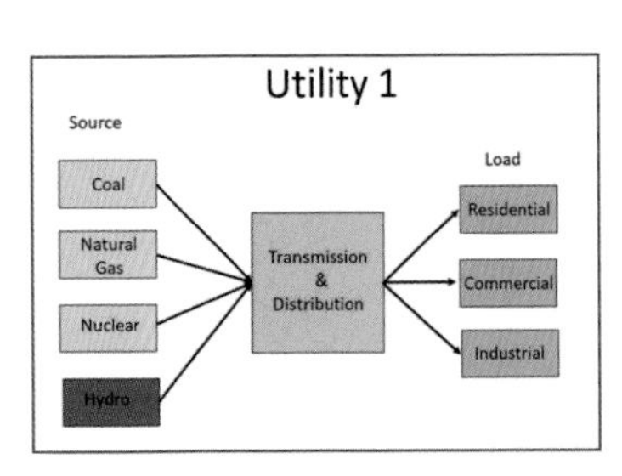

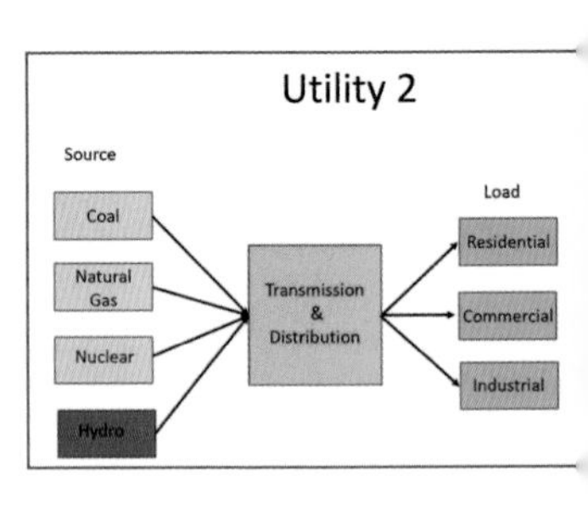

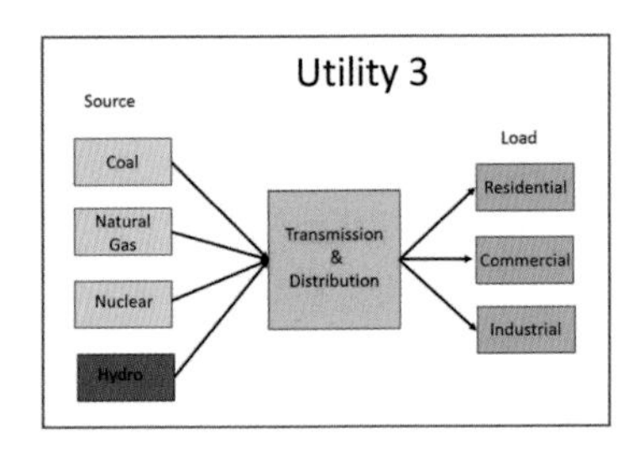

Figure 7-2. Several Utilities in Proximity to Each Other.

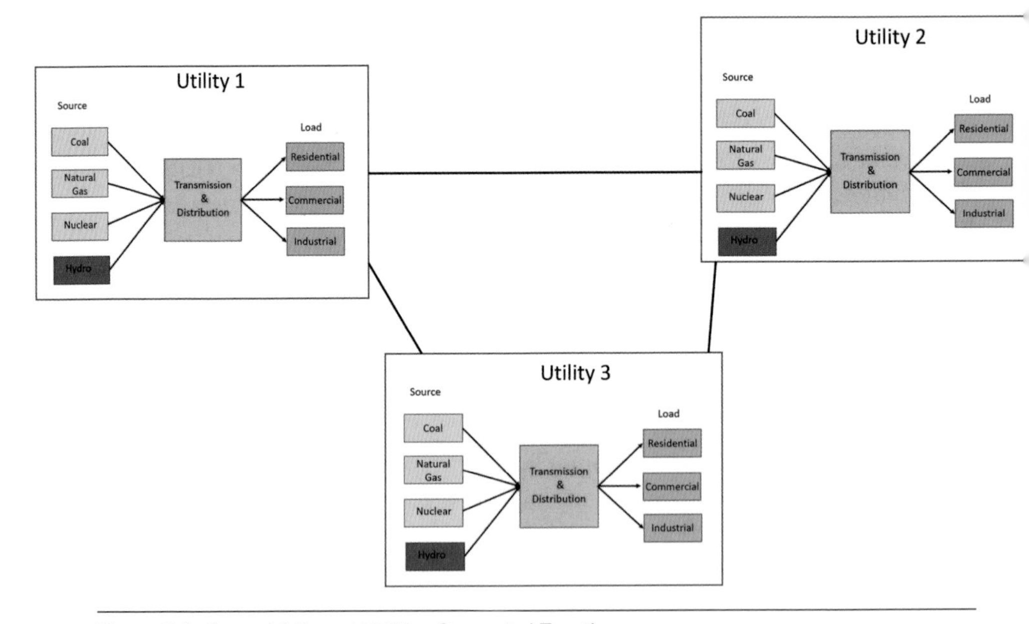

Figure 7-3. Several Adjacent Utilities Connected Together

Often, separate utilities join together in a *regional geographic area* overseen by a *regional transmission organization* (RTO) or *independent system operator* (ISO) as shown in Figure 7-4. More detail on those entities appears later in this chapter.

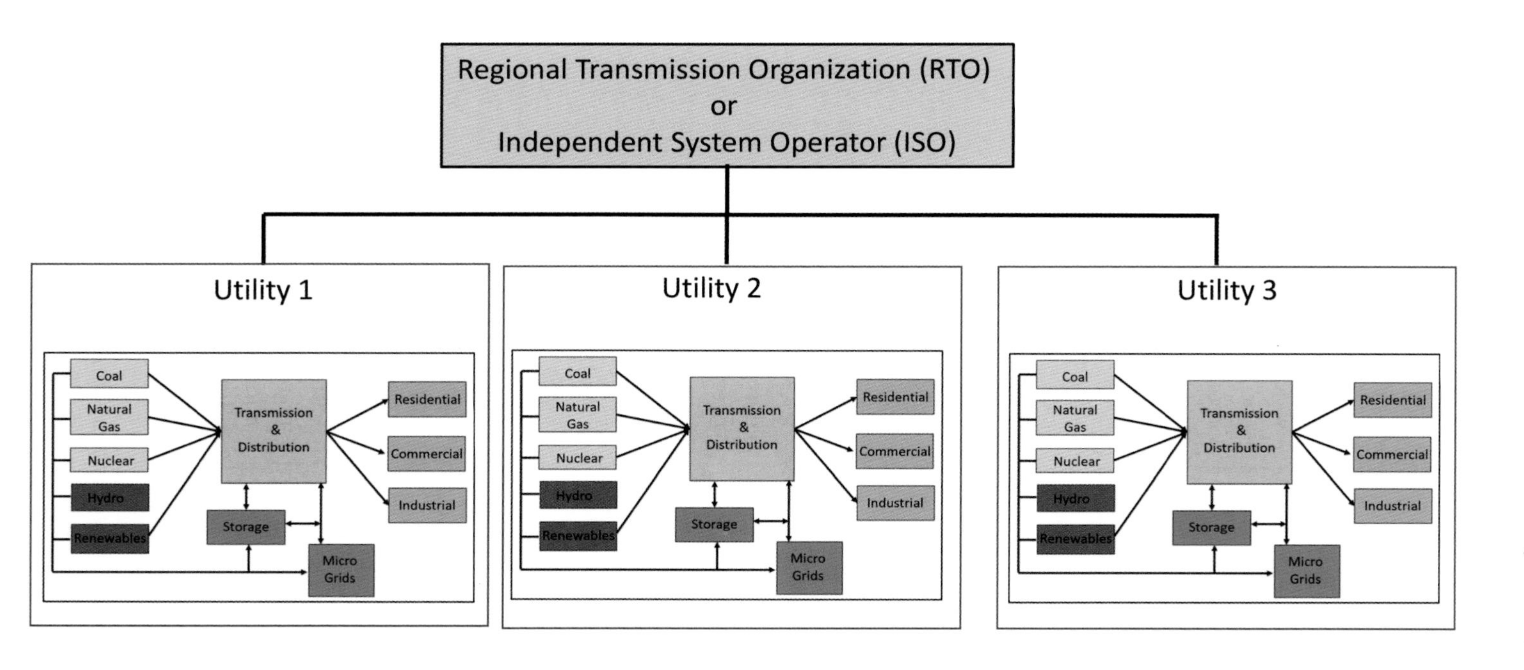

Figure 7-4. Several Utilities in a Geographic Area Joined Under an RTO or ISO

Reliability Coordinator Area

The generation, transmission, and loads within the boundaries of the RC, which coincides with one or more BAAs.

Balancing Authority Area

The collection of generation, transmission and loads in the metered boundaries of the BA. The BA maintains load-resource balance within this area.

No entity has a solo part – all work together to maintain grid stability. In North America, an entity called a *reliability coordinator* (RC) has the highest level of authority for reliably operating a part of the electric grid. An RC has the widest view of a large portion of the electric grid called a *reliability coordinator area* (RCA). Beneath the RC is the *balancing authority* (BA), which integrates generator output plans ahead of time, maintains demand and resource balance in its *balancing authority area* (BAA), and supports grid frequency in real time. The individual utilities described above are usually the entities serving as BAs.

Reliability Coordinator (RC)

RCs have the highest level of authority for reliably operating the electric grid because they have the widest view of the system. They have the operating tools, processes, procedures, and data to operate large portions of the overall grid, including the authority to prevent or mitigate emergency operating situations in next-day analyses and real-time operations. The RC has a broad enough view of the grid to calculate *interconnection reliability operating limits* (IROLs) for equipment, which it may base on the operating parameters of transmission systems beyond any one BA's operational vision. The RC maintains generation-demand balance within its RCA to ensure frequency remains within acceptable limits.

SOLs and IROLs

A *system operating limit* (SOL) applies to a particular part of the overall grid (a specified "system") with a goal of ensuring equipment operates within acceptable reliability criteria.

An IROL is an equipment operating limit that, if violated, could lead to overall grid instability or cascading outages adversely impacting reliability. An IROL covers an entire *interconnection* of systems.

The RC may also direct one or more BAs within its RCA to take certain actions to maintain grid reliability. North America has numerous RCs.[50] Next in importance for operating the electric grid is the BA.

Balancing Authority

Each BA monitors and controls a portion of the overall electric grid (its BAA). Often, a utility that owns and operates the equipment in its part of the grid serves as a BA for its BAA. However, a BA *may* be an ISO or RTO (which does not actually *own* the transmission facilities and equipment). (See Figure 7-4). A BA performs the following functions:

- Integrates resource plans ahead of real-time
- Maintains load-generation balance (hence the BA designation)
- Supports interconnection frequency in real time
- Receives real-time operating information from transmission operators in its BAA, adjacent BAs, and generator operators in its BAA
- Provides real-time operational information for RC monitoring
- Receives reliability alerts from its RC
- Complies with reliability-related requirements (e.g., reactive requirements, location of operating reserves) specified by its RC
- Verifies implementation of emergency procedures to its RC
- Informs the RC of confirmed equipment changes (e.g., due to generation or load interruptions) in its BAA
- Dispatches generation resources to ensure generation-load balance in real time
- Directs transmission operators (or distribution providers) in its BAA to reduce voltage or shed load, if needed to maintain load-generation balance
- Redispatches generation facilities to manage congestion as directed by the RC
- Takes corrective and emergency actions as directed by the RC
- Implements system restoration plans as needed

Independent System Operators

Because no electric utility wants another electric utility operating the grid (due to the commercial advantage the grid operator would hold), commercial and competitive concerns gave rise to the concept of a single, commercially unrelated entity called an *independent system operator* (ISO) to control (operate) parts of the grid covered by several (or many) utilities.

ISOs generally do not *directly* operate generation or transmission facilities. Instead, operators at (or at least connected to) the individual utilities control those assets. In the example shown in Figure 7-4, the ISO sends operating instructions to the separate utilities in its operating area (Utilities 1, 2, and 3). The ISO will also send *dispatch instructions* to generation resources, meaning the ISO instructs generators to supply an amount of electricity at a specific output level and time period, based on factors like price, grid congestion, and system *topology*. Likewise, distribution systems generally have

System Topology

The electric grid's *topology* means the general layout of system elements and their capacity to transmit electricity, considering wires, switches, transformers, protection systems, and other equipment.

their own control rooms and interface directly with the ISO's system operators to ensure the reliable supply of electricity.

Figure 7-5 depicts the relationship among RTOs, ISOs, and BAs (all responsible for discrete parts of the larger, overall grid) to show the complexity of the *bulk power system* (BPS).

The entity operating the overall grid and the name of that entity vary by country (and sometimes region) depending on the type of government and the particular industry structure of the country or region. Some examples follow:

China

China passed its *Electric Power Law* in 1996 to diversify ownership and control of electrical assets; protect the legal rights of investors, managers, and consumers; and regulate generation, distribution, and consumption. China has two wide area synchronous grids – the *State Grid* and the *China Southern Power Grid.*

Continental Europe

Europe, a continent consisting of many independent nations, has the *European Network of Transmission System Operators*, representing more than forty electricity transmission system operators (TSOs) from over thirty countries across Europe.

United States

The U.S. provides an example of how and why nomenclature differs between countries and is driven by regulations. The ISO concept arose in connection with electric industry deregulation. ISOs grew out of Federal Energy Regulatory Commission (FERC) Orders, No. 888 and 889, in which the FERC suggested the concept to allow existing power pools to satisfy the requirement of providing nondiscriminatory access to transmission systems.

Later, in Order No. 2000, the FERC encouraged the voluntary formation of *regional transmission organizations* (RTOs) to monitor and control the transmission grid on a regional basis throughout North America. According to the FERC, RTO regions serve load in two-thirds of the country. The U.S. contains the following ISOs and RTOs:

- California ISO
- Midcontinent ISO
- New England ISO
- New York ISO
- PJM Interconnection
- Southwest Power Pool
- Electric Reliability Council of Texas, Inc. (ERCOT)

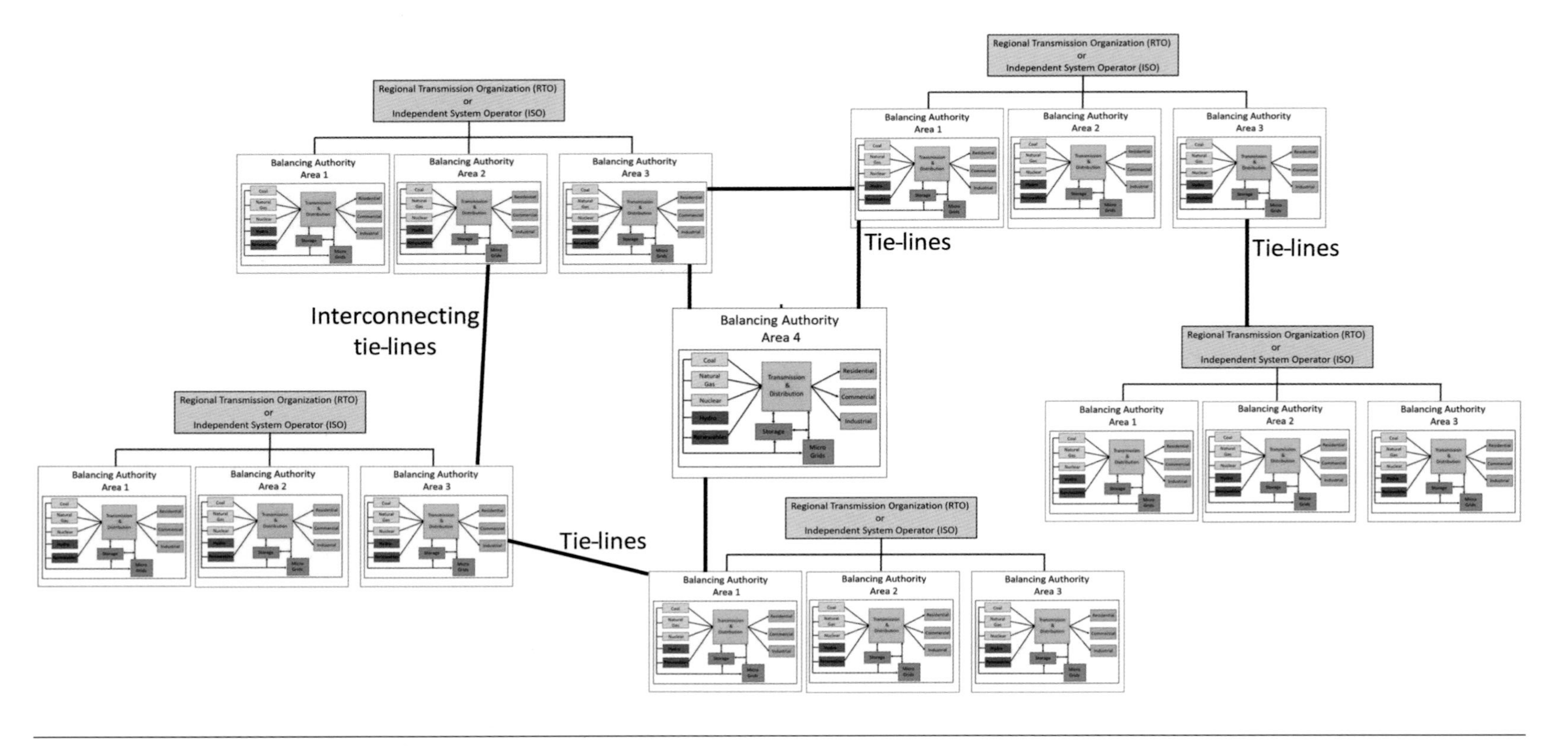

Figure 7-5. Depiction of the BPS in the U.S.

System Operations

Control rooms are common in process industries, including oil and gas, brewing, distilling, chemical, and electric transmission and distribution. While this chapter deals with control rooms in the electric industry, many of the processes, hardware, software, and skills apply across a variety of industries.

> **Reliability**
> Electrical power is delivered in a manner consistent with customer expectations.
>
> **Resiliency**
> The grid's ability to recover after it experiences an upset condition.

Control rooms in the electric industry differ from those of most other industries because they control facilities far removed from the physical control room. When an issue arises, the system operator cannot just walk over and check it out. Instead, the operator must determine remotely what happened (or is about to happen) and take corrective action. If the operator cannot address the issue remotely, the company dispatches a crew to the site.

Control room operators typically work rotating shifts around the clock as they keep operations safe, reliable, and efficient. Their titles usually start with the system they operate and then add Transmission System Operator, Distribution System Operator, Electric System Operator, or T&D System Operator. Figure 7-6 shows three electric system operators at work in the control room of ERCOT.

Figure 7-6. Electric System Operators.

Control rooms contain the "brains" of the grid. As in human bodies, nerves (the control system) stretch out to the far reaches of the grid, collecting data from instruments and equipment along the way. Centralized servers receive, process, and store the data in databases. Continuing with the human analogy, like the autonomic nervous system's control of breathing, heartbeat, and digestive processes, many control decisions occur without human intervention, based on data analysis conducted by algorithms. Other data appears on displays for review and action by the operator.

Human Factors

The field of study called *human factors* examines the relationship between human beings and the systems with which they interact, to minimize errors by improving efficiency, creativity, productivity, and job satisfaction. Control room designers consider human factors (including fatigue) when designing control rooms. Fundamental to control room operations are two key concepts: *situational awareness* and *permission to operate*, both influenced by human factors.

Situational Awareness

Situational awareness derives from work performed in connection with pilots during World War II. The Army Air Corps lost many pilots when they moved from one kind of airplane to another. They studied this problem and discovered each plane had a different cockpit setup. As pilots moved between planes, the extra time required as they looked for controls and instruments located in different places caused them to lose awareness of the situation around them. Consequently, the Army identified the most important instruments and controls and placed them in the same location in every type of airplane. This practice continues to this day, and the setup of the most critical flight instrumentation remains the same in airplane cockpits. (A familiar situational awareness condition for most readers involves talking on a cellular telephone while driving and, as a result, losing track of surrounding traffic.)

Permission to Operate

The second concept – *permission to operate* – states that a person has permission to operate equipment so long as that person maintains situational awareness. When the operator loses situational awareness (i.e., does not know what is going on in the system), permission to operate ceases, and the operator must return the system to a safe operating mode, which could require shutting down equipment or changing operating parameters. In modern operations, control room systems use *alarms* or *alerts* to assist the system operator in maintaining situational awareness.

Alarms and Alerts

When part of the system passes a predetermined threshold (e.g., a thermal limit or voltage limit) and enters an abnormal state, the operator receives an automated alarm (usually visual and audio) meaning something needs attention. The operator then determines how to address the abnormal situation. Trained system operators work at each desk and have specialized expertise in their respective areas.

Improving Operations with Technology

Utilities increasingly leverage computer networking technology for video, audio, and control data. That technology offers flexibility and scalability for centralized or distributed control room operations. Moving from direct-connected systems to networks allows utilities to configure connectivity architecture and the overall size of system installations based on specific requirements and designs, to accommodate large distances, high performance, and varying levels of system redundancy and resiliency.[51]

Networking solutions allow system operators to monitor, access, and troubleshoot utility assets from any networked computer. While multiple displays at operator workstations help them view information to improve operations, these new systems also allow the display of data from anywhere in the network on a large video wall (think NASA), enabling a better overview of system status and faster response times to extraordinary events. The combination of audiovisual (AV) and IT continues to develop. Combining technologies allows utilities to connect AV electric system components, integrating information to facilitate responsive operations, command and control, and physical security.

These sophisticated systems must be secure and reliable on a 24/7 basis. Consequently, those systems use scalable authority levels, encrypt communications, require individualized logins, and have failover/backup support. Technology professionals can often log in to a system remotely (e.g., from home in the middle of the night) to troubleshoot and fix problems or issues.

Control Room Processes

Control rooms execute five main processes sequentially and repeatedly as shown in Figure 7-7.

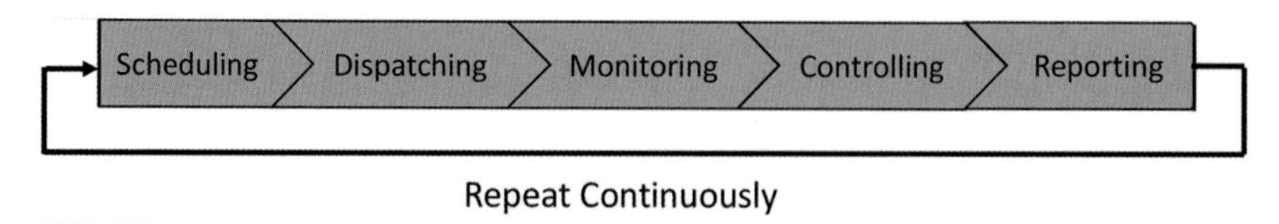

Figure 7-7. Control Room Processes.

Scheduling and Dispatching

Scheduling involves several activities. Control room personnel must coordinate outages of equipment with BAs or an RC to ensure grid stability. This chapter will not discuss the situation when a BA is a single utility because that entity need not – technically – *schedule* its generation (with itself). Typically, those types of BAs simply perform *load following* (i.e., the utility generates only the amount of electricity needed to serve its own native load).

Scheduling and dispatch are closely related processes. Scheduling happens before dispatching and serves as its basis. Scheduling involves comparing future supply to future demand and developing a plan to make them match. Dispatching turns that plan into periodic dispatch instructions. These plans are developed one day at a time and updated as needed (generally multiple times per day).

Each day, generation companies inform their BA of the amount of generation they have contracted to sell for each hour of the next day (based on bilateral sales contracts). At the same time, the BA determines the amount of *load* it expects for each hour of the next day based on historical usage, weather forecast, and other relevant factors. Figure 7-8 shows a sample supply/demand forecast for the ERCOT region in Texas.

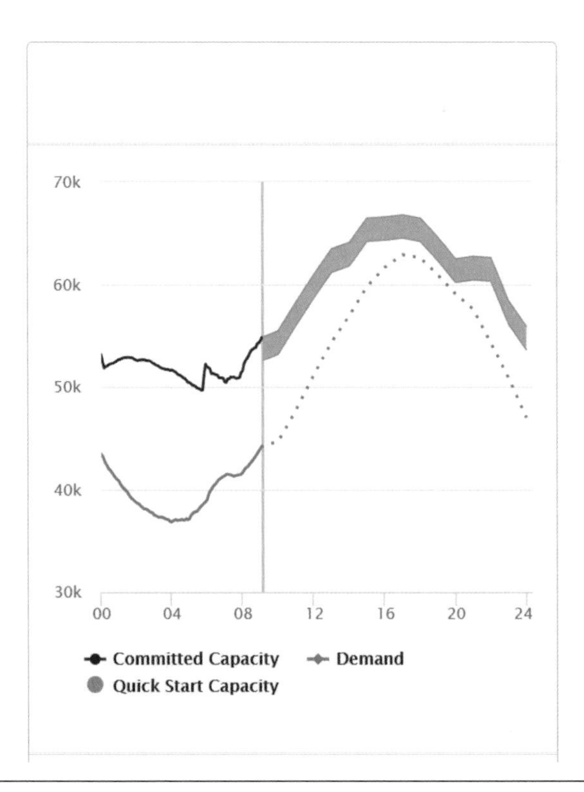

Figure 7-8. Sample Supply/Demand Forecast.

Figure 7-8 shows that ERCOT will have available *capacity* of approximately 65,000 MWs at 4:00 p.m. (shown as "16" on the horizontal axis), and ERCOT anticipates *demand* of approximately 63,000 MWs at that time. That means ERCOT will have a reserve capacity of 2,000 MWs at that time. If, in real time, ERCOT experiences higher demand than forecasted (for example, if the day is hotter than anticipated and customers use more electricity), ERCOT can call on this 2,000 MWs of idle generation units to cover the additional demand.

Because utilities cannot (yet) effectively store large amounts of electricity, the BA must continuously adjust power plant outputs to meet electricity demand in a process called *dispatching* the power plants. Dispatching usually has two stages occurring over different time horizons.

Unit Commitment: The first stage of scheduling (unit commitment) typically occurs one day ahead of *real time* and involves the BAs taking scheduling data and, from it, developing a preliminary plan listing the power plants (units) to run the next day and the levels at which they should run to meet expected load. The BA then continually performs studies of the preliminary plan (through sophisticated software) to optimize dispatching of generation plants based on economic factors and transmission system constraints. When the BA plans for a plant to run during this process, it considers the plant *committed* or *scheduled* (hence the title unit commitment).

Dispatch: The second stage, or *dispatch*, occurs when the BA provides schedules to generation plants for specific amounts of electricity to generate (a *set point*) during each time period. BAs make dispatch decisions based on economic factors and transmission system constraints, such as whether the transmission system can accommodate the levels of electricity demanded in various geographic areas. BAs also consider other operational factors such as a unit's ramp rate (time in which it takes a generation unit to reach a certain output level) and minimum run times (length of time a unit must run before shutting down per unit design or environmental regulations).

Example: Figure 7-9 depicts a scheduling example. The *base load* line shows the minimum load level anticipated for the day (approximately 38,000 MWs). The BA must ensure that it has *at least* that amount of generation scheduled to operate all day. The *peak demand* line shows the highest level of anticipated load for the day (approximately 63,000 MWs). The BA must ensure it has at least that amount of generation available at that specific time (shortly after 4:00 p.m.) and all other times throughout the day. The power plants the BA dispatches at peak demand times are those that can start and stop quickly and tend to be expensive to operate. The BA must, however, also have additional generation in reserve in case load exceeds expectations or a scheduled generation unit cannot run for some unanticipated reason.

Unit commitment and dispatch for the BA in this example are:

- The BA commits base load plants – typically those with long minimum run times and low marginal costs.

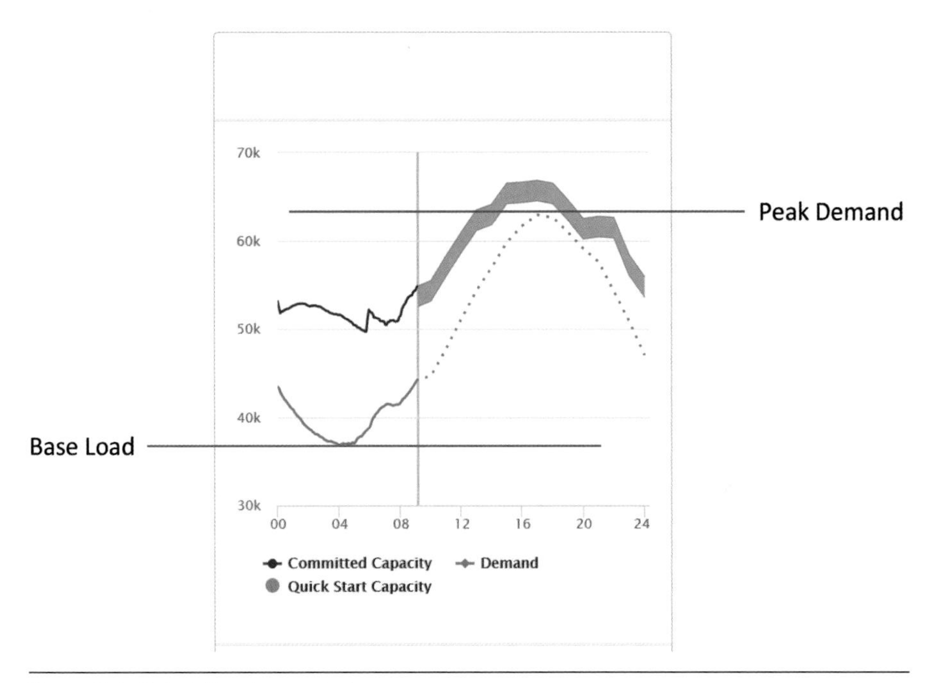

Figure 7-9. Scheduling Demonstration.

- Each hour, the BA uses economic factors and system topology data to determine the set point for each dispatched plant. The BA also commits plants that can quickly start up during the day, as needed, and makes hourly decisions about how much electricity each plant should generate.
- Finally, the BA commits *peaking plants* it can dispatch within minutes if needed at peak load times.

All these tasks, taken together, ensure BPS reliability and security.

Monitoring and Controlling

Like scheduling and dispatching, monitoring and controlling are closely linked, with monitoring providing the input for controlling the grid.

Before modern technology allowed remote system monitoring and control, changing equipment settings occurred by telephone or radio communication, with personnel physically located at strategic locations throughout the system such as generation plants and substations. Control room operators would contact field personnel and tell them what to do. For example, a control room operator would contact a person at a generation plant and orally dispatch (inform) that plant operator to run the plant at a specific output level (in MWs). The plant operator would then manually change the plant's output (typically through analog controls). Alternatively, the control room operator might contact a field hand at a substation and instruct him to take a particular line out of service or place a particular line into service. The person at the substation would manually change the status of switches from open to closed (or *vice versa*) to fulfill the control room operator's instruction. All the while, the people at the generating plants and substations monitored system operating data to ensure that equipment remained within safe operating limits. Over time and with the advance of computer and communications technology, utilities began remotely monitoring the system and controlling equipment using SCADA systems, as discussed in the previous chapter.

Operating Tasks

System operators in control rooms (whether for an ISO, RTO, or particular utility) assist with electric system maintenance and perform specific operational activities established by their companies. The basic goal is *safely* keeping their portion of the electric system under control and working properly. Their duties include monitoring, troubleshooting, and managing the system and field crews working around the utility, with an emphasis on safety and efficiency.

Whether for monitoring, decision-making, responding, controlling, collaborating, or communicating, control room operational tools and the reports generated by them serve as an essential part of a company's day-to-day operations. System operators benefit from automated solutions for situational awareness, surveillance, and command/control. In electric industry control centers, automated systems continuously monitor a never-ending flow of data involving infrastructure and equipment statuses. Information systems provide operators an integrated real-time overview of power flow and equipment status for situational awareness and decision-making (permission to operate). Information integration and visualization are important – from network/distribution information to topological service area overviews – to ensure continuous operations.

System Operators

As mentioned above, control room operators (or *system operators*) perform many functions, including:

- Monitoring the overall electric grid in their service area, interconnections with other electric utilities (to determine correct generation needs), electric system voltage, and instrumentation control panels.
- Remotely operating field-based equipment.
- Operating the computerized electric control system.
- Monitoring weather conditions for system operation needs.
- Operating radio and pager systems to communicate with work crews to resolve outages.
- Dispatching service and trouble calls to field personnel.

- Reviewing and assigning switching operations for real-time field crew work assignments.
- Troubleshooting and solving problems in electric transmission and distribution systems.
- Analyzing data to discover patterns or discrepancies and determining appropriate response actions.
- Responding to customer complaints by troubleshooting outages and determining whether to dispatch a field crew.
- Recording data on appropriate forms and logs for tracking crew and work assignments and electric utility grid systems monitoring.
- Reading maps and schematics to determine system problems.
- Resolving routine problems in generation dispatch.

System operators typically work on shifts because they must monitor and operate the grid on a 24/7 basis. They have demanding jobs. Errors can lead to injury or death. Consequently, optimal ergonomics is important to minimize human error and maximize performance and effectiveness. A better ergonomic design for the control room and operating systems leads to better workflow and equipment management. Equipment must be located optimally and displays located for comfort and easy viewing. Controls access must be intuitive to allow operators to remain alert so they can instantly learn of system events and react appropriately (operators typically work eight-hour to twelve-hour shifts).

Utilities need experienced operators to reliably orchestrate grid operations, respond to system disturbances, and, in emergencies, restore the system. A typical control room team consists of several operators trained on performing tasks to ensure the reliable operation of the BPS.

Coordinating with Field Crews

Over the course of a day, system operators coordinate with field crews to take equipment out of service to allow maintenance or repairs. System operators look at the work schedule for the next day and develop plans to isolate the affected equipment. The next day – in real time – the system operators coordinate with the field crews opening and closing switches to isolate the equipment on which the field crew will work. Communications generally occur via radio (and sometimes by cell phone).

Occasionally, due to weather events, equipment malfunctions, or other extraordinary events, system operators dispatch field crews to address problems or issues on the electric grid. For example, when severe weather passes through an area, it may affect power lines (e.g., tree limbs take down lines). System operators learn about problems either through customers reporting outages or through sensing devices located throughout the system. They respond to those issues by dispatching field crews to perform restoration activities. Figure 7-10 shows an example of an outage map (from Austin Energy in Austin, Texas). At the time of the screen capture, Austin Energy had three outages on its system, and crews were on-site at two of the locations (denoted by the round dots with an icon of a person in a hard hat).

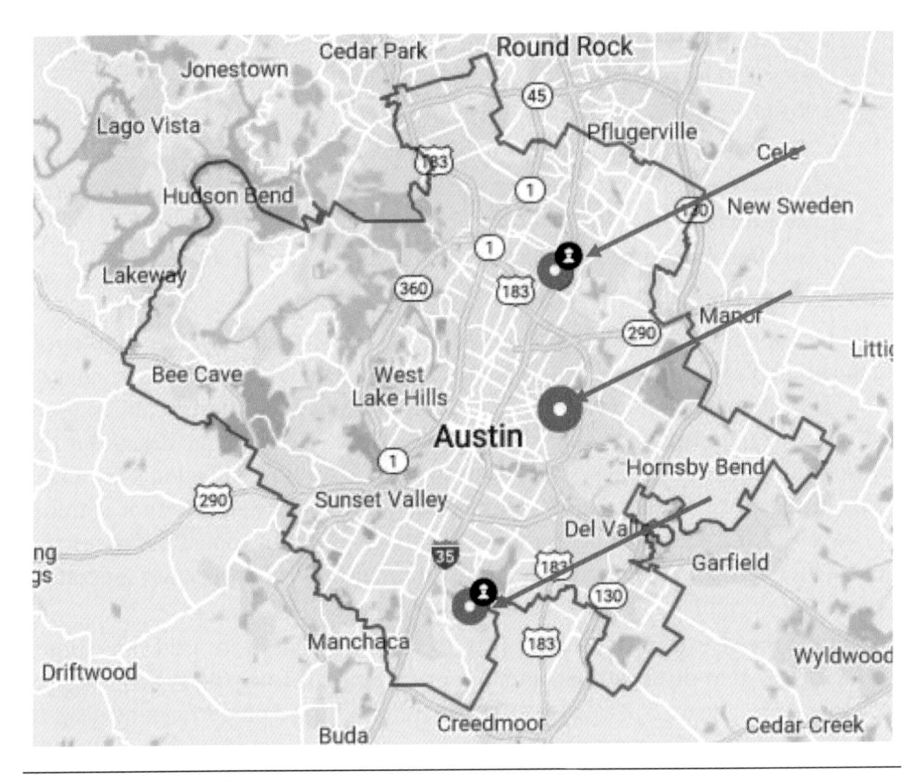

Figure 7-10. Outage Map.

Switching Operations

Switching operations involve opening and closing switches to isolate equipment, thereby ensuring electrical work associated with high-voltage equipment takes place in accordance with safe work practices. Switching operations apply to all electrical switchgear from the point of generation to the load center disconnect switch, including transformers, main breakers, and tie breakers. Switching operations apply during repairs, maintenance, modifications, and installations.

System controllers generally follow priorities in switching operations for power grid equipment:

1. Physical safety of employees and public
2. Integrity and reliability of the power system
3. Protecting equipment
4. Serving customers

When it comes to switching operations, operators must open and close switches in the right order. Accordingly, system operators perform system studies to ensure the work will not negatively affect system reliability (such as SOLs or IROLs) before they begin switching operations. Based on these studies, operators prepare, review, and revise (if necessary) switching procedures showing the sequence of required switching or operations. Before performing switching operations, a system operator typically ensures a second qualified system operator reviews the switching procedure to check it for accuracy. Next, the operator notifies other operating entities and the RC of planned work and real-time status changes of equipment and work with potential operational impact. Finally, they authorize, direct, and log switching procedures on equipment affecting the power grid.

Field crews, in turn, inform the control room operator when they arrive at a facility and provide specific details of their work that may affect grid operations or reliability. The system operator records the visit and reasons for the visit in the control room log. Each on-site employee associated with switching activities generally does the following:

- Understands rules and switching procedures.
- Analyzes the switching instructions in advance.
- Resolves any questions about the completeness or correctness of the switching order before starting work.
- Performs the operations in the listed sequence or as directed by the system operator.
- Ensures the switching orders are: (a) understood by all involved; (b) properly applied and (c) strictly observed.
- Correctly identifies all equipment.

Responding to System Disruptions

One of system operators' most important jobs is responding to system events. Those events include inclement weather, equipment malfunctions, and even things like snakes or squirrels causing equipment faults. When one author worked for an electric utility, he routinely received emails from field workers who, when arriving at a substation to address an equipment outage, would find fried snakes, squirrels, or other critters who got into a piece of equipment and caused an electrical arc by getting too close to live circuits.

To prevent damage, utility companies protect substations with interruption devices (circuit breakers or fuses) to shut off power if a short circuit occurs. In overhead distribution networks, most faults are temporary, like lightning strikes, surges, or foreign objects (tree limbs, mylar balloons, and animals) contacting distribution lines. For example, a tree limb may land on a power line and cause a short circuit, but the fault can quickly clear without intervention if the limb falls to the ground. If a circuit breaker is the only protection system, such a transient event could cause part of the distribution network to black out while the utility sends a repair crew to reset the breaker. Consequently, utilities developed approaches to avoid the need to send a work crew for transient events. The first line of defense against such transient faults are *automatic circuit reclosers* (ACRs) and *sectionalizers,* as discussed in Chapter 5.

If the circuit does not contain an ACR, when the system operator receives an alarm indicating a system fault, the operator responds by remotely opening and closing breakers to isolate the fault and direct the flow of electricity around the fault. Doing so involves using data from the SCADA system as well as predefined switching operations or personal experience developed during the controller's career. If the system operator cannot solve the problem remotely from the control room, the operator dispatches a field crew to the problem area to manually perform switching operations and repair or replace the equipment causing the fault. In a worst-case scenario, the utility may have to take equipment out of service for an extended period to repair, replace, or reconfigure the system.

System Operator Training

The North American Electric Reliability Corporation (NERC) has Reliability Standards and other rules requiring system operator certification to operate Bulk Power System (BPS) equipment. NERC's system operator certification program promotes BPS reliability by ensuring system operators meet minimum qualifications. The system operator certification program provides a framework to obtain initial certification in one of four credentials: Reliability Operator; Balancing, Interchange, and Transmission Operator; Transmission Operator; and Balancing and Interchange Operator.

NERC administers a system operator certification exam to test specific knowledge of operator job skills and Reliability Standards and to prepare operators to work on the BPS during normal and emergency operations. NERC issues a "system operator" credential to a person who successfully passes a NERC system operator certification exam. System operators maintain their certification through NERC-approved continuing education courses and activities. NERC monitors system operators to ensure they maintain their credentials to work in control centers. NERC's credential maintenance program oversees learning activities by approving providers who meet NERC guidelines and standards.

Passing an exam earns candidates a credential and a certificate valid for three years. To maintain certification, operators must obtain at least thirty continuing education hours (CEHs) from NERC-approved learning activities during a three-year period. The CEHs must include simulations (i.e., tabletop exercises, training simulators, emergency drills or practicing emergency procedures, system restoration, blackstart or other reliability-related activities).

Managing the Grid in Emergencies

Most of the time, system operators spend their workdays performing routine operational tasks. Occasionally, however, events occur requiring them to address emergent situations. Events, including hurricanes, tornadoes, wind or snowstorms, and fires require drastic actions to protect the reliability of the power grid. Because generation and load must always remain equal, operators must respond quickly when they lose either large amounts of load or generation. Most often, operators address very serious system emergencies by *load shedding*.

Load Shedding

The concept of *load shedding* involves opening circuit breakers or switches serving end-use customers to preserve the overall integrity of the power grid. When energy demand exceeds supply, if system operators do not react quickly to shed load, frequency on the grid drops and protection systems automatically activate to protect equipment from damage. In a recent example, ERCOT operators shed a significant amount of load in response to a freak winter storm in Texas (winter storm Uri). The storm had a significant impact throughout Texas, bringing snowfall, ice, and the coldest temperatures in decades.

As temperatures plummeted, Texans' electric use rose, forcing ERCOT to call for *rolling outages* intended to last forty-five minutes per circuit. For some, however, the *rolling* outages did not "roll" and, instead, lasted several days. According to a study from the University of Houston, 69% of Texans had no power between February 14 and February 20, 2021, for an average of forty-two hours. At the storm's worst point, 48.6% of all generation capacity in the ERCOT region was forced offline.[52] At the same time, ERCOT experienced its highest winter electricity usage ever (69,222 MWs).[53] Beginning at 1:20 a.m. on February 15, 2021, ERCOT began load shed operations that progressed as follows:

Time	Load Shed
1:23 a.m. –	1,000 MWs
1:47 a.m. –	1,000 MWs (2,000 MWs total)
1:51 a.m. –	3,000 MWs (5,000 MWs total)
1:55 a.m. –	3,500 MWs (8,500 MWs total)
2:00 a.m. –	2,000 MWs (10,500 MWs total)[54]

At its highest points on February 15 and 16, ERCOT shed approximately 20,000 MWs of load.[55]

Blackstart

If grid stability deteriorates sufficiently and protection systems automatically isolate generation plant equipment, the grid can experience a total collapse, causing a widespread blackout. After such an event, the grid must go through a *blackstart*. Blackstart events involve restoring electric generation stations or a part of the electric grid without relying on the *external* electric power grid to recover from the shutdown. Normally, the electric power used in a generation plant comes from the plant's own generators. If the plant's generators shut down due to an emergency, power comes from the grid through the plant's service line. During a large-scale outage, however, power from the grid is not available.

To perform a blackstart operation, some generation plants have diesel generators (or other support resources) to start larger generators, which, in turn, start the main power station generators. A typical blackstart scenario might occur as follows:

1. A small diesel generator starts at a generating station.
2. The diesel generator brings the generating station into operation.
3. System operators energize key transmission lines between the generation plant and other areas of the grid (those transmission lines are called *blackstart corridors*).
4. Operators use the power from *the first* generation facility to start *another* generation facility (referred to as a *next start* unit).
5. The combined power from the restarted power plants restarts other power plants in the system.

Ultimately, power is restored to the overall electricity network. Because starting the entire grid at once is not feasible, blackstarts occur gradually, area-by-area (called *blackstart islands*). In a widespread blackout, blackstart involves starting multiple generation islands (each supplying its local load area) and synchronizing/reconnecting the islands to re-establish the complete grid (like a patchwork quilt). System operators must closely coordinate with each other (usually through the RC or BA) throughout the blackstart event.

Generation Plant Interface and Control Room

Virtually every generation plant also has a control room. Depending on the size of the facility, the control room may be very modest. The operators in generation plant control rooms start and stop the generation units and open and close switches to direct the transmission of electricity. At a power plant, the interconnection between the plant and the transmission system is referred to as a *switchyard* (instead of substation). In the switchyard, a *generator step-up transformer* takes the electricity from generator voltage to transmission system voltage and connects the power plant to the transmission system.

Summary

- Historically, vertically integrated utilities generated, transmitted, and distributed electricity to their end-use customers.
- Over time, separate utilities interconnected to make broad-ranging electric grids to improve overall system reliability.
- The FERC issued several orders affecting the structure of the electric markets in the U.S. and giving rise to ISOs and RTOs.
- ISOs and RTOs administer large portions of the electric grid.
- A BA/RC schedules and dispatches generation units based on economic factors and system topology.
- System operators control equipment on the power grid.
- Control room designers account for human factors when designing control rooms.
- Control room processes include scheduling, dispatching, monitoring, controlling, logging, coordinating with field crews, and responding to system disruptions.
- System operators must pass a certification exam and obtain minimum levels of continuing education credits.
- Operators coordinate with field crews to address system outages.
- Power plants connect to the transmission grid at a switchyard.
- Power plant operators control the equipment at a power plant.
- In extreme emergencies, system operators respond by shedding load to avoid blackstart events.
- Operators coordinate with field crews to bring the system back from a blackstart event.

8

Electrical Energy Storage

Energy can neither be created nor destroyed – only converted from one form of energy to another.

—First law of thermodynamics

"Grandpa Tom, my flashlight quit working," said Luke.

"Try these new batteries," the grandfather said, handing Luke two batteries.

Luke replaced the batteries and his face lit up along with the light. "Thanks, Grandpa Tom!" Luke exclaimed. But then he said, with a quizzical frown, "How does light get into the batteries?"

"Light is not in the batteries; rather, chemicals in the batteries produce electricity when the switch is closed, completing the circuit. That is why battery storage is referred to as *electrochemical*," Grandpa Tom explained.

"What are rechargeable batteries, and how do they work?" asked Luke.

"Rechargeable batteries also contain chemicals and generate electricity through a chemical process. When rechargeable batteries are exhausted, they are put into a charger and electricity from the charger drives the chemical process in reverse. When the batteries are removed from the charger, they can generate electricity again. So, electricity is not really stored as electricity; it is stored as chemical energy in the batteries," explained Grandpa Tom.

"Cool," said Luke.

Introduction

This chapter's title is misleading – electricity energy, with a few exceptions, cannot be "stored." Rather, it is generated from primary energy sources, converted into another energy form for "storage" and regenerated (converted back) into electricity when needed. Converting electricity to another form of energy, storing it, and then regenerating it creates heat. This means the amount of electricity going into storage is always less than the amount of electricity coming out of storage. The total efficiency of generating electricity, converting it into another form of energy, and converting it back into electricity again is called round-trip efficiency and is calculated as:

$$\text{Electricity Out/Electricity In} = \text{Round Trip Efficiency}$$

All electrical energy storage solutions suffer losses during this round-trip process.

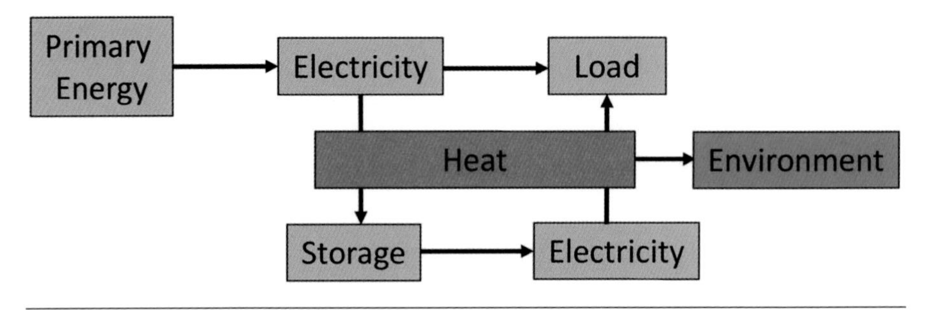

Figure 8-1. From Primary Energy to Electricity, to Storage, and Back to Electricity.

Historically, electricity was generated from primary energy sources and immediately consumed without ever involving storage. In fact, use of storage was so infrequent in 2006, when the previous edition of this text was published, that storing electrical power for use on the grid was not discussed in any detail – let alone, having a whole chapter devoted to it. Over the past fifteen or so years, the need for electrical energy storage emerged – driven primarily by the move to wind and solar generation resources.

The next section steps back to the broader topic of energy (not just electrical) storage. Then the chapter leaves other storage techniques behind, focusing only on electrical storage.

Primary Energy Storage

Chapter 1 discussed primary energy sources including fossil fuels (coal, natural gas, and oil), nuclear, and renewables (wind, water, biomass, geothermal, and sunlight). This section explores the role of energy from the sun as it relates to creating and then "storing" primary energy before using it to generate electricity. Many people hear the term *energy* from the sun and think about the solar panels on their neighbors' house or, perhaps, a community solar field. Solar energy, however, consists of much more than these common images. Some scientists say *all* primary energy sources connect back to the sun, as discussed in the following paragraphs.

Solar Energy

The most widely recognized solar electricity generation process involves *direct* capture of solar energy in photovoltaic (PV) cells, converting photons directly into electrons and yielding megawatt-scale power plants on par with traditional utility-scale electrical power generation (oil, coal, natural gas, nuclear, hydroelectric).

Indirect capture of solar energy can be witnessed by anyone playing outside with a magnifying glass on a sunny day. The solar energy can be concentrated, captured, and used to generate heat. Various common, utility-scale solar energy concentrating processes, including trough, linear Fresnel, and power tower (collectively called concentrated solar power or CSP), capture heat and use it to generate steam. This steam spins a turbine/generator and produces utility-scale electricity.

Neither direct nor indirect solar energy processes provide for an energy storage component, however.

Fossil Fuels

Scientists say solar energy captured by plants through photosynthesis and serving in the diet of the local marine or land animals became fossil fuels (coal in the case of plants and crude oil and natural gas in the case of animals). In other words, fossil fuels are stored solar energy.

Biomass and Biofuels

All the energy stored in biomass and biofuels produced today easily links to plants grown with the direct input of energy from the sun (sunlight).

> **Biomass or Biofuel**
> Plant-based material that serves as fuel for producing heat or electricity (e.g., wood, crops, industrial or household waste).

Water

Electricity generated from water results from the hydrologic cycle (evapotranspiration, condensation, precipitation, infiltration, and collection) driven by sunshine. In the past, power from water drove grist mills, sawmills, industrial plants, and hydroelectric power plants. The energy stored in reservoirs for hydroelectric plants continues as an important electricity generator.

Wind

Local environmental heating (driven by the sun) and subsequent cooling, cause high- and low-pressure zones in the atmosphere. Air moving from higher pressure to lower pressure produces wind. Wind energy has been, and still is, used for transportation, agriculture, and recreation. Increasingly, solar energy in the form of wind generates gigawatts of power from onshore and offshore wind turbines, but does not provide energy storage.

Other primary energy forms may also connect to the sun, although perhaps in a more circuitous fashion:

- Nuclear energy – formation of radioactive heavy elements in the gravity of the sun
- Wave and tidal energy – Sun, Moon and Earth gravitational interplay
- Geothermal energy – earth's mass (and gravity) formed as a result of the sun's gravity

Energy Density

Energy density is the *amount* of energy a given *volume* can store ("volumetric energy density") expressed as megajoules per liter (MJ/L), or in a given mass ("gravimetric energy density," also called "specific energy") expressed as megajoules per kilogram (MJ/kg). The higher the energy density of a fuel source, the greater the amount of energy it contains.

The corollary of "power density" is the amount of power per unit volume. Higher-power density means energy can be released at a faster rate. Ideal storage provides both high energy and high power. Figure 8-2 provides the energy densities of relevant primary fuels (including cow and camel dung), as compared to energy densities of batteries, hydrogen, flywheels, and water.

Material	Energy Density (MJ/L)	Specific Energy (MJ/kg)
Uranium	1,539,842,000	80,620,000
Natural Gas	0.0364	53.6
Coal	26-49	24 - 35
Diesel Fuel	38.6	45.6
Cow and Camel Dung	NA	15.5
Rechargeable Li Battery	NA	9
Batteries	0.56 – 6.02	0.17 - 9
Hydrogen (gas)	0.0101 – 0.0911	120 - 142
Flywheel	5.3	0.36 – 0.5
Water (100 m dam height)	0.000978	0.000981

Figure 8-2. Comparison of Relevant Primary Energy and Storage Options.

The energy density of uranium compared to that of coal or natural gas is why nuclear power plants do not have uranium continually fed into them like coal- and natural gas-fired generation plants. Lithium-ion battery energy density, compared to that of coal and natural gas, is an indication of just how much battery storage will be needed when lower-capacity-factor generation resources replace higher-capacity resources.

Certain primary energy sources (solar, wind, wave, and tidal) do not fit into the standard energy density definition and do not appear in this table.

Storage and the Grid

The burgeoning focus on electrical power generation from renewable energy sources, primarily wind and PV solar, has caused a shift from storing energy *before* the point of first electricity generation (as with uranium, natural gas, coal, and water) to storing energy *after* the point of first electricity generation. Accordingly, the balance of this chapter considers only this *electrical* energy storage.

Nondispatchable

Wind and solar (without storage) generate electricity when the wind blows strongly enough or the sun shines brightly enough to cross a certain minimum threshold. Electrical power industry vernacular refers to generation resources that cannot supply electricity "on demand" as *nondispatchable.* Non-dispatchable generation resources create problems for system operators because they cannot necessarily rely on the generation resources when consumers want electricity.

Dispatchable

Conversely, sources that can supply power on demand (like fossil fuels, nuclear, geothermal, and hydro) are called *dispatchable.* The term dispatchable goes back to the notion that system operators send *dispatch instructions* to generators and the generators respond by adding more – immediately available – electricity to the grid. Each of these conventional energy sources can provide instantly dispatchable energy because they bring along their own integrated storage mechanism to provide for this on-demand functionality. Renewable primary energy sources (like solar and wind) can become dispatchable with the addition of an appropriate energy storage technology after their first point of electricity generation.

Capacity Factor

Dispatchability links to another concept – capacity factor (discussed in Chapter 3) and to storage. A simple example follows:

- A six-gigawatt-capacity pulverized coal plant operating at an 85% capacity factor (see Chapter 3) can produce a maximum of 3,672 ($6 \times 0.85 \times 24 \times 30 = 3{,}672$) gigawatt-hours (GWh) of electricity over one month.
- Six gigawatts of wind generation resources operating at a 30% capacity factor (see Chapter 3) can produce a maximum of 1,296 ($6 \times 0.30 \times 24 \times 30 = 1{,}296$) GWh over the same month, considerably less than that generated by the equivalent coal capacity.
- To produce the same 3,672 GWh as the coal plant, a wind resource must have a capacity of almost seventeen gigawatts ($3{,}672/1{,}296 = 2.83$ and $2.83 \times 6 = 16.98$).

Storage Capacity

Utility-scale electrical storage capacity is measured in terms of instantaneous power – megawatts or gigawatts (also called nameplate power) – and how long that power can be delivered, in terms of megawatt-hours (MWh) or gigawatt-hours (GWh).

Going back to the coal and wind capacity factor example – assuming the six-gigawatt coal plant serves a constant five-gigawatt load and operates 85% of each day, it could generate a maximum of 122.4 GWh ($6 \times 0.85 \times 24$). While the plant operates, the grid load would consume 102 GWh ($5 \times 0.85 \times 24$). While the plant does not operate, the grid load would demand 18 GWh ($5 \times 0.15 \times 24$). In this case, storage of five gigawatt and eighteen gigawatt-hours capacity could reliably supply the grid.

Assuming the seventeen-gigawatt wind resources serve the same constant five-gigawatt load and operate 30% of each day, they could generate a maximum of 122.4 GWh ($17 \times 0.30 \times 24$). While the resources operate, the grid load would consume 36 GWh ($5 \times 0.30 \times 24$). While the resources do not operate, the grid load would demand 84 GWh ($5 \times 0.7 \times 24$). In this case, storage of five gigawatt and eighty-four gigawatt-hours capacity could reliably supply the grid. The eighty-four gigawatt-hours of storage would need to charge at a rate of 11.6 ($84/(24 \times 0.3)$) gigawatts and discharge at five gigawatts to reliably supply the grid.

Storage Required

In reality, no grid has only one generation source, and this scenario would never happen; it merely demonstrates how storage requirements increase as capacity factors decrease. Suffice to say, the lower the capacity factor, the longer the time between generation (storage recharge) cycles, and the larger the variability in load, the higher the amount of storage required to maintain grid stability.

Figure 8-3 presents a comparison of storage required in this simplistic example, graphically depicting how much more storage is required.

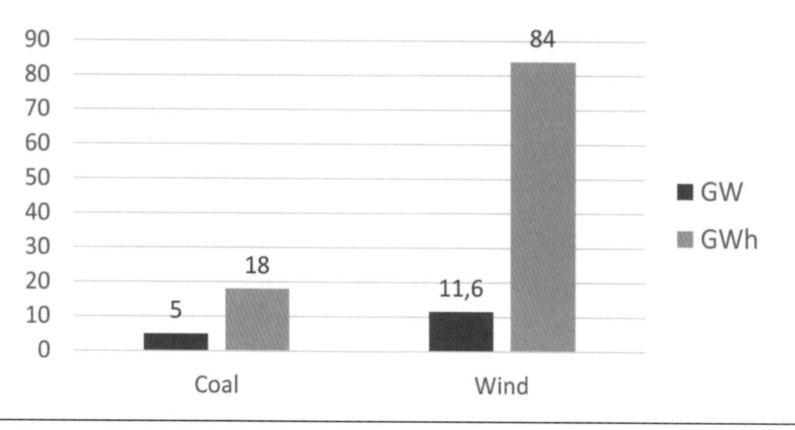

Figure 8-3. Comparison of Storage Capacity Required Between Coal and Wind.

Electrical Energy Storage Technologies and Uses

Multiple different storage technologies exist, with some better suited for certain applications and others more appropriate for other applications.

Storage Technologies

The U.S. Department of Energy (DOE) identifies three primary energy storage technologies:

- Electromechanical
- Thermal
- Electrochemical, electrostatic, and electromagnetic

This chapter discusses each technology type along with its primary sub-types. While several well-established energy storage technologies are currently in use, pumped hydro is the largest by far.

The rapid increase of utility-scale electrochemical storage (batteries), driven by wind and solar generation resources, means Figure 8-4 provides a historic view at the time of publication. Electrochemical storage should continue growing at a faster rate than pumped hydro. Other technologies are also emerging in pilot-scale operations, and even more technologies are in the research and development stage.

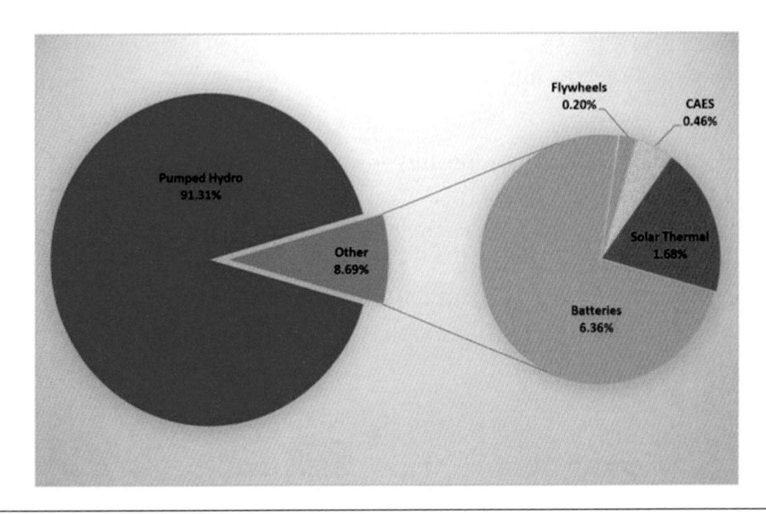

Figure 8-4. Percent Electricity Storage Capacity in the United States by Technology.

Figure 8-5 shows energy storage sub-technologies. The vertical axis is the amount of time over which the technology discharges, the horizontal axis is the system power rating, and the top axis lists the technology type. This text deals with utility-scale storage, which requires various performance characteristics. Bulk energy storage with high capacities and long discharge times appears on the right side and towards the top, with pumped hydro and compressed air energy storage (CAES). High-speed, shorter-duration storage for ancillary services/power quality, including batteries and superconductors, appears at the left and lower on the chart. System power ratings in Figure 8-5 do not necessarily come from one unit – various modules may aggregate for larger capacities and durations.

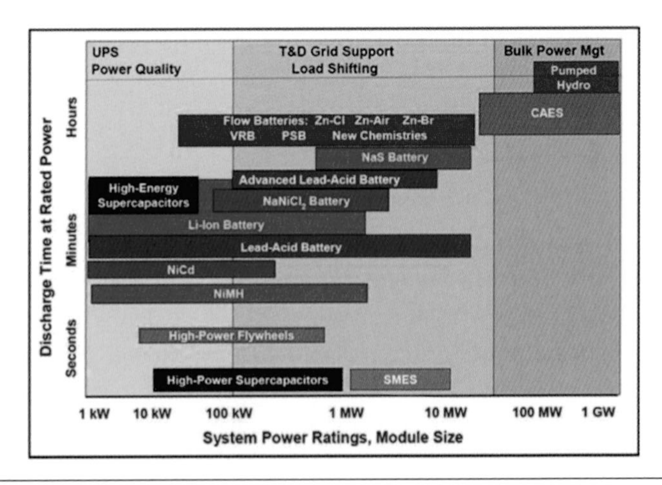

Figure 8-5. Energy Storage Applications Based upon Technology, Rated Power Capacity, and Discharge Duration.

Storage Uses

The United States DOE classifies storage applications into the five categories and seventeen sub-categories shown in Figure 8-6.

The required discharge duration and power capacity – the four left columns of Figure 8-6 – vary by the five different categories. Categories 1 (Electrical Supply), 2 (Ancillary Services), and 3 (Grid System) are the most important for grid stability. Category 4 (End User) provides customers flexibility to manage how they use electricity and service quality and reliability. Finally, Category 5 (Renewables Integration) is the fastest-growing category.

#	Benefit Type	Discharge Duration*		Capacity (Power: kW, MW)	
		Low	High	Low	High
Category 1 - Electric Supply					
1	Electric Energy Time-shift	2	8	1 MW	500 MW
2	Electric Supply Capacity	4	6	1 MW	500 MW
Category 2 - Ancillary Services					
3	Load Following	2	4	1 MW	500 MW
4	Area Regulation	15 min.	30 min.	1 MW	40 MW
5	Electric Supply Reserve Capacity	1	2	1 MW	500 MW
6	Voltage Support	15 min.	1	1 MW	10 MW
Category 3 - Grid System					
7	Transmission Support	2 sec.	5 sec.	10 MW	100 MW
8	Transmission Congestion Relief	3	6	1 MW	100 MW
9.1	T&D Upgrade Deferral 50th percentile***	3	6	250 kW	5 MW
9.2	T&D Upgrade Deferral 90th percentile***	3	6	250 kW	2 MW
10	Substation On-site Power	8	16	1.5 kW	5 kW
Category 4 - End User/Utility Customer					
11	Time-of-use Energy Cost Management	4	6	1 kW	1 MW
12	Demand Charge Management	5	11	50 kW	10 MW
13	Electric Service Reliability	5 min.	1	0.2 kW	10 MW
14	Electric Service Power Quality	10 sec.	1 min.	0.2 kW	10 MW
Category 5 - Renewables Integration					
15	Renewables Energy Time-shift	3	5	1 kW	500 MW
16	Renewables Capacity Firming	2	4	1 kW	500 MW
17.1	Wind Generation Grid Integration, Short Duration	10 sec.	15 min.	0.2 kW	500 MW
17.2	Wind Generation Grid Integration, Long Duration	1	6	0.2 kW	500 MW

*Hours unless indicated otherwise. min. = minutes. sec. = seconds.

Figure 8-6. Energy Storage Uses.

Capacity is expressed in *megawatts* (MWs) and discharge duration in *hours, minutes,* and *seconds.* Because total energy stored depends on capacity (MW) and duration (hr), these are critical attributes required for understanding energy storage potential. This is particularly important when storing substantial amounts of electrical energy for Categories 1, 2, and 5. Category 5 is the newest and fastest-growing storage use owing to the increase in wind and solar with their relatively low-capacity factors.

The maximum energy storage as shown in Figure 8-6 (on line 1) – 500 MW at 8 hours (or 4,000 MWh) – demonstrates one of energy storage's challenges: how to store substantial amounts of electrical energy for prolonged periods of time. *Long duration* or *seasonal storage* is discussed in more detail at the end of this chapter.

Electromechanical Energy Storage

An electromechanical energy storage system (ESS) converts electrical energy into kinetic or potential energy used to regenerate electricity as needed.

Pumped Storage Hydropower

As shown in Figure 8-4, pumped storage hydro (PSH) currently accounts for about 91% of electrical storage in the U.S. PSH stores energy not as electricity but as potential energy and generates electricity in the same way as regular hydro (discussed in Chapter 3). Water in a reservoir (stored energy) falls in a controlled manner through a tunnel (penstock) to a turbine driving an electric generator. Regular hydroelectric generation is a one-directional process, whereas PSH is two-directional.

With regular hydroelectric generation, falling water transfers its stored (potential) energy (represented by the *head* – elevation of the water above the turbine) to a turbine, converting this potential energy to kinetic energy. The turbine then transfers kinetic energy to a generator, and the generator converts kinetic energy into electrical energy.

PSH has one significant difference from regular hydro – the ability to use it *to push water back uphill.* When the grid has *surplus energy,* that energy is used to pump water from a lower body (reservoir or river) up to a higher elevation reservoir. When dispatched, the water flows down to the lower body through a turbine to generate electricity. The largest pumped hydro energy storage plant in the U.S. is the Bath County Pumped Storage Station in Virginia with a power rating of 3,003 MW and an energy storage capacity of 30,930 MWh.

Figure 8-7 shows the pumped storage upper reservoir of another PSH facility – Taum Sauk – located in the St. Francois Mountain region of Missouri, about 90 miles (140 km) south of St. Louis. The upper reservoir can hold approximately 1.5 billion gallons of water.

Figure 8-7. Taum Sauk Upper Reservoir.

PSH uses more electricity as it pumps water up than it generates as the water flows back down, which, on the surface, does not seem to add value. But PSH generates electricity during times of high use (and high prices) and pumps water at times of lower electricity use (and lower prices). The first known uses of PSH occurred in Italy and Switzerland in the 1890s and first appeared in the U.S. in 1930.[56] PSH facilities now appear all over the world and account for 95% of all utility-scale energy storage in the U.S.[57]

Closed Loop

In some instances, the same water that came downhill through the turbine/generator gets pumped back uphill and reused repeatedly. This design is called *closed-loop* PSH and represents approximately 25% (22 GW) of all PSH capacity worldwide.[58] Figure 8-8 shows a closed-loop system.

Open Loop

Also shown in Figure 8-8, is an open-loop PSH system, which takes water from a flowing river or stream, pumps it uphill to the upper reservoir, and discharges it back into the stream to generate electricity – using the water once. The world's largest PSH plant is the Fengning Pumped Storage Power Station in China,[59] which will provide 6,612 gigawatt-hours of energy storage per year (~18 GWh/day).

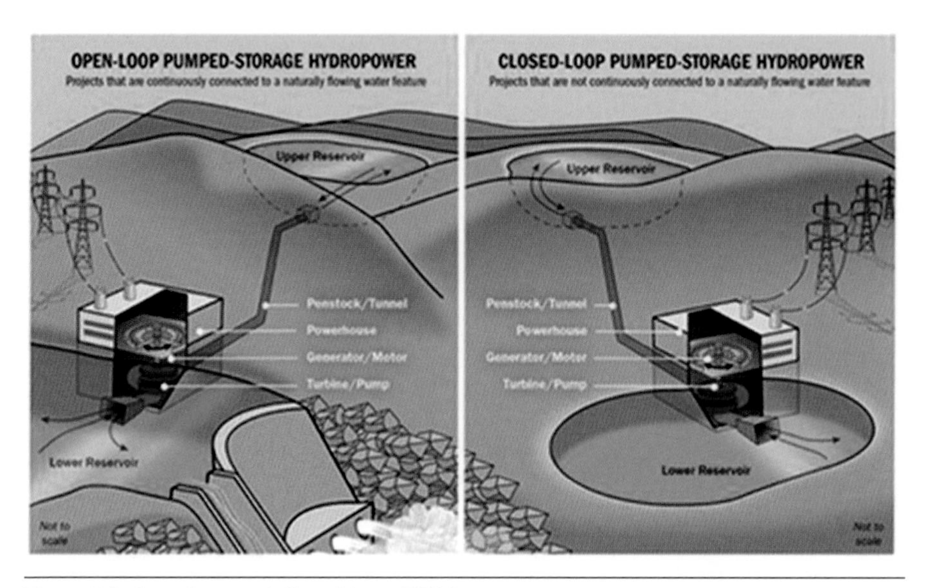

Figure 8-8. Open-Loop and Closed-Loop PSH Systems.

Emerging Technology – Gravity Compressed Hydroelectric Storage

A new and creative variant on traditional PSH turns the idea of hydraulic potential energy on its head. *Gravity storage* uses hydraulic lifting of an extremely large mass of rock by water pumps providing potential energy subsequently released when the pressurized water discharges through a turbine.[60] This creative solution (not yet proven at large scale) claims to potentially yield a very high-capacity (GWh-scale) storage solution at 80% efficiency and lends itself to the many locations not offering appropriate topography for a naturally-elevated PSH facility.

Compressed Air Energy Storage (CAES)

Like PSH, CAES takes electricity from the grid during periods of low demand and puts electricity back onto the grid during periods of high demand. A motor/compressor combination compresses atmospheric air into a confined space (pressure vessel) that holds the compressed air and releases it as needed. As it expands, it drives a turbine/generator supplying electricity to the grid. Hybrid systems combine compressed air with natural gas-fired turbines to deliver greater power. The larger the pressure vessel and the higher the pressure, the more energy it can store.

Compressibility

Gases are compressible – forcing more molecules into a pressure vessel requires energy that increases the pressure inside the vessel. As molecules get released to the lower pressure, the volume they occupy expands as they release the energy that forced them into the vessel.

In some cases, CAES uses underground salt domes or caverns as large pressure vessels.[61] Combining this large volume of compressed air, with pressures that can exceed 1000 pounds per square inch (PSI), provides for both high-energy and high-power storage solutions. CAES advantages include a readily available and free working fluid (air) and storage enclosures generally provided by nature. Figure 8-9 shows the CAES process flow.

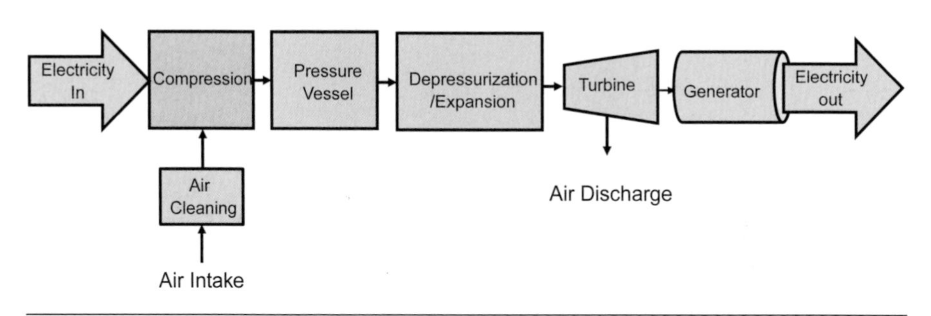

Figure 8-9. CAES Process Flow.

The largest operational CAES project in the world is the McIntosh CAES Plant in Alabama with a rated power capacity of 110 MW. It can run for 26 hours at full power and has an energy storage capacity of 2,860 MWh. The oldest CAES plant, built in 1978 in Huntorf, Germany, has a higher-rated power capacity of 290 MW but can run for only two to three hours and so has a lower energy storage capacity. Both plants represent a hybrid technology combining natural gas with compressed air storage. Both plants store air underground in excavated salt caverns produced by solution mining.[62]

On the other end of the scale, small-scale CAES systems provide a few tens of kilowatts for less than a minute as an uninterruptible power supply (UPS) bridge until a motor/generator comes online.

Kinetic Energy Storage Systems (KESS)

A *kinetic energy storage system* (KESS) converts electrical energy into kinetic energy and then back into electrical energy as needed. The two most common systems are rotating flywheels and linear energy gravity systems.

Rotating Flywheels: Electric motors add energy to flywheels, making them spin. Flywheel energy depends on the flywheel's mass, diameter, and rotary speed. Larger-diameter, heavier flywheels, turning faster, contain more energy than smaller, lighter flywheels turning slower. A motor-generator with appropriate controls provides kinetic energy to the system and electrical energy out of it, as needed.

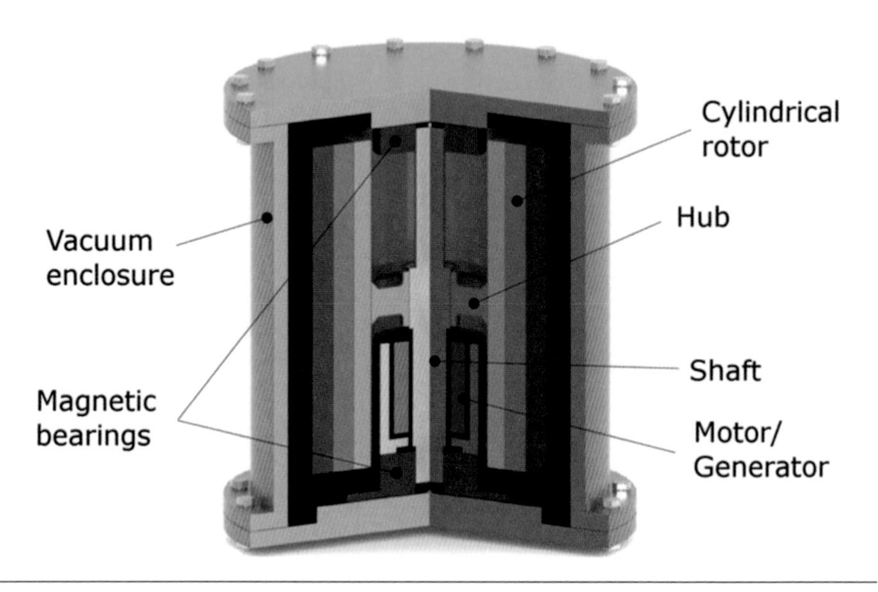

Figure 8-10. Typical Cylindrical Flywheel Rotor.

Flywheel capacities are in the tens of kilowatt-hours, meaning individual flywheels cannot provide grid-scale storage. Combining units in series or parallel, however, may successfully create meaningful grid-scale storage.

While individual flywheels cannot supply grid-scale storage, rotating flywheels, spinning at 10,000–20,000 revolutions per minute (RPM) work well for rapid response to grid demands by providing short-term electrical energy until other, higher-power systems can start up and serve the load. One common example is a flywheel paired with a diesel-powered generator. The flywheel responds seamlessly to a grid outage and bridges the twenty to thirty seconds needed to start a backup diesel generator.

As with every power system, but especially given the flywheel's modest capacity, efficiency is important. Common design elements incorporated in the service of greater efficiency include magnetic (non-contact) bearings to levitate the flywheel and reduce friction losses, vacuum chamber housing (to reduce air drag losses), and motor/generator design.

Emerging Technology – Linear Gravity Energy Storage (LGES)

While linear mechanical movement may not seem beneficial when compared to high-speed rotary motion, large distance and large weights make LGES feasible. Like flywheels, LGES directs electricity to a motor. The motor lifts a large mass vertically, converting electrical energy into potential energy. Electricity is generated by dropping the weight, allowing gravity to work. The cable holding the weight rotates a drum and drives a motor-generator, typically through a gearbox.

One advantage of LGES over rotary flywheels is the potential for greater capacities. The size/weight of a single working mass does not limit the system and the mass can be duplicated many times in a single system. Additionally, the system is simple, with no centrifugal force limits, vacuum systems, or sophisticated magnetic bearings. Another advantage is the distance the mass travels can be long, translating into higher-capacity energy storage. Some developers design their own vertical structures, while others take advantage of pre-existing features such as cliffs or mine bores to reduce cost. An example of a gravity energy storage system using working mass weights appears in Figure 8-11.

Figure 8-11. Energy Vault Commercial Demonstration Unit (CDU).

The picture to the right in Figure 8-11, shows the tower to which various weights connect, and the picture to the left is a detail of the individual weights. More weights and a longer vertical drop mean more stored energy.

Thermal Energy Storage

A thermal energy storage (TES) system stores energy as heat (or cold) in various materials and releases them for future use. Some systems, like concentrated solar power (CSP), store heat *before* the point of electrical generation during the day and release it during the night, allowing continued electricity generation when the sun is not shining.

In other systems, like liquid air energy storage (LAES), electricity is stored as cryogenic air and regenerated into electricity. Still other systems use energy to run refrigeration equipment at night and support daytime air-conditioning requirements without ever regenerating into electricity again.

Concentrated Solar Power with Storage

All CSP technologies (parabolic trough, linear Fresnel, and solar power tower) concentrate heat from the sun into a working fluid – commonly water, oil, or molten salt. The thermal medium goes through a heat exchanger, producing high-temperature steam to drive a turbine/generator. When CSP incorporates energy storage, the energy *storage* mediums may include all the above but may also extend to other high-specific-heat-capacity materials such as sand, rocks, or concrete blocks. Because the sun shines only during the day, some CSP systems incorporate energy storage, allowing for higher-capacity factors. Figure 8-12 is a picture of a solar parabolic trough. The trough concentrates sunlight on a pipe carrying the working fluid and capturing the solar energy as heat.

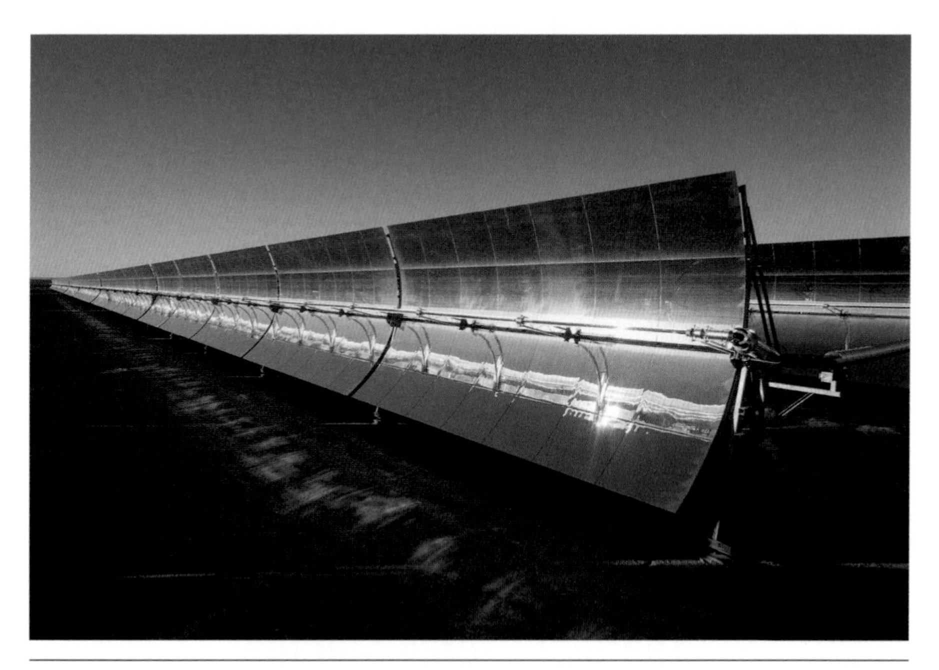

Figure 8-12. Picture of a Solar Parabolic Trough.

One example of a CSP plant with storage is the 280 MW Solana Generating Station in Gila Bend, Arizona. Since 2013 it has successfully captured solar energy as heat during the day and used it later, in the peak evening hours, to time-shift the solar energy and satisfy demand in a directly dispatchable manner.[63] The plant uses synthetic oil as the heat transfer fluid and molten salt as the heat storage medium to provide six hours of nameplate power production. Figure 8-13 is a block diagram of the CSP process.

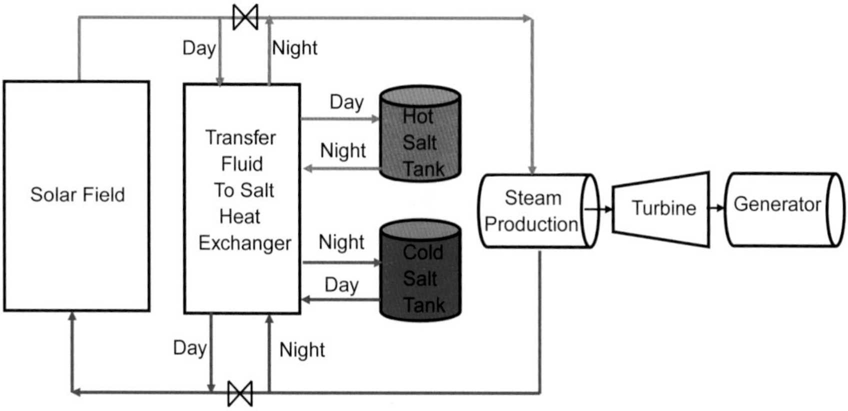

Figure 8-13. Concentrated Solar Power (CSP) with Storage Flow Diagram.

Solar Field

During the day, the solar field intake receives cold transfer fluid from the heat exchanger, and steam production discharges, heats the transfer fluid, and discharges it to both the heat exchanger and steam production. The solar field shuts down at night.

Heat Exchanger

During the day, the heat exchanger receives hot transfer fluid from the solar concentrators and cold salt from the cold salt tank. It heats the cold salt and discharges it to the hot salt tank. During the night, the heat exchanger receives hot salt from the hot salt tank and uses it to heat the cold transfer fluid, which is discharged to steam production. The now cold salt goes back to the cold salt tank.

Salt Tanks

The hot salt tank receives hot salt from the heat exchanger during the day, stores it, and sends it back to the heat exchanger during the night. The cold salt tank sends cold salt to the heat exchanger during the day and receives cold salt from the heat exchanger during the night.

Steam Production

Hot transfer fluid enters the steam production intake, heats water to generate steam, and exits the steam production discharge. The produced steam drives the turbine connected to the generator.

Ice Energy Storage

Thermal energy storage, as discussed above, captures high-temperature energy in a working fluid, or other medium, used to generate steam and spin a turbine. Alternatively,

Phase Change

Substances have four phases, solid, liquid, plasma and gas. The phase they are in depends on temperature and pressure. They change from one phase to another when heat is added or removed (at constant pressure).

energy can also be stored at *lower* temperatures using a medium such as water or ice. This ice stores electrical energy which may then be used later to displace the electrical energy used in refrigeration equipment for cooling.

The key to ice energy storage is a property called "latent heat of fusion," which relates to phase change. In simple terms, it takes a lot of energy to convert liquid water to a solid (ice). The amount of energy depends on the "latent heat of fusion" and, because water has a high latent heat of fusion, ice can store a lot of energy. This energy is "regenerated" as cooling capacity when the ice melts.

Ice energy storage is common in urban core areas where large ice energy storage facilities create copious amounts of ice (generally at night) using surplus or low-cost energy. Through a series of heat exchangers and pipes, the ice provides chilled water for cooling urban buildings the following day. This efficient energy storage technique satisfies air-conditioning demand from the buildings and moderates electrical power consumption by time-shifting cooling loads from the day to the night. Figure 8-14 is a picture of a building housing an ice energy storage plant.

Figure 8-14. Austin Energy District Cooling Plant Number 3 – Downtown Austin, Texas.

Emerging Technology – Liquid-Air Energy Storage

Liquid-air energy storage (LAES), also called cryogenic energy storage (CES), drives a turbine with expanding air (like CAES). However, LAES takes CAES another step – through a series of compression and refrigeration cycles. LAES removes enough heat to make the air change phase from gas to liquid. The liquid is stored at low pressure in insulated tanks at temperatures less than -196°C (the boiling point of nitrogen). When electricity is needed, heat is added to the liquid air causing it to evaporate (change back into a gas), expand, and run through a turbine/generator.

This energy storage technology relies on well-established industrial processes of air liquefaction and enjoys the benefit of high-energy density, which translates to efficient storage volumes, allowing LAES to work almost anywhere. This technology best supports large-scale energy storage due to the significant industrial equipment required. The U.S. DOE Global Energy Storage Database indicates about 30 MW of LAES operates today. Larger plants (like the Highview Power 50MW/250MWh CRYOBattery in Greater Manchester, U.K.) are under development.[64]

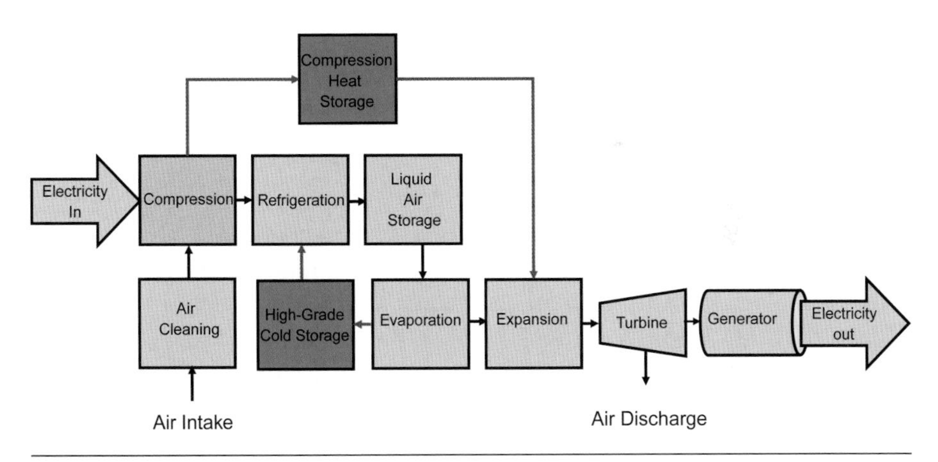

Figure 8-15. Liquid Air Energy Storage – Schematic Flow Diagram.

The gray boxes in Figure 8-15 show the process flow. The top green box shows capturing heat generated during compression and using it for expansion. The lower green box shows cold temperatures generated in the expansion process directed to the refrigeration cycle.

Electrochemical, Electrostatic, and Electromagnetic Energy Storage

Three words in the title of this section all have the same root – *electro* – meaning they relate to electricity. The second part of those words, *-chemical, -static,* and *–magnetic,* indicate the form into which the electricity is converted for storage.

Electrochemical (Battery) Energy Storage

Batteries are self-contained, chemical power packs. Many people remember the periodic table of elements from high school chemistry with all the known elements arranged in order of their atomic mass and atomic number indicating each element's *reactivity* – an important feature when selecting chemical combinations for battery design.

Battery Components

Batteries consist of four main components: two electrodes (cathode and anode), an electrolyte, and a separator that separates the cathode and anode. The anode and cathode each contains a different mixture of chemicals based on their reactivity and energy density.

Electrochemical Capacity

Electrochemical capacity is rated in terms of amp-hours (Ah) – the average amount of current released over a period of time under normal use. For example, a battery discharging twenty amps for eight hours has a rating of 160 Ah.

Ever since Alessandro Volta first discovered electrochemical potential (called *voltage* in his honor), battery scientists have searched for combinations of chemicals in nearly infinite permutations to find the "perfect" combination.

Nonrechargeable and Rechargeable Batteries

Batteries bought at a store come "charged" – they produce electricity through chemical reactions. When inserted into a circuit, batteries produce electricity while the chemical reactions continue. Once the reactions are exhausted, the battery is "dead" – unless it is a *rechargeable* battery. Chargers connected to rechargeable batteries drive the reactions in the opposite direction recharging the batteries.

The common lead acid battery (used in cars) provides an example. They have a cathode of lead-dioxide and an anode of metallic lead. The two electrodes are placed into a container with an electrolyte (sulfuric acid) with a separator between them. When the two electrodes connect to a load, the chemical reaction begins and strips electrons from some atoms. When an atom loses an electron, it becomes an *ion*. Chemists consider ions "electrically unstable" – they want that electron back. Electrons exit the battery through the top of the anode and travel to the load. From the load, the electrons travel to the top of the cathode and enter the battery. In the meantime, the ion travels through the electrolyte to the cathode, where it rejoins the electron. The separator prevents physical contact between the anode and cathode, while facilitating ion transport in the cell.

When the driver completes the circuit by pushing the starter or turning the key, a chemical reaction in the battery sends electricity to the starter (and all the other electrical components). Once the car starts, its alternator generates electricity for the entire car and sends some electricity back to the battery reversing the chemical reaction and recharging the battery. Other battery types follow the same basic approach but use different cathode, anode, and electrolyte formulations and different separators and packaging schemes.

Figure 8-16 depicts a simple lead acid battery of the type commonly used in cars.

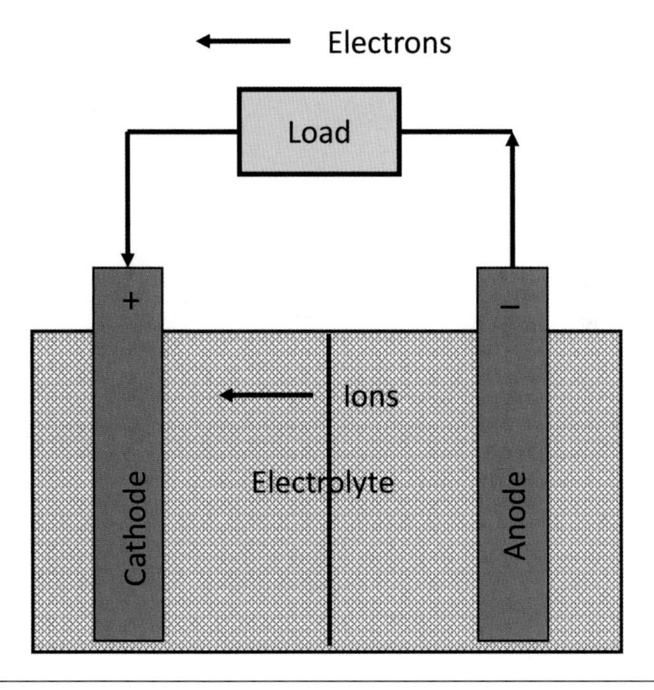

Figure 8-16. Battery Schematic Showing the Cathode, Anode, Electrolytes, and Electron Flow.

Battery Chemistries

Some of the most common battery materials include lead acid, nickel cadmium, nickel metal hydride, and alkaline. The Nobel Prize-winning chemistry of the lithium-ion battery (LIB), however, currently holds the most meaningful place in grid-scale electrochemical storage discussions.

Lithium-Ion Batteries

The formats of LIBs are similar to many other battery formats – "cells" made of electrodes (anode and cathode), separators (separating the electrodes), electrolyte (providing a conduction path for the lithium ions), with all contained in a housing – and are the building block of today's most common battery energy storage systems.

These cells come in various formats (cylindrical, prismatic, and pouch being the most common), capacities (mAh to multiple Ah), and chemistries. Each format offers advantages in cost, packaging, and thermal management.

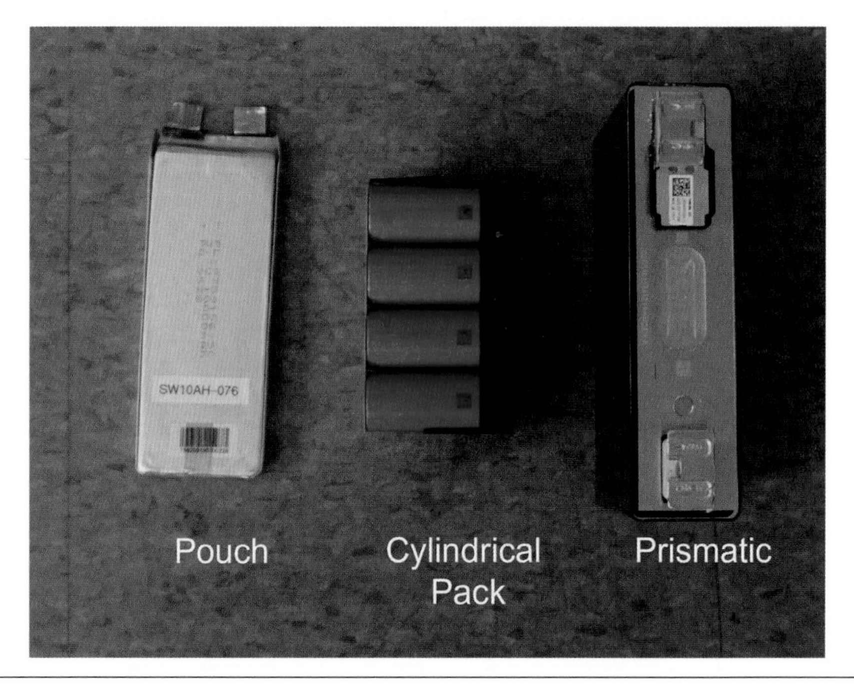

Figure 8-17. Lithium-Ion Cell Format Types; left to right: Pouch, Cylindrical (in a pack), and Prismatic.

The most common commercially available LIB chemistries today include:

- Lithium Iron Phosphate (LFP)
- Nickel Manganese Cobalt (NMC)
- Nickel Cobalt Aluminum (NCA)

In each case, one of the three metal oxides is coated on an aluminum cathode substrate, and the anode is most commonly graphite film coated onto a copper substrate. Cells are connected to each other in serial and/or parallel fashion, creating "modules," and then these modules are connected to achieve the system voltage and capacity (Ah or kWh) as required.

Modules are packaged into enclosures and provided with a battery management system (BMS) and a thermal management system (TMS). The BMS provides charge and discharge control of voltage and current and monitors all parameters to assure safe operation. The TMS is critical as cells/modules/packs dissipate heat under both charge and discharge operations. If all the cell temperatures are not maintained within a prescribed temperature band, the system can suffer reduced cycle life, or worse, battery fires. As a frame of reference, it takes over 50,000 LIB cells (similar in size to a AA battery) connected to make a single MWh of storage.

Materials of Construction

The discussion above uses words like lithium and cobalt, both elements in short supply and mined in not necessarily the most responsible manner or in the most secure parts of the world. Governments, industry, and academia around the globe seek to address supply concerns, look for more eco-friendly formulations, and try to find secure supplies of the elements needed to support the use of electrochemical storage for wind and solar power.

Emerging Technology – Advanced Battery Technologies

The field of electrochemical energy storage is not limited to the three lithium-ion chemistries already discussed. Other technologies intent on addressing current shortcomings are emerging. One promising area of lithium-ion-based advancement is "solid-state" batteries in which a solid-glass electrolyte or solid-ceramic separator replaces the liquid electrolyte, providing a safer battery with greater cycle life and energy density, yielding faster charge times.

Emerging Technology – Flow Battery

Another class of electrochemical storage technology is the flow battery (also called a redox flow battery). Unlike the "cell" configuration prevalent in the section above, flow batteries have a different architecture and are specialized for grid-scale applications. They are comprised of two electrolyte tanks and a cell stack. One tank contains a liquid anolyte (negative) and the other contains a liquid catholyte (positive). Each tank has its own pump that supplies one side of the cell stack. The cell stack contains a membrane separating the liquid anolyte from the liquid catholyte and provides the selective ion transfer with the resulting electric current.

The volume of the tanks correlates to battery energy capacity, and the aggregate size of the membrane(s) correlates to system power. There are several dozen demonstration projects worldwide, generally in the kW scale, commonly with several hours of storage each. Similar to the cell-based batteries discussed above, there are numerous flow battery chemistries in contention, with vanadium redox representing 82.8% of commissioned projects worldwide, and zinc-bromine representing 16.8% per the U.S. DOE Global Energy Storage Database. Benefits include long cycle life, energy efficiency and ability to scale, while challenges, as always, include cost.

Electrostatic – Capacitors

Capacitors are an exception to the statement "electricity cannot be stored as electricity." They consist of two metal plates connected to an electrical circuit. The plates are insulated from each other by air or some other insulation material. Electricity *wants* to transmit from one plate to the other, but the insulation prevents transmission. Scientists say positive charges coalesce (build up) on one plate and negative charges coalesce on the other plate to store electricity in an electrical field. Static electricity provides a useful example – an excess electrical charge accumulates as a person walks across a wool rug and discharges when the person touches a metal object.

Supercapacitors

Supercapacitors, also referred to as ultracapacitors (the two terms are interchangeable), store energy using electrochemistry (similar to batteries) combined with electrostatics (an electrical charge is "stored" in an electric field – similar to traditional capacitors) or as a hybrid of both.

Supercapacitors complement batteries. They provide higher-*power density* in contrast to a battery's higher-*energy density*. Supercapacitors and batteries both respond in milliseconds, but supercapacitors deliver their entire "charge" of stored energy in a matter of seconds while batteries typically deliver their energy in minutes to hours. This combination of high power, long life and "super-fast" charge/discharge capability of supercapacitors provides grid-integration opportunities, especially in the rapid response ancillary services area.

Emerging Technology – Superconducting Magnetic Energy Storage (SMES)

SMESs convert electricity into an electromagnetic force (emf) and store it in a coil. They are the most efficient energy storage technology because they have no mechanical, thermal, or chemical processes

Superconductivity

Ability to conduct DC current without energy loss *when cooled below* a critical temperature (referred to as T_c).

First proposed in the 1970s, SMES suffers no losses in the storage coil due to resistance because it incorporates the benefits of superconductivity. Electricity (converted to DC) charges the coil, which stores it until needed. In addition to the coil, SMES contains control, power conversion, and cryogenic cooling. Systems tend to be modest in capacity (tens of MW and less) but GWh-scale systems are in discussion. Limited demonstration and commercial systems exist, but the U.S. DOE Global Energy Storage Database does not currently list any SMES in operation or proposed.[65]

Hydrogen

Even a cursory glance at today's energy-related news identifies an article or two about hydrogen rewriting the global energy playbook and giving rise (hopefully) to *the hydrogen economy*. The smallest and most abundant of all elements, nearly all the hydrogen in the earth (and in the atmosphere) is locked up as molecules with other atoms (including water and organic compounds). Free hydrogen in the atmosphere is so light gravity cannot hold it and it floats out to space.

Hydrogen Colors

Using hydrogen for electrical "storage" requires understanding the colors of hydrogen. These colors, shown in Figure 8-18, describe the hydrogen's source – how it is produced. Each color depends on the raw material feedstock, processing method, and primary process energy source. Also important are the other process outputs (beyond hydrogen), for example, carbon dioxide versus oxygen – an important differentiator as the grid decarbonizes. All processes yield an identical hydrogen (H2) molecule.

Color	Hydrogen Source	Process Used to Extract Hydrogen	Emissions
Black and Brown	Coal	Coal gasification with steam methane reforming.	Carbon dioxide (CO_2) and heat released to atmosphere.
Gray	Natural Gas (CH_4)	Steam methane reforming.	Carbon dioxide (CO_2) and heat released to atmosphere.
Blue	Natural Gas (CH_4)	Steam methane reforming.	Carbon dioxide (CO_2) sequestered with heat released to atmosphere.
Turquoise	Natural Gas (CH_4)	Methane pyrolysis using various energy sources.	Solid carbon with heat released to atmosphere
Green	Water (H_2O)	Electrolysis using electricity from renewable resources.	Oxygen (O_2) and heat released to atmosphere.
Pink, Purple, and Red	Water (H_2O)	Electrolysis using electricity from nuclear energy.	Oxygen (O_2) and heat released to atmosphere.
Yellow	Water (H_2O)	Electrolysis using electricity from solar energy.	Oxygen (O_2) and heat released to atmosphere.
White	Natural Hydrogen (H_2)	In the research stage.	Unknown

Figure 8-18. Hydrogen Source "Color" Listing.

Electrical Storage

Green, pink, purple, red, and yellow hydrogen use electricity to produce hydrogen via a process called electrolysis. Using hydrogen to store electricity, like most other storage processes, involves converting electricity into another energy form (in this case, hydrogen) and converting it back into electricity as needed – in the case of hydrogen, either using a fuel cell or turbine generator.

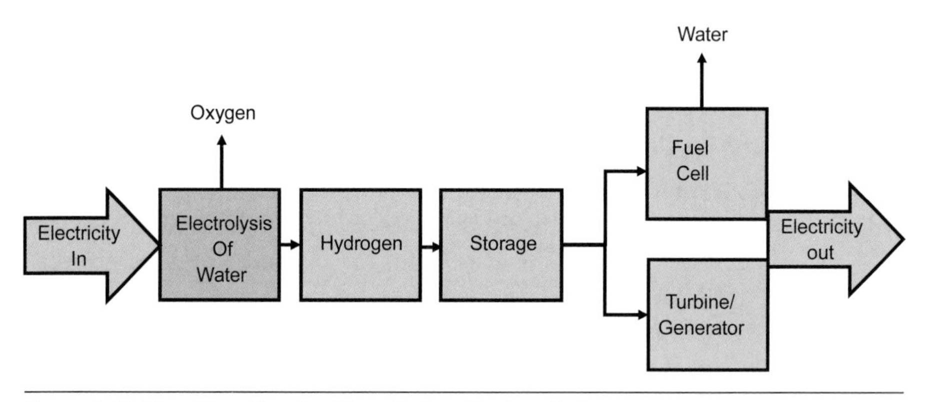

Figure 8-19. Flow Chart Showing Electrolysis to Produce Hydrogen.

Hydrogen Economy

The prospect of a hydrogen economy generates excitement because hydrogen could serve as either a clean fuel or storage medium, replacing nearly all fuel in use today. Hydrogen processing infrastructure already exists at enormous scale worldwide. Over 70 million tons of hydrogen were produced globally in 2018[66] for industrial processes (not for energy storage), including oil refining, ammonia production, and steel production.

Perhaps the most important takeaway regarding hydrogen energy storage – more than any other energy storage technology – is that hydrogen holds the broad-based potential of a *strategic* differentiator, creating an entirely new energy paradigm: the *New Hydrogen Economy.*

Long-Duration Storage

The need for long-duration storage (days, weeks, or months) may become acute as low-capacity wind and solar generators continue to penetrate the generation mix or seasonal disruptions of water flows affect hydro-power production. There are currently no viable long-term storage solutions, although some technologies, including redox flow batteries, LAES, and hydrogen storage, are angling for market positions with active demonstration projects. One such example is the 840 MW

Intermountain Power Project in Utah (supporting the Los Angeles Department of Water and Power) based on *green hydrogen,* which uses multiple salt domes with massive storage potential.

Summary

- Electrical energy, with a few exceptions, cannot be "stored."
- Electrical energy is generated from primary energy sources, converted into another energy form for "storage," and regenerated (converted back) into electricity when needed.
- Converting electricity to another form of energy, storing it, and then regenerating it creates heat.
- The amount of electricity going into storage is always less than the amount of electricity coming out of storage.
- Over the past approximately fifteen years, the need for electrical energy storage emerged – driven primarily by the move to wind and solar generation.
- Energy density is the *amount* of energy that can be stored in a given *volume.*
- The U.S. DOE identifies energy storage technologies as electromechanical, thermal, electrochemical, electrostatic, and electromagnetic energy storage.
- An electromechanical energy storage system (ESS) converts electrical energy into kinetic or potential energy used to regenerate electricity as needed.
- Thermal energy storage (TES) stores energy as heat (or cold) in various materials and releases them for future use.
- Batteries are self-contained, chemical power packs.
- Batteries consist of four main components: two electrodes (cathode and anode), an electrolyte, and a separator that separates the cathode and anode.
- Even a glance at today's energy-related news identifies an article or two about hydrogen rewriting the global energy playbook and giving rise (hopefully) to *the hydrogen economy.*
- The need for long-duration storage (days, weeks, or months) may become acute as low-capacity wind and solar generators continue to penetrate the generation mix or seasonal disruptions of water flows affect hydro-power production.

9

Gas-Electric Coordination

Love thy neighbor as thyself.

King James Bible, Matthew 22:39

"Grandpa Tom, our stove at home has rings that get red, but your stove has fire. Why is that?" asked Luke.

"Your stove uses electricity, and mine uses natural gas," answered the grandfather.

"Is that why ours doesn't work when the power is off and yours does?" asked Luke.

"That's right," replied Grandpa Tom. "Electrical power companies and natural gas distribution companies have many of the same customers and are customers of each other," he added.

"If electricity companies and gas companies have the same customers and are customers of each other, how do they decide who gets what if there is not enough natural gas or electricity to go around?" wondered Luke quizzically.

"Good question. Everybody expects all the natural gas and electricity they want when they want it, but they don't want to pay a lot for it – and guaranteeing we can get all we want when we want it costs a lot of money," stated Grandpa Tom. "The electrical power industry, the natural gas industry, and governmental regulators work together trying to decide how to allocate electricity and natural gas between the customer groups without costing too much," added Grandpa Tom. "But they have a tough challenge," Grandpa Tom concluded.

"I hope they figure it out; I don't like when we lose power," Luke sighed.

The Challenge of Gas-Electric Coordination

Figure 9-1 from the BP Statistical Review of World Energy 2021 shows the percentage of world electricity by primary energy. The percentage of electricity generated from coal (currently the largest percentage) is falling as the percentages from natural gas and renewables rise.

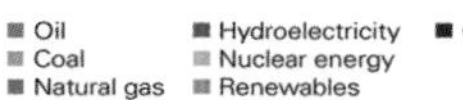

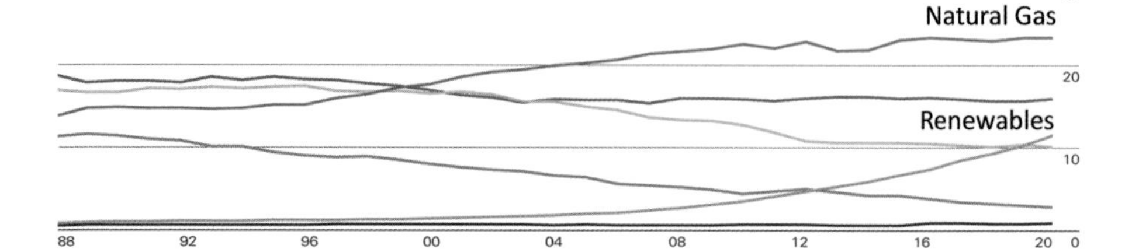

Figure 9-1. Global Electrical Generation by Fuel Source.

Assuming coal use continues declining, these lines and their directions imply natural gas (the "least polluting" fossil fuel) will continue as one of the primary energy sources for several decades to come. Extrapolating the three lines implies the time frame might be one or two decades, but the amount of energy demanded by the peoples around the globe continues increasing. Thus, based on the trend lines, eliminating all use of fossil fuel will probably take quite a long time.

The growing and continuing role of natural gas in generating global electrical energy is why a chapter regarding Gas-Electric Coordination – how natural gas moves from production point to generation – appears in this text. This chapter provides background on the natural gas value chain for perspective and then describes how regulators, the gas industry, and the electrical power industry attempt to coordinate the (usually competing) needs of gas and electricity customers to meet the needs of both.

Natural Gas and Electricity

Although gas and electricity both provide heat and light, they are very different types of energy. Gas is a *primary* energy source – a fuel. Electricity is a *secondary* energy source produced from fuels and does not occur in nature. Gas is easily and cheaply stored for later use. Electricity is not.

Dispatchable Electrical Power

Dispatchable electrical energy and *nondispatchable electric energy* – discussed previously in this text – are important for understanding gas – electrical coordination. Consequently, they are discussed in this chapter as well. Dispatchable electrical energy can be turned on and off when and as needed.

Currently, it comes from electric generation plants powered by fossil fuels, nuclear fuel, geothermal energy, water, and, to a small extent, batteries and other storage technologies.

Nondispatchable Electrical Power

Most renewable energy sources (including wind and solar) generate electricity only intermittently. Engineers say wind and solar have low-capacity factors. Electricity generated from them is called *nondispatchable* because electric system operators cannot turn it on or off at will. Reliably keeping peoples' lights on depends on having enough *dispatchable* power connected to the grid.

> **Storage**
>
> Storage makes nondispatchable electrical generation types pseudo-dispatchable.

Competing Customers

In addition to powering electric generators, natural gas also heats homes, businesses, schools, hospitals, and industrial plants; powers vehicles; and serves as a feedstock to manufacture chemicals.

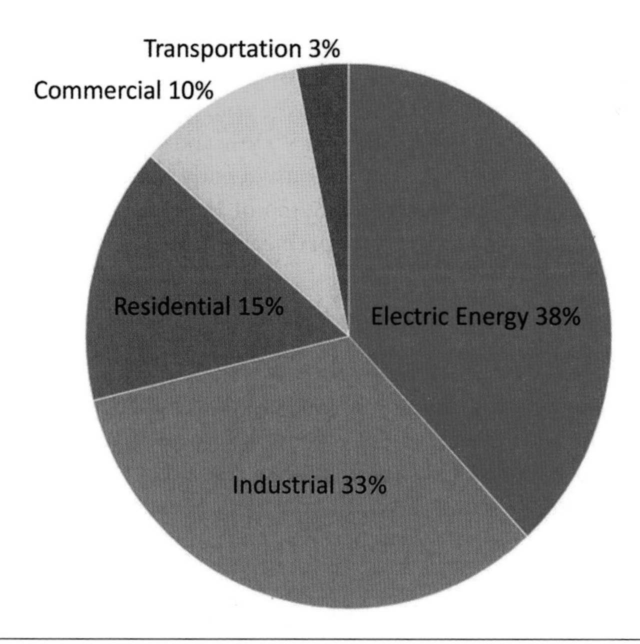

Figure 9-2. U.S. Natural Gas Usage by Customer Group.

One key challenge for coordinating gas and electric is prioritizing natural gas usage among customers in times of high demand or short supply. High demand can result, for example, from a snowstorm when everyone wants heat – some directly from burning gas and some from electricity generated from gas. Supply shortages can occur for various reasons – equipment malfunctions, weather changes, natural disaster, or civil unrest. In a worst-case scenario, high demand occurs at the same time as a supply shortage. Such an event occurred in Texas in February 2021 when Winter Storm Uri froze natural gas equipment while gas and electric demand skyrocketed due to unprecedented cold temperatures.

The Natural Gas Industry

Generally accepted theory provides that hydrocarbons formed over eons from decomposing organic materials (once living microorganisms and vegetation). Ancient rivers washed those remains along with soil, gravel, and sand downstream, depositing them in geological basins. Successive layers of "overburden" from the rivers covered the remains along with the soil, gravel, and sand.

> **Permeability**
> Property of geologic structures determining how easily fluids can flow from or through it.

Weight of the overburden exerted pressure on the mixture of organic remains, soil, gravel, and sand, and the depth raised the temperature, essentially "cooking" the remains and converting them into hydrocarbon fluids and converting the soil, gravel and sand into rock formations – shale and sandstone for example.

The rock surrounding the remains (and consequently the hydrocarbon fluids produced in them) is called "source rock." Some of the source rock formations from which hydrocarbons are produced are permeable, while other formations are impermeable.

Oil and Gas Reservoirs

The U.S. EIA defines reservoir as "a porous and permeable underground formation containing an individual and separate natural accumulation of producible hydrocarbons (oil and/or gas) which is confined by impermeable rock or water barriers and is characterized by a single natural pressure system."[67] The rock formations under the impermeable rock or water barriers have minuscule pores (voids) containing hydrocarbons.

Hydrocarbon Production from Reservoir Rock

Hydrocarbons created in permeable source rock flowed towards the earth's surface. In some cases, the hydrocarbons reached the surface and dissipated into the atmosphere or onto the ground. In other cases, the hydrocarbon fluids flowing towards the surface were trapped under layers of impervious materials. The hydrocarbons trapped under impervious rock layers remained there until geophysicists located them, drillers drilled through them, and production staff extracted the hydrocarbon fluids from the reservoir rock.

Hydrocarbon Production from Source Rock

Oil and gas that formed in less permeable source rock could not flow and waited until drillers bored into the formation and injected water, chemicals, and sand (i.e. proppant), cracking the rock and increasing its permeability. The oil and gas industry refers to this process as "hydraulic fracturing" or simply "fracking." Fracking source rock goes back to at least the 1970s, although it has become controversial in recent years. Figure 9-3 graphically recaps the previous paragraphs.

> **Proppant**
> Sand forced into the fractured rock to prop open the fissures created from a "frac job" is called, not surprisingly, proppant.

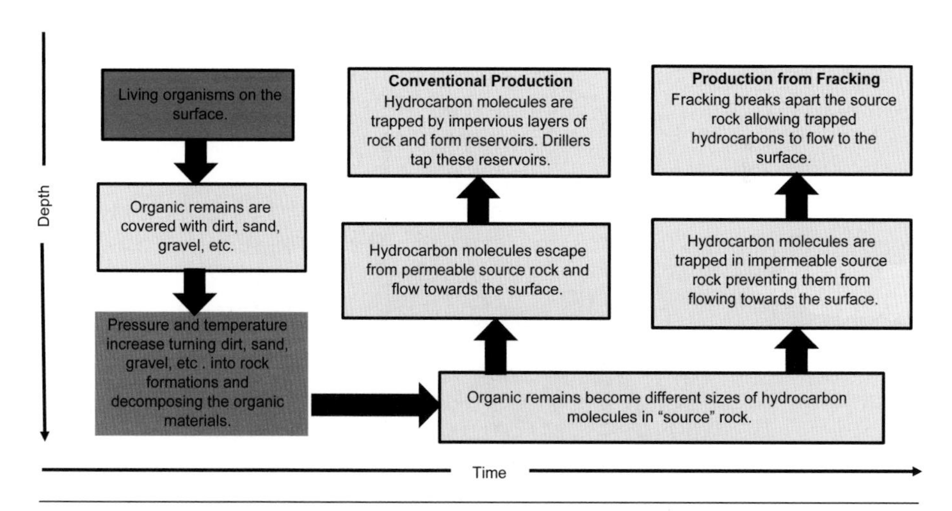

Figure 9-3. Oil and Gas Formation Flow Chart.

Production Stream

The hydrocarbon fluids flowing out of reservoir rock and fracked source rock contain mixtures of different size hydrocarbon molecules in liquid or gaseous form. This mixture of hydrocarbon gases and liquids exiting the wellhead is typically called the "production stream." The size and type of hydrocarbon molecules contained in the production stream vary by location, temperature, and pressure

– all related to source rock depth. Every producing field yields a mixture of various hydrocarbon molecules. Hydrocarbons from producing zones at different depths in the same field may differ in the combination of hydrocarbon molecules they hold.

The smallest hydrocarbon contained in the production stream is methane (CH_4). As already discussed, different production streams contain concentrations of the various sizes of hydrocarbon molecules. The Marcellus Shale in Pennsylvania, for example, contains much more gas than oil. In southwestern Pennsylvania, producers consider the gas "wet" because it contains ethane (C_2H_6), propane (C_3H_8), and butane (C_4H_{10}) in addition to the methane. The natural gas in north-central and northeast Pennsylvania is considered "dry" because it contains very little (if any) ethane, propane, or butane. The oil sands of Alberta fall on the other end of the spectrum. They contain large, complicated molecules like asphalt, because they were not buried very deep nor exposed to high pressures and temperatures.

Production

Oil and gas production from different wells contains different mixtures of molecules. Whatever the molecular composition, they exit the ground through a wellhead at the top of a well bore.

Figure 9-4. Natural Gas Wellhead. Manifolds (often called a "Christmas tree" or just "tree") are located atop wellbores. Production field workers adjust the various valves to control pressures and flows.

Wells are connected to production processing pads by pipes called flow or gathering lines. At the processing pad, equipment separates the stream into its constituent parts (crude oil, gases, and water) and removes impurities. The number of wells connected to a specific processing pad varies by factors such as well spacing, production rates, and economics. The gaseous stream normally contains a mixture of methane, ethane, butane, propane, and other heavier hydrocarbons. It leaves the facility through a meter into a pipeline that gathers gases from surrounding production pads and aggregates them for transport to gas processing plants.

Figure 9-5. Orifice Meter. The meter at the production processing area is typically the first component in the natural gas gathering system.

Natural Gas Pipelines

Natural gas systems consist of small-diameter lines that gather the gas and larger-diameter lines that move the gas to local distribution company (LDC) lines.

In gas pipeline nomenclature, gathering lines extend from oil and gas processing pads to gas treating and processing plants and then to transmission pipelines. At the delivery end, transmission lines connect to distribution lines at facilities called "city gates." From there, distribution lines carry natural gas to end-use customers. Electric generation facilities receive natural gas from either transmission or distribution lines depending on facility size and location.

Regulations in many countries, including the U.S., treat natural gas transmission pipelines as *transporters*, meaning the pipeline companies take *custody*, but not *ownership*, of the gas. They transport for shippers who sell to large users (such as hospitals, schools, factories, malls, and electrical generation plants), LDCs, and gas marketers. Alternatively, large users or LDCs may ship the gas on transmission pipelines themselves, purchasing the gas directly from producers before it enters the transmission system.

LDCs sell to individual homes and businesses and to many of the same customers served directly by transmission pipelines. They receive gas from natural gas transmission pipelines, liquefied natural gas (LNG) gasification plants, and storage – their own or those owned by others. Figure 9-6 shows the flow of natural gas from well to consumers.

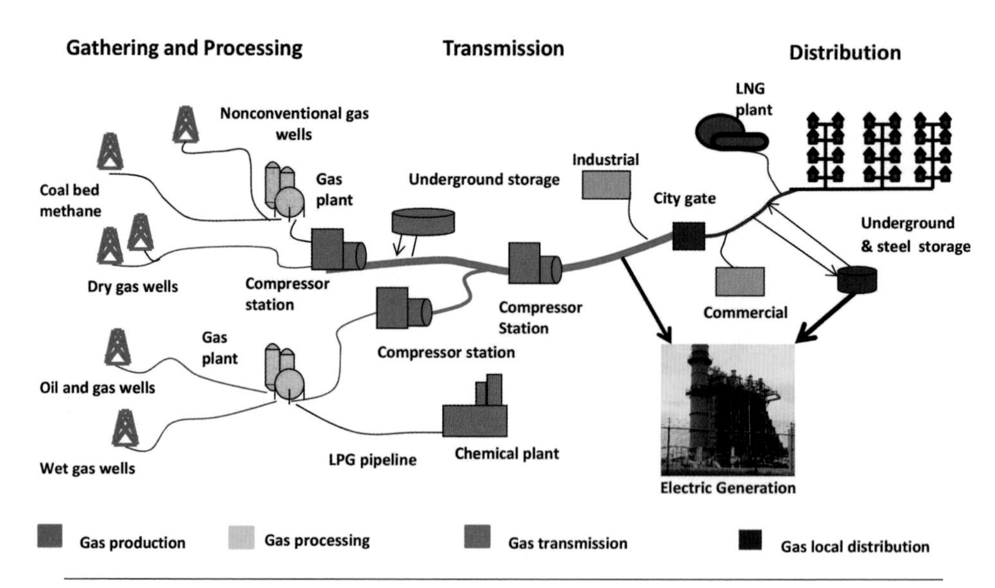

Figure 9-6. Natural Gas Flow from the Production Area to Customers.

The same company may own production, gathering, transmission, and local distribution lines through different subsidiaries. Pipeline companies (or their parent) may own gas processing plants or have a trading company buying and selling gas between related and non-related entities.

The distinction between natural gas gathering, transmission, and local distribution lines has more to do with history, regulations, traditional customers, line size, and operating pressures than hydraulics or operating principles. Natural gas, regardless of the country, obeys the same universal laws of physics. Accordingly, natural gas (and other) pipelines operate largely the same around the globe regardless of ownership structures.

Gathering and Processing

Lines used to gather the gas are usually between four and ten inches diameter, with size determined by the amount of gas available to move. Along its way, the gaseous stream may undergo additional processing to remove water, carbon dioxide, nitrogen, or other contaminants. Finally, the hydrocarbon stream goes to a processing plant to separate methane from the rest of the stream and the natural gas liquids (NGLs) (hydrocarbon molecules larger than methane). The methane stream then enters the transmission line.

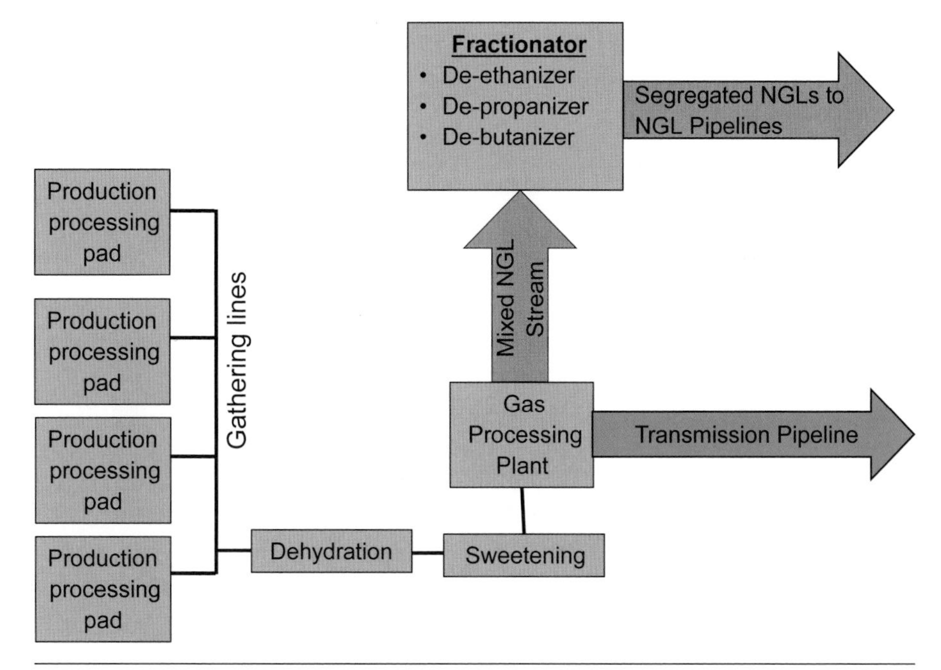

Figure 9-7. Gathering Operations Extend from Production Processing Pads to Gas Plants. Transmission, shown by the green arrow, moves the natural gas to the marketplace or long-term storage.

The lines from the production processing pad to the transmission lines are like neighborhood streets that lead to progressively larger thoroughfares and, eventually, the freeway system – the transmission lines in this analogy.

NGLs from the gas processing plant serve as either fuel or feedstocks for the chemical industry. In either case, the NGLs are separated into their individual components – ethane, propane, butane, and other components – at processing plants called *fractionators*. Sometimes fractionation happens at the same location as the gas processing plant; other times the mixed stream moves by pipeline to other locations, sometimes many miles away, for fractionation.

Transmission Lines

Transmission lines range from a few miles to thousands of miles long. They serve as the *autobahn* of the natural gas industry, hauling massive amounts of energy from gas plants, LNG facilities, other pipelines, and storage fields for most (and in some cases all) its journey to homes, schools, commercial establishments, manufacturing plants, and electric generation plants. Transmission lines may have diameters as small as six inches or as large as forty-eight inches or more. Some pipelines transport natural gas from just a few origination points to a limited number of destinations, but many have several hundred receipt and delivery points.

According to the CIA Fact Book, the world has more than 1.2 million kilometers of natural gas transmission lines. Figure 9-8 shows the distribution of these lines among countries.

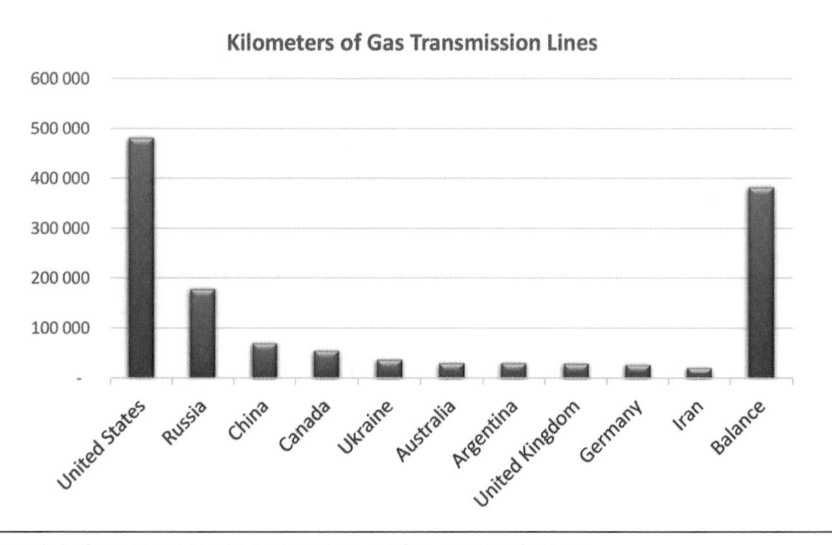

Figure 9-8. Countries with the Largest Natural Gas Pipeline Systems.

Liquefied Natural Gas (LNG) Facilities

LNG is methane, the same colorless, odorless hydrocarbon molecule in natural gas pipelines, cooled to approximately -260° F (or -162° C) transforming it from gas to liquid. In liquid form, the gas takes up approximately 1/600th the volume of its gaseous state at 60° F and atmospheric pressure. The smaller volume allows for moving methane in specially designed oceangoing tankers from producing countries to consuming countries. Figure 9-9 shows an LNG export facility.

Figure 9-9. LNG Export Facility. The facilities in the foreground liquefy the gas by refrigeration. The domed tank stores the LNG until loaded on the ship in the background.

At the delivery end, LNG import terminals temporarily store the LNG cargo until they re-gasify it by warming it up to normal pipeline temperature and deliver it into transmission or distribution lines.

Compressor Stations

Origination compressor stations sit at the beginning of transmission lines. Figure 9-10 shows the outside of an origination station, and Figure 9-11 a partial view of the engines and compressors inside the station.

Further along the line, *booster* stations "boost" the gas pressure (energy) previously lost due to friction.

Engines, like car motors, require cooling and lubrication. Cars use radiators and motor oil for those functions. Compressor stations also have cooling and lubricating systems. Compressor cooling systems sometimes also cool the just-compressed discharge gas. Control equipment monitors the station discharge temperatures of gas pushed into the transmission line to ensure it does not exceed safety limits. Station operators must deal with ancillary equipment (such as cooling and lubrication systems) to keep the station operating.

Figure 9-10. Natural Gas Origination Compressor Station.

Figure 9-11. Inside a Natural Gas Origination Station.

Figure 9-12. Inside a Natural Gas Booster Station.

Normally, compressor stations have local control rooms – small versions of the large, remote central control rooms. Figure 9-13 shows a local control room computer screen connected to the local SCADA system.

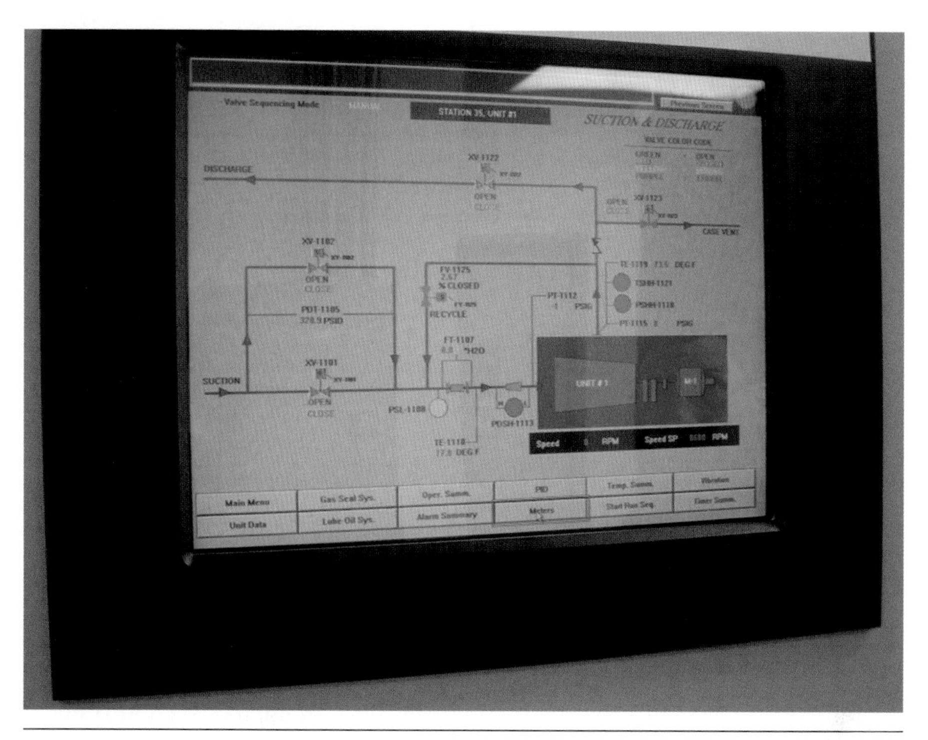

Figure 9-13. Local Control Screen.

Pipeline Hubs

Supply flexibility and reliability relies on the ability to receive from and deliver to other pipelines. Over time, regional pipeline hubs – where many pipelines interconnect – developed. Hubs provide physical interconnects allowing gas to move efficiently from one pipeline to another. They also serve as clearing points for gas trading. Examples of hubs include the National Balancing Point (NBP) in the UK, Zeebrugge in Belgium, Baumgarten in Austria, and the Henry Hub in the U.S.

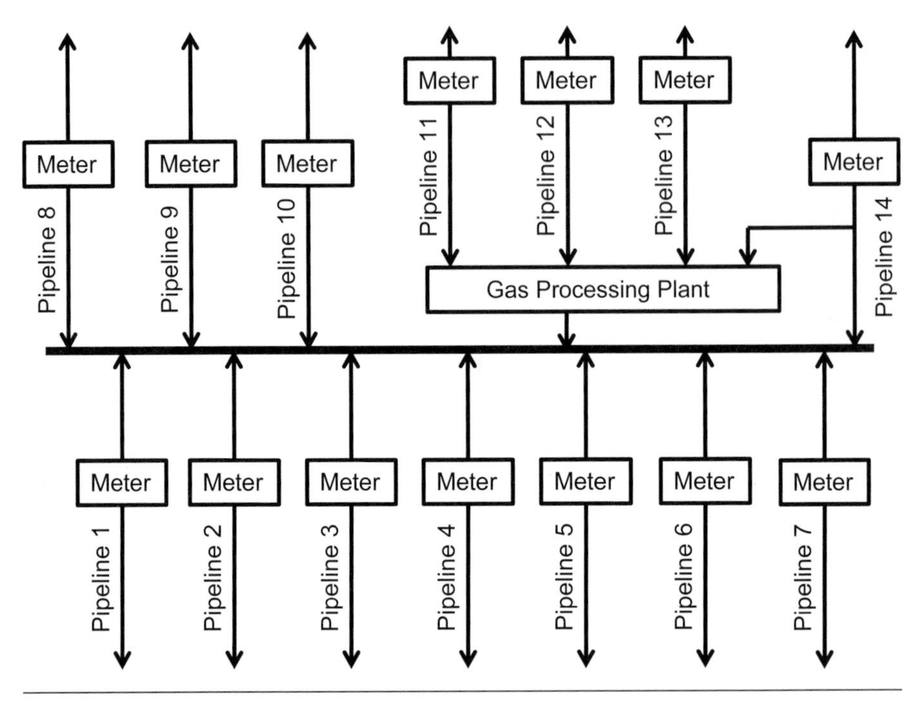

Figure 9-14. Schematic of Natural Gas Hub.

Pipeline Interconnects

When two pipelines cross, they sometimes have an interconnection containing valves to redirect flow and meters to measure the flow. Interconnects provide flexibility and increase access to multiple sources and markets.

Figure 9-15. Natural Gas Interconnect Between Two Gas Transmission Lines.

Pipeline City Gates

Transfer of gas to a local distribution company (LDC) or other customer takes place at facilities referred to as *city gates* or *town gates.*

Figure 9-16. Natural Gas City Gate.

City gate stations control pressure, reducing it from the higher transmission pipeline pressures to the lower pressures at which LDCs operate. They also measure and ensure the quality of the gas and often serve as the point for adding odorant – the material that gives natural gas its distinctive smell. (In its natural state, gas has no odor, and the industry adds odorant to make it easier for consumers to notice a leak.)

Natural Gas Storage

The reservoirs from which gas was produced are often used as gas storage, especially when near consumption points. Gas companies inject gas into the reservoir when supply exceeds demand and withdraw it as needed, particularly in the winter or during periods of peak electricity demand in the summer.

These underground storage facilities are often structurally fragile and have limitations on injection and withdrawal rate. Depleted reservoirs (D in Figure 9-17) account for more than 85% of total gas stored. Salt caverns (A in Figure 9-17) account for less than 5%.

When natural gas-powered generation plants start up, they need a lot of gas quickly. Because salt caverns are geologically more stable than reservoirs, they generally support faster injection and withdrawal rates, an important feature for dispatchability. Over a season, caverns fill and empty many more times than

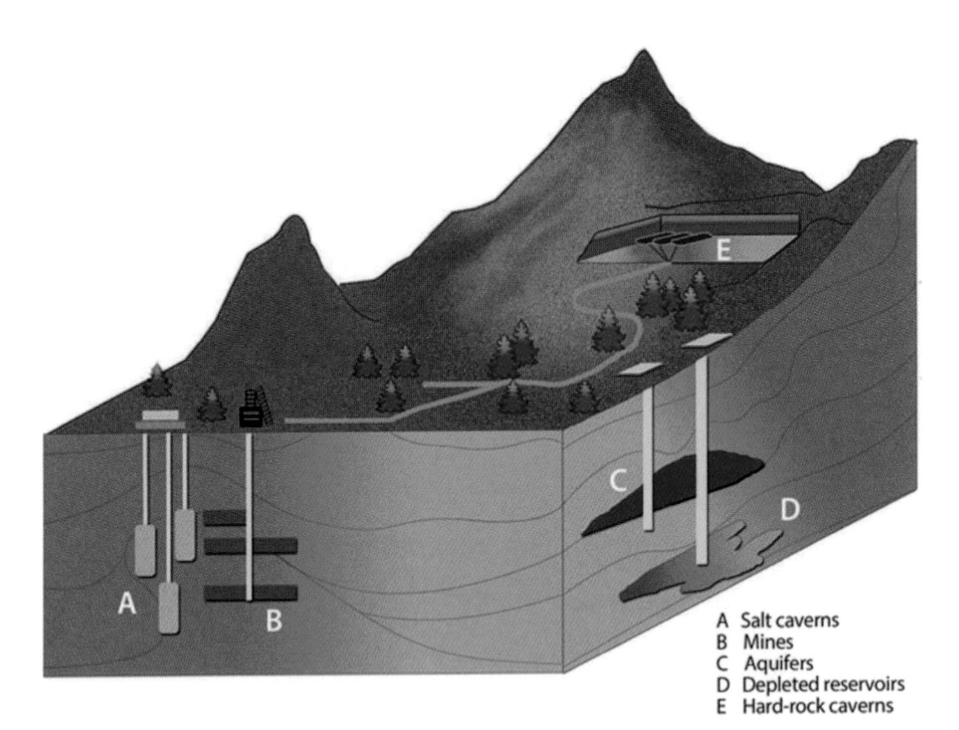

Figure 9-17. Types of Underground Gas Storage.

depleted reservoirs. They may even fill one day and empty the next. In other words, salt caverns typically serve as short-term storage and reservoirs long-term and inter-seasonal storage.

Natural Gas Pipeline Operations

In the early days of the gas pipeline industry, no central control rooms existed. Instead, operators in the field along the pipeline performed all operating activities.

As early as the late-1800s, the idea of centralized dispatch emerged. From a central office, a few dispatchers communicated with operators at each station, first by telegraph and then by telephone, telling them which valves to open and close, which compressors to start and stop and how to conduct other functions to move the natural gas. Some operators dispatched the lines centrally, but other operators and technicians along the lines still physically operated the equipment. Often, the field staff, who worked round-the-clock shifts, lived in company towns around the facilities.

Over time, communication systems, instruments, smart devices, and automation equipment developed and matured, enabling remote monitoring and control of process variables – pressures, flow rates, and others from control rooms hundreds of miles away. Central staff (now called controllers) could operate valves, compressors, and other equipment. Centralized dispatch evolved into centralized control, reducing the operating tasks physically performed by field staff.

Direct Operating Tasks

At a high level, operating tasks, whether performed in the field or centrally, focus on safely, reliably, and efficiently moving natural gas from its origin point to its destination, with due regard for the environment. Direct operating tasks subdivide into:

- Controlling
  - Pressures
  - Flow rates
- Directing
  - Receipts
  - Deliveries
- Measuring
- Assuring
  - Quality
  - Flow

Pressure Control and Rate Control: Line pressures must remain low enough to accept gas into the line in the amount and at the rate desired and high enough to deliver it out of the line in the amount and at the rate desired. Controlling pressures and flow rates calls for starting and stopping compressors and modulating control valves, variable speed drives, engines, and turbines.

Directing Receipts and Deliveries: Controllers open and close valves, allowing or stopping natural gas from flowing into or out of the pipeline at designated locations, quantities, and pressures. Those valves require frequent opening and closing and commonly include an actuator allowing remote operation from the central control room. Technicians physically at the site must manually open or close valves without actuators.

Measuring Quantities: Pipeline companies meter gas receipts into the line and deliveries out of the line for custody transfer, ownership, inventory control, and release management. Meter readings provide essential data to the accounting and release management systems. Field technicians periodically calibrate meters and other instruments to ensure accuracy.

Ensuring Quality: Samplers and other instruments closely monitor the gas stream's energy content and contaminants. Energy content can vary depending on the concentration and types of non-methane molecules in the pipeline. At lower-volume connection points along the line, technicians attach small pressure vessels or bottles to a sampler device to extract gas from the line. The technicians periodically collect the bottles and take them to a lab for quality analysis.

Figure 9-18. Remote Receipt and Delivery Facility. Some of the valves have actuators (red cylinders) and some, such as the one at the far right, do not. Technicians come to the station to open or close this valve by turning the hand wheel.

Figure 9-19. Metering and Sampling Location. The sample container is in the center foreground, and an orifice meter is on the right with two tubes connected.

At higher-volume custody transfer locations, gas chromatographs measure and report to the control room the quality of the gas on a real-time basis.

Assuring flow: Water and hydrocarbon dew points are two important quality measures. Water dew point measures the amount of water in the line.

Hydrocarbon dew point measures the concentration of heavier molecules in the line. Operators regularly monitor both dew points for quality, flow assurance, and integrity purposes.

Water and/or heavier hydrocarbons condensing into liquids and collecting in low points may temporarily block flow.

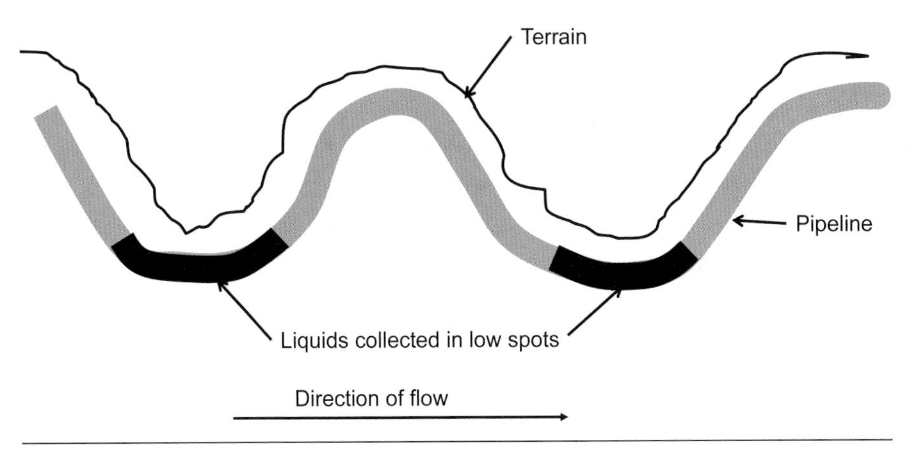

Figure 9-20. Liquids Collected in Low Spots Blocking Flow.

Pressure builds behind (upstream of) the blockage, and the force behind the blockage eventually increases until it forces the liquid forward. The liquid surges forward and flows to the next low spot, forming another block. The pipeline industry refers to this blocking and surging as "slug flow." Water and other liquids can severely damage compressor stations if not removed before reaching the compressor. In addition to causing slug flow, collections of water can cause internal pipeline corrosion.

"Pigging" a Line: Periodically, pipeline field technicians send devices through lines to remove liquids and solid debris. Technicians insert these devices, called *pigs*, into the line from *pig launchers*. The fluids in the line behind the pigs push them along until they reach a receiver. At that point, they are pushed into the pig receiver and removed from the line. Both launchers and receivers are sometimes called *pig traps*.

> **Why are they called pigs?**
> Early pipeliners wrapped leather around rags and other materials and sent them through the line to clean it. They thought the sound made by these contraptions as they inched through the line driven by positive displacement pumps sounded like pigs oinking.

The generic term *pig* has many subsets, including spheres, polly pigs, scrapers, brush pigs, and even gel pigs. The names provide a clue to their functions. Spheres generally just push out liquids; scraper pigs and brush pigs clean the inside of the pipe. An industry organization, the Pigging Products and Services Association, dedicates itself to the art and science of pigging as a service to the pipeline industry. Figure 9-21 shows loading of one type of cleaning pig into a pig launcher.

Figure 9-21. Loading a Pig into a Pig Launcher.

Field operators and technicians work closely with employees in the central control room to ensure they complete all these operating tasks safely, reliably, efficiently, and with due concern for the environment and the landowners where pipelines cross private property.

Indirect Operating Tasks

The field technician job involves a lot more than operating the pipeline. They also interface with stakeholders including:

- Landowners
- Regulators
  - safety
  - environmental
  - land use
- Local emergency response officials
- Operating personnel at receipt and delivery points
- Others

Maintenance and Integrity

Field technicians serve as "boots on the ground" when it comes to:

- inspecting;
- calibrating;

- maintaining; and
- repairing equipment.

They also work with pipeline integrity management professionals to understand pipe condition and make repairs – hopefully well in advance of any failures. Technical advances provide the ability to run intelligent pigs, called in-line inspection (ILI) tools inside the pipeline.

Figure 9-22. Magnetic Flux Leakage (MFL) ILI Tool. This tool uses the principles of magnetism to detect metal wall loss.

ILI tools come in a variety of technologies, with some better at detecting wall loss, others cracks, and still others deformations. Gas pushes a tool along as it looks for potential defects. All potential defects are mapped to Global Positioning System coordinates and analyzed. Some require immediate repair. Those not posing an immediate risk are monitored during subsequent tool runs.

Central Control Room Operations

Natural gas control rooms and electric industry control rooms have similar functions and processes. One controls molecules while the other controls electricity. Both use process data supplied (data acquisition) to understand and then send commands (supervisory control) so the pipeline or power line performs its intended function.

The gas pipeline grid must move enough natural gas molecules from the left to the right of Figure 9-23, so every gas user receives gas in the pressure range required to support combustion – not too much pressure nor too little pressure.

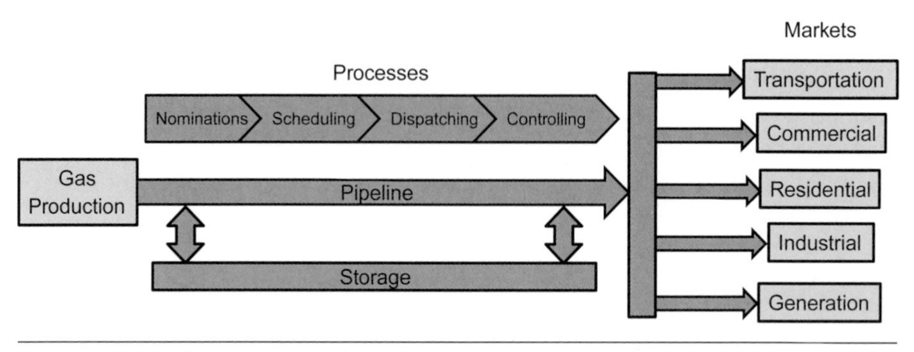

Figure 9-23. Gas Flow and Control Room Processes.

> **Line Pack**
>
> The quantity of gas in the pipeline. Line volume remains constant but the quantity of gas in the line varies depending on pressure.

Figure 9-23 looks simple and straightforward but there are thousands of market participants and many thousands of receipt and delivery points along the lines.

Receipts into the system and deliveries out of the system must balance within the *line pack* constraints. The challenge for pipeline controllers is ensuring gas remains constantly available in sufficient quantities and in the pressure range supporting combustion for customers on both the coldest and hottest days. That challenge presents a wide-ranging and quickly changing demand profile.

Control Room Processes

Figure 9-23 shows four control room processes: nominations, scheduling, dispatching, and controlling. Of course, a lot more activity takes place, but those four are keys to understanding gas-electric coordination.

Nominations

The first pipeline process, nominations, involves various gas shippers in the five market groups in Figure 9-23 "nominating" (informing) pipeline companies of the natural gas quantities they want to ship and the origination and destination points for each day.

> **NAESB**
>
> An industry forum for the development and promotion of standards for a marketplace for wholesale and retail natural gas and electricity.

Pipeline companies and their regulators dictate exactly how these nominations occur. In the U.S., the Federal Energy Regulatory Commission (FERC) has adopted various North American Energy Standards Board (NAESB) recommendations allowing shippers to change their nominations five times each day. More information on this topic appears on the FERC and NAESB websites. Other regulators or authorities and independent system operators in other countries or regions adopt their own procedures. The important takeaway is natural gas nominations and scheduling processes must be fluid and dynamic and be consistent with and support the unique needs of each customer group.

Scheduling

Next, schedulers compare nominations to available capacity and contractual obligations. When available pipeline capacity exceeds demand, the pipeline accepts the nomination. If nominations exceed capacity, pipelines allocate space to shippers based on *transportation service agreements* (TSAs). Shippers with TSAs for *firm transportation* get priority. Those with *interruptible transportation* get next priority, and the pipeline may or may not have available capacity to move all the gas the shipper needs.

Firm Contracts: In a *firm transportation* contract, the pipeline company promises to reserve a certain amount of capacity for the shipper, and the shipper in turn, promises to pay for that capacity whether used or not. Pipeline companies charge more per unit for firm transportation than interruptible transportation contracts.

Interruptible Contracts: The astute reader has probably already deduced pipeline companies can interrupt the gas flow in an interruptible contract; meaning the gas may not move when the shipper wants it moved. However, the shipper saves money when the pipeline is not full because it pays less per unit for interruptible transportation than for firm transportation.

Dispatching and Controlling

Prior to automation, different people performed these two functions – dispatchers in centralized locations communicated by telegraph or telephone with operators at remote stations, instructing them to stop and start compressors and open and close valves. The remote operators then controlled the gas flow by changing equipment settings accordingly.

Now, typically, the same position (commonly called a controller) performs both functions. Controllers receive the daily schedule (electronically) from schedulers and manage receipts into and withdrawals from the pipeline to best meet the schedule at the lowest cost. They continually monitor the line with SCADA systems as described in Chapter 7.

Figure 9-24. SCADA Displays at a Gas Controller's Console.

The Gas Pipeline – Electric Generators Dilemma

Electrical power grid participants attempt to build sufficient capacity to transport power to every customer safely, reliably, and efficiently, with due concern for the environment during even the highest demand periods. They rely on power generators for electrical power. When it comes to natural gas supply, electric generators face two dilemmas – one related to contracts and the other related to customer classes.

Balancing Firm and Interruptible Rates

Natural gas electric generators must decide how much gas to transport at firm transportation rates (more expensive) versus interruptible transportation (less expensive) rates. If they guess wrong and do not contract for enough firm capacity to meet their needs, they may not have sufficient gas supplies to generate enough electricity to meet customer needs. If they contract for firm transportation to meet peak demands, they will pay more and pay for capacity they may need only infrequently.

Figure 9-25 graphically represents the dilemma for three of the major U.S. natural gas users by season.

If electric power generators using natural gas sign *firm* transportation agreements for gas (the black line in the middle graph), they will pay for more capacity than they need most of the time. If they contract for firm volumes at the level of the red line and depend on summer downturn in the residential and commercial market to provide enough space to handle summer peak, they may not have enough capacity most of the time.

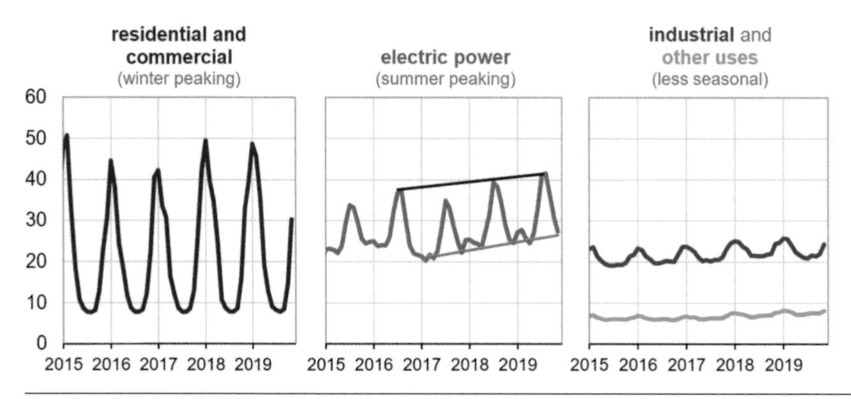

Figure 9-25. U S. Natural Gas Consumption by Sector.

Customer Priority

Natural gas customers, including electric generation plants, want their needs met twenty-four hours per day, 365 days per year. Electric generation plant customers, many of whom are also natural gas customers, also want their needs met at all times.

Given world energy forecasts, ensuring customers have all the energy they want, when they want it, is not a simple task. In periods of insufficient gas supply to address all customer demands, meeting electric generation needs might mean natural gas pipelines stop delivering to other customer classes to serve electric generation needs.

The essence of gas-electric coordination involves natural gas controllers delivering gas for electric generation at exactly the right time so generators can deliver electricity to the grid at exactly the right time to serve the electric customers.

Looking to the Future

Gas-electricity coordination is the natural gas system delivering molecules to electric energy generators so they can supply electric energy to the grid exactly when needed.

Ensuring all customers have all the energy they want, when they want it, now and in the future, is a key task for the electrical power industry, the natural gas industry, and regulators.

Summary

- Natural gas (the "least polluting" fossil fuel) will continue as one of the primary energy sources for decades to come.
- Although gas and electricity both provide heat and light, they are very different types of energy.
- Gas is a *primary* energy source – a fuel. It is easily and cheaply stored for later use.
- Electricity is a *secondary* energy source produced from fuels and does not occur in nature. Electricity is not easily and cheaply stored for later use.
- In addition to powering electric generators, natural gas also heats homes, businesses, schools, hospitals, and industrial plants; powers vehicles; provides fuel for cooking, lighting, and other activities; and serves as a feedstock to manufacture chemicals.
- One key challenge for coordinating gas and electric is prioritizing natural gas usage among customers in times of high demand or short supply.
- The reservoirs from which gas was produced are often used to store gas, especially when near consumption points, by injecting gas into the reservoir when supply exceeds demand and withdrawing it as needed, particularly during periods of peak electricity demand.
- Natural gas control rooms and electric industry control rooms have similar functions and processes.
- The essence of gas-electric coordination involves natural gas controllers delivering gas for electric generation at exactly the right time to allow generators to deliver electricity to the grid at exactly the right time to serve electric customers.

10

Bulk Electrical System Reliability

We should remember that good fortune often happens when opportunity meets with preparation.

—Thomas Edison (1887 – 1931)

"Grandpa Tom, our lights went out last night! My friend next door said we had a blackout," exclaimed Luke.

"Well, it wasn't really a blackout," said the grandfather. "That only happens when all the electricity to a large area goes out," he added.

"Why did our electricity go out last night?" asked Luke.

"Your dad told me the transformer serving your street blew out, and you were without electricity until the local distribution company replaced it," answered Grandpa Tom.

"So, would a blackout be a really big transformer blowing out?" asked Luke.

"Blackouts require more than just one transformer blowing out," replied Grandpa Tom. "My friend Andrew told me there is a group called NERC, which I think he said stands for the North American Electrical Reliability Corporation, that works to ensure we don't have widespread blackouts," the grandfather added.

"Let's look them up!" shouted Luke, jumping up and heading to the computer. "Here it is. NERC develops and enforces Reliability Standards; annually assesses seasonal and long-term reliability; monitors the bulk electrical system through system awareness; and educates, trains, and certifies industry personnel," read Luke. "Sounds like they do a lot," he added.

"Well, I do like to have electricity," chuckled Grandpa Tom.

Introduction

Generation, transmission, and distribution of electricity are regulated in various ways around the world by federal, state, and local regulations. Some of these regulations are the same as those pertaining to other industries, but two regulatory areas, reliability and economics, have their own sets of regulations. This chapter discusses reliability and associated regulations, and the next discusses electricity markets and their regulations. Before moving to reliability, this chapter briefly discusses regulations in general.

Regulations

Standards usually start as industry-recommended practices. As they gain greater acceptance, industry associations elevate recommended practices to *standards*. Recommended practices and standards are voluntary unless and until governmental authorities incorporate them into laws (codes) or regulations, making them mandatory.

Regulators interpret laws passed by governmental authorities and turn their interpretations into regulations. For many industries, regulators also adopt voluntary industry standards into regulations. The U.S. standards for the bulk electric system are not called *regulations* – they are called *Reliability Standards.* Generation and transmission operators and other affected parties use Reliability Standards to develop policies and procedures for compliance and maintain records demonstrating compliance with the Standards. As in many industries, electric industry regulators monitor compliance by auditing those records and taking enforcement actions when an audit reveals a violation.

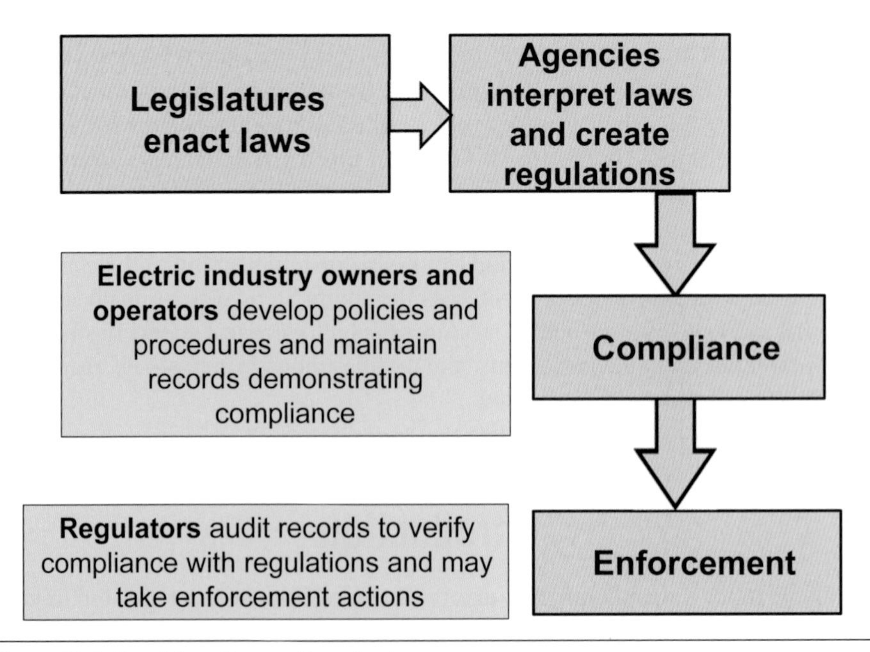

Figure 10-1. Regulatory Process Flow Chart.

In the case of the U.S. electric power industry, however, the regulator (the Federal Energy Regulatory Commission or FERC) delegates audits and enforcement to an industry organization (North American Electric Reliability Corporation or NERC). The authors consider the terms *regulations* and *Reliability Standards* as synonymous for the purpose of this text.

Regulatory Categories

Broadly speaking, regulations cover the following categories:

- Public safety
- Employee safety
- Environmental protection
- Land use
- Reliability and security
- Markets

The first four categories apply to all industries, not just the electrical power industry, and are not discussed in this text. The final two, reliability and security and markets are unique and are discussed in this chapter and the next.

North American Electrical Reliability Standards

Currently the primary U.S. reliability regulator is the FERC. The primary industry association involved with electrical reliability in North America is NERC, originally founded as the "National Electric Reliability Council" in 1968 by representatives of the electric power industry in response to governmental pressure after a large blackout in the northeast U.S. in November 1965. At that time, NERC developed and promoted *voluntary* compliance with best practices – or standards – to reliably operate the bulk transmission systems of North America.[68]

NERC Definition of Reliability

NERC defines a reliable Bulk Power System (BPS) as one able to meet the electricity needs of end-use customers even when unexpected equipment failures or other factors reduce the amount of available electricity. NERC divides reliability into two categories: adequacy and security.

Adequacy

"Adequacy means having sufficient resources to provide customers with a continuous supply of electricity at the proper voltage and frequency, virtually all of the time. Resources refer to a combination of electricity generating and transmission facilities that produce and deliver electricity, and demand-response programs that reduce customer demand for electricity. Maintaining adequacy requires system operators and planners to take into account scheduled and reasonably expected unscheduled outages of equipment, while maintaining a constant balance between supply and demand".[69]

Security

For many years, the electrical power industry defined system security as the ability of the BPS to withstand sudden, unexpected disturbances like short circuits or unanticipated loss of equipment due to natural causes. In today's world, security includes withstanding disturbances caused by manmade physical or cyberattacks. The BPS must be planned, designed, built, and operated taking into account these modern threats as well as more traditional security risks.[70]

Definition of Bulk Power System and Bulk Electrical System

NERC defines the BPS as, "the electricity power generation facilities combined with the high-voltage transmission system which together create and transport electricity around North America." The BPS does not include the distribution lines delivering electricity directly to homes and businesses. A local utility generally provides that service under the jurisdiction of state, provincial, or local utility regulatory agencies.[71]

BPS or BES

As currently used, BES comprises the entire BPS and includes some lower-voltage resources not included in BPS. This text uses BES except when specifically intending to refer to the BPS.

FERC Jurisdiction

From a practical standpoint, NERC oversees large generators (>20 MWs and connected to the grid at >100 kV) and most transmission systems (>100 kV) engaged in interstate (between states and not solely within a state) transmission of electricity. State and local regulators oversee small generators (< 20 MWs), distribution (< 100 kV), and microgrids.

Evolution of Reliability Regulations in the United States

Chapter 5 discussed the global evolution from local generation and distribution to connected grids – and then to microgrids. This section uses the United

States as a case study of the circumstances leading up to current Reliability Standards.

In the early days of the industry, no regulations existed. Subsequently, cities and states began establishing rules and regulations for the industry and, ultimately, the federal government got involved in 2005 with passage of the Energy Policy Act of 2005.

Incidents and Response

Most reliability-related regulation resulted from significant *outages*. For example, in November, 1965, the northeast experienced a significant blackout when more than 30,000,000 people lost electricity for up to 13 hours.

Figure 10-2. Life Magazine Cover – November 1965.

The blackout started when a protective relay on a transmission line in Ontario near Niagara Falls acted as (incorrectly) programmed tripping (opening) because other equipment failed to operate properly.

Power demand at the time was high due to unseasonably cold weather, overstressing transmission lines into Ontario. When the incorrectly set relay opened to isolate that line, power was redirected to other lines, causing them to overload. Protective equipment on those lines acted to isolate them from overloading.

The power intended for the loads on the isolated lines then flowed over lines connected to New York, causing them to overload. As a result, several power plants ceased generating as they disconnected from the grid. The overloads, coupled with lost generation plants, caused widespread power imbalances on the grid. More and more lines and plants isolated themselves (referred to in the industry as *cascading* outages and *islanding*). Starting with the first failure near the New York-Canada border, the blackout moved east until, around 5:30 p.m., the lights in most of New York City went out (as shown in Figure 10-3). By 7:00 a.m. the next day, most power was restored.

Response to the 1965 Blackout

In 1967, largely as a result of the 1965 incident, the U.S. Congress considered passing the Electric Power Reliability Act to create a body to oversee electric reliability coordination. To avoid government regulation, however, the electric industry formed the first electric reliability *council.*

Shortly thereafter, the Federal Power Commission (FPC), predecessor to the FERC, recommended a group consisting of members from regional coordinating organizations in the U.S. to share information and review, discuss, and help resolve inter-regional coordination issues.

In 1968, the electric industry formed the *National Electric Reliability Council* (NERC) with twelve regional entities. In 1969, NERC created its first (unenforceable) regional coordination *guidelines.* Each regional entity developed a wide-ranging program to coordinate planning and operation to optimize reliability of the interconnected system, with committees coordinating activities of member systems.

1977 Incident

The next significant development occurred in July 1977 when New York City experienced another major blackout. According to a report prepared for the U.S. Dept. of Energy (DOE), the event caused a complete shutdown of the Consolidated Edison (ConEd) system in New York and resulted from a combination of lightning, equipment malfunction, questionable system design, and operating errors.

This incident started when lightning struck two high-voltage lines at the northern end of ConEd's system. About twenty minutes later, lightning struck two more lines, and protection equipment operated improperly, causing circuit breakers on three of the four remaining lines to trip (open), which removed a large generator and other important transmission lines from the grid.

Transmission ties to other utilities covered the power loss but eventually overloaded. The situation continued deteriorating until all ties to external sources opened through either protective equipment operating or system operator (human) actions to protect neighboring systems. ConEd's customer demand exceeded its available electricity supply and, at approximately 9:30 p.m., the ConEd system completely shut down.

Power restoration took a long time. The first attempt at restoring service caused significant equipment damage. ConEd then developed a comprehensive restoration plan but experienced setbacks due to lack of emergency power and equipment malfunctions and did not begin restoring power until at about 2:00 a.m. the next day, finishing shortly after 11:00 p.m. that day.

Response to 1977 Incident

In 1980, the North American Power Systems Interconnection Committee (NAPSIC) became part of NERC as its *Operating Committee* and brought industry operations and planning together in one organization. NERC adopted NAPSIC operations criteria and guides. Shortly thereafter, NERC changed its name to *North American Electric Reliability Council,* recognizing Canadian participation.

1992 Action Plan

In 1992, the NERC Board opined that complying with NERC and regional reliability policies, criteria, and guides should be mandatory. Shortly thereafter, NERC published *NERC 2000* – an action plan recommending mandatory compliance with NERC policies, criteria, and guides and a way to address violations. The plan included strategies for interconnected systems operation, planning reliable bulk power systems, membership, and resolving disputes.

1996 Incident

The next major developments occurred in 1996 with *two* blackouts in the western U.S. First, in July, fourteen states, parts of Alberta and British Columbia in Canada, and Baja, Mexico, lost 11,850 MW of generation, causing a widespread blackout for more than 2,000,000 people. Then, in August, the same area lost 28,000 MWs of generation, causing a blackout for more than 7,500,000 people. Those events caused the Western Systems Coordinating Council (WSCC) to develop its *reliability management system*, through which members voluntarily agreed to pay fines for violating reliability standards. WSCC was one of NERC's regional reliability organizations and later became the Western Electricity Coordinating Council (WECC).

Early Reliability Standards

In 1997, the DOE Electric System Reliability Task Force created an *Electric Reliability Panel* that, with NERC, argued the industry needed mandatory and enforceable reliability rules. It recommended creating an independent, audited, self-regulating *Electric Reliability Organization* to create and enforce reliability standards for North America. The panel believed the plan needed federal legislation for adoption and, subsequently, NERC began transforming its policies, criteria, and guides into *reliability standards*.

In 2015, NERC changed the ES-ISAC's name to the Electricity Information Sharing and Analysis Center (E-ISAC).

In 1999, a group of industry, state, and consumer organizations proposed legislation to create an Electric Reliability Organization to draft and enforce mandatory Reliability Standards and have FERC oversee the organization. Contemporaneously, NERC established the Electricity Sector Information Sharing and Analysis Center (ES-ISAC) and, in 2000, agreed to serve as the industry's primary point of contact for the government on national security and critical infrastructure protection issues. In 2001, the province of Ontario, Canada, made NERC's operating policies and planning standards mandatory and enforceable.

2003 Incident

In August, 2003, North America experienced its worst blackout when more than 50,000,000 people lost power in the Northeast and Midwest U.S. and Ontario, Canada. The blackout affected 9,300 square miles of land, 508 generating units (at 256 plants), and caused more than $6 billion in economic losses.

Ultimately, experts determined the blackout resulted from three primary causes:

- Inadequate situational awareness by FirstEnergy Corporation (FEC).
- FEC's failure to properly manage tree growth in its rights-of-way.
- Failure of the regional reliability organizations to provide effective diagnostic support.

In a November 2003 report, a joint U.S.-Canada task force cited several violations of NERC guidelines as contributing to the event and made forty-six recommendations to prevent future blackouts.

"All the News That's Fit to Print"

The New York Times

Late Edition

ONE DOLLAR

POWER SURGE BLACKS OUT NORTHEAST, HITTING CITIES IN CANADA AND 8 STATES; MIDDAY SHUTDOWNS DISRUPT MILLIONS

OVERLOADED GRID

Buildings Are Evacuated and Hospitals Fill — Bush Plans Review

Figure 10-3. New York Times Front Page – August 15, 2003.

FERC and NERC

In the aftermath of the 2003 blackout, the U.S. Secretary of Energy stated he would not seek to penalize FEC because the U.S. had no enforceable electric reliability standards. The U.S. Congress responded by passing the Energy Policy Act of 2005 which, among other things, gave FERC new and broad jurisdiction over the BPS. The operative section of the Act gave FERC jurisdiction over:

> all users, owners and operators of the *bulk-power system*...for purposes of *approving reliability standards* established under this section and *enforcing compliance with this section.* All users, owners and operators of the [BPS] *shall comply* with reliability standards....

FERC Certifies NERC as Electrical Reliability Organization

The Act also gave FERC authority to certify an *Electric Reliability Organization* (ERO) to oversee development and enforcement of reliability standards. Shortly thereafter, NERC changed its name to *North American Electric*

> **NERC Regional Entities**
>
> Currently NERC has six regional reliability entities – Northeast Power Coordinating Council (NPCC), SERC Reliability Corporation (SERC), Midwest Reliability Organization (MRO), ReliabilityFirst (RF), Texas Reliability Entity (TRE) and Western Electricity Coordinating Council (WECC).

Reliability Corporation and, in July 2006, FERC certified NERC as the statutory ERO. At that time, NERC delegated its monitoring and enforcement authority to Regional Entities while retaining oversight of the process.

Meanwhile, in 2004, NERC converted its operating policies, planning standards, and compliance requirements into a set of ninety standards called *Version 0 Reliability Standards*.

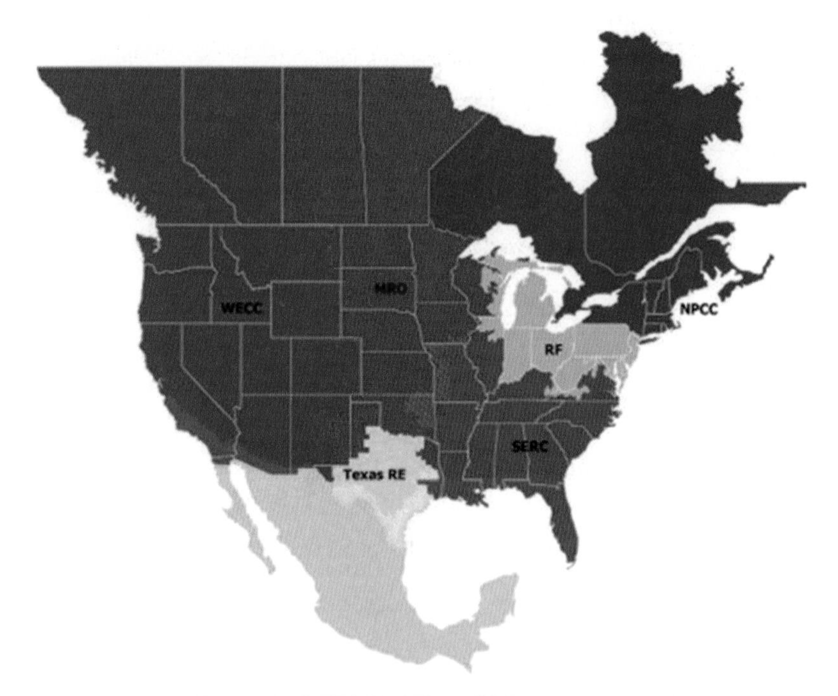

MRO	Midwest Reliability Organization
NPCC	Northeast Power Coordinating Council
RF	ReliabilityFirst
SERC	SERC Reliability Corporation
Texas RE	Texas Reliability Entity
WECC	Western Electricity Coordinating Council

Figure 10-4. NERC Regional Reliability Entities.

Canada Recognizes NERC as International ERO

Contemporaneously, the Canadian federal and provincial governments acknowledged NERC's role as the ERO for *North America* (because the U.S. and Canadian grids interconnect). Canada's National Energy Board (NEB) recognized NERC as the international ERO through a memo of understanding in 2006. Official recognition followed from all other interconnected provinces.

FERC Approves NERC Reliability Standards.

In April 2006, NERC submitted 102 Reliability Standards to FERC, Canadian provincial authorities, and the Canadian NEB. In March 2007, FERC approved eighty-three of those Reliability Standards which, on June 18, 2007, became the first legally enforceable standards for the U.S. BES.

At the same time, FERC gave NERC the authority to levy fines of up to $1 million per day, per violation, against companies violating those standards. Between March and June 2017, NERC gave electric utilities a *grace period* during which they could self-report deviations from the new Reliability Standards and create *mitigation plans* to address the deviations without incurring fines. In January 2008, FERC approved the first Critical Infrastructure Protection (CIP) Reliability Standards.

Mexico Recognizes NERC

In September 2010, WECC, the Comisión Federal de Electricidad (CFE), the Centro Nacional de Control de Energía (CENACE), and Area de Control Baja California (ACBC) signed a Membership and Operating Agreement for compliance with Reliability Standards in the Mexican state of Baja California. Thus, NERC now oversaw reliability standards covering the U.S., Canada, and parts of Mexico.

NERC Primary Duties and Powers

Pursuant to the Energy Policy Act of 2005, NERC and its regional entities have several primary duties:

- Entity registration and certification
- New or revised Reliability Standards
- Event analysis
- Compliance monitoring and enforcement

Entity Registration and Certification

All entities with responsibilities pursuant to Reliability Standards must register with NERC. These entities – users, owners, and operators of the BES – register by their functional responsibility:

- Reliability Coordinator (RC)
- Transmission Operator (TOP)
- Balancing Authority (BA)
- Planning Authority/Planning Coordinator (PA/PC)
- Transmission Planner (TP)
- Transmission Service Provider (TSP)
- Transmission Owner (TO)
- Resource Planner (RP)
- Distribution Provider (DP)
- Generator Owner (GO)
- Generator Operator (GOP)
- Reserve Sharing Group (RSG)
- Frequency Response Sharing Group (FRSG)
- Regulation Reserve Sharing Group (RRSG)

NERC oversees registration activities performed by the six Regional Entities, who collect data and track the location of facilities having a material impact on the BES within the Regional Entities' geographic boundaries. The Regional Entities also approve registration applications and review entity certification applications. If an entity does not agree with a registration determination, it may request a NERC-led registration review panel evaluation.

In addition to registration, NERC requires certain entities to be *certified* for specific roles. Prospective and existing registered entities intending to perform or performing the RC, TOP, and/or BA functions must be certified to operate one or more RC, TOP, and/or BA *areas.*

Each area must have a certified RC, TOP, and BA to perform tasks required by the Reliability Standards. A NERC Certification/Review Team recommends certification based on a review and assessment of an entity's ability to perform the tasks of the certifiable function.

New or Revised Reliability Standards Development

FERC must approve all new or revised national and regional Reliability Standards. Drafting new standards or revising existing standards follows NERC's Standards drafting process, which follows the American National Standards Institute (ANSI) model and begins with a Standards Authorization Request (SAR). After receiving a SAR, NERC forms a *SAR drafting team* consisting of Subject Matter Experts (SMEs) to review the SAR language and a *Standard Drafting Team* (SDT) to draft the language for the new or revised standard. Those two teams typically consist of the same people.

Drafting

Based on the expertise of the SMEs on the SDT and informal feedback the team collected, the SDT drafts proposed standard language, conducts field tests (if necessary), performs quality reviews, and submits draft standards to the NERC Standards Committee (SC). If the SC approves the draft standard, NERC posts it on its website for industry comment and forms a *ballot pool* consisting of industry participants and stakeholders to vote on the proposed Standard.

Simultaneously, NERC conducts a non-binding poll on proposed *Violation Risk Factors* (VRFs) and *Violation Severity Levels* (VSLs) for the Reliability Standard.

Violation Risk Factors

Each Reliability Standard has assigned VRFs through the NERC Reliability Standards development process. VRFs are assigned to requirements to provide clear, concise, and comparative association between the violation of a requirement and the expected or potential impact of the violation to BES reliability. NERC assigns one of three defined levels of VRF to each Reliability Standards requirement: Lower; Medium; or High.[72]

Violation Severity Levels

VSLs are defined levels of the *degree* to which an entity violated a Reliability Standard. Whereas NERC establishes VRFs *pre*-violation, and they indicate the relative potential impacts violations of each Reliability Standard could pose to BES reliability, NERC assesses VSLs *post*-violation, and they indicate the severity of the actual violation of the Reliability Standard in question. VSLs are designated as: Lower, Moderate, High, and Severe, and each Reliability Standard specifies its VSLs.[73]

If a ballot for a new or revised standard fails (ballots must receive at least two-thirds positive votes to pass) or if the draft Standard requires "significant" changes based on comments received from the industry, the SDT reviews and revises the draft language and resubmits it for an additional ballot. Once a ballot receives the required positive votes, NERC submits the Standard to its Board of Trustees (BOT) for approval and, finally, to FERC for approval.

FERC may approve, reject, or remand Reliability Standards. Often, when FERC remands a Standard to NERC, it includes *directives* describing items NERC must address before resubmitting the Standard to FERC for approval. Figure 10-5 shows a flow chart of the NERC ballot process.

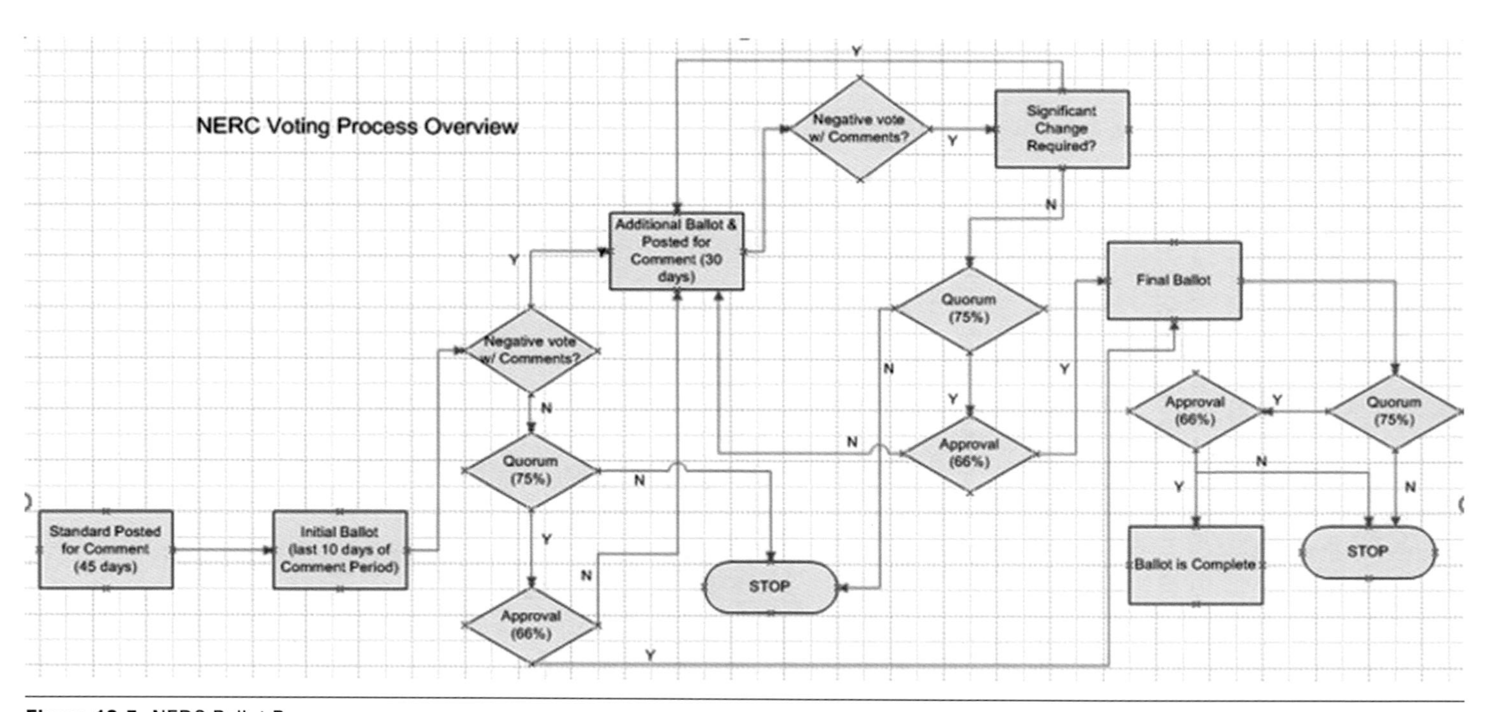

Figure 10-5. NERC Ballot Process.

Event Analysis

In the quest to continually improve reliability, NERC analyzes events impacting reliability and that potentially impact reliability, using the NERC event analysis process. The process provides a structured, consistent approach to analyzing system events such as unexpected outages, intended and controlled system separation, failure or misoperation of a protection system, system-wide voltage reduction, etc.[74] NERC promotes review and analysis of operations, planning, and CIP processes as a learning opportunity by providing insight and guidance. Based on the analysis, NERC identifies and disseminates information to BES owners, operators, and users for use in continuous improvement and making operations more reliable.

Compliance Monitoring and Enforcement

As show in Figure 10-1, entities develop policies and procedures based on standards and regulations, and regulators monitor those procedures and entity performance.

Monitoring

Through monitoring compliance with Standards, NERC seeks to ensure BES reliability and security. To that end, NERC developed and refined a *Compliance Monitoring and Enforcement Program* (CMEP). Compliance Enforcement Authorities (CEAs), typically the six regional entities, administer the CMEP on behalf of NERC. Each year, CEAs develop CMEP schedules for their area for that year based on entity *risk* to BES reliability and security.

Risk: Risk depends on the probability of an event happening and the severity of the consequences of that event. In electric reliability terms, if the *probability* of a transformer failure is the same regardless of the area served, the consequences for a transformer serving a large area are greater than the consequences for a transformer serving a smaller area. Thus, the *risk* of failure is greater for the transformer serving the larger area than for the one serving the smaller area. Larger risks warrant more mitigation measures than smaller risks. Reducing risk involves understand and mitigating it.

Inherent Risk: Each entity has risks associated with it by virtue of its size, complexity and a variety of other factors. Inherent risk is the risk associated with a system before additional risk mitigation or control measures are instituted. Complex systems and poorly performing systems generally have higher inherent risk than simpler systems with excellent performance history.

Inherent Risk Analysis

Using the risk-based approach, the CEA performs an *Inherent Risk Assessment* (IRA) for each registered entity in its region to determine the level of compliance monitoring for that entity. The IRA focuses on *entity-specific risk* (described in detail below). Figure 10-6 shows a flow chart of the IRA process.

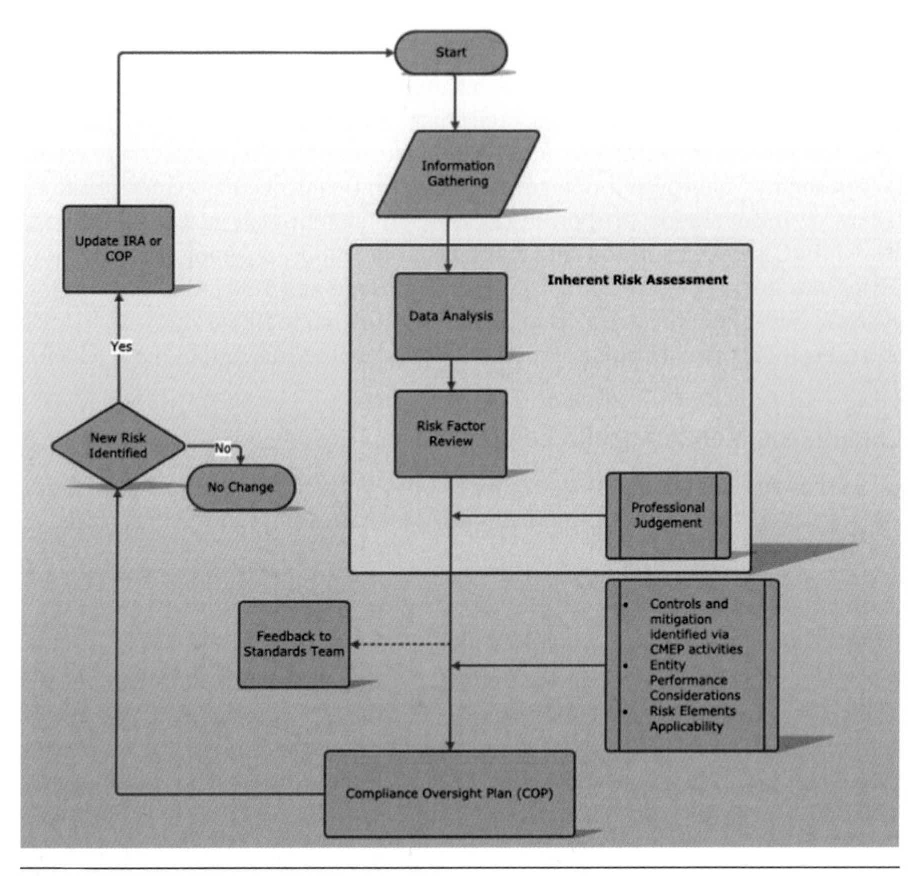

Figure 10-6. IRA Process.

To begin an IRA, the CEA considers specific NERC risk elements, the functions performed by the registered entity, and the Reliability Standards applicable to that entity.

2020 RISK ELEMENTS		
ELEMENT	STDS	REQS
Management of Access and Access Controls	11	13
Insufficient Long-Term and Operations Planning Due to Inadequate Models	4	8
Loss of Major Transmission Equipment with Extended Lead Times	2	2
Inadequate Real-time Analysis During Tool and Data Outages	2	2
Improper Determination of Misoperations	1	2
Gaps in Program Execution	6	10
Texas RE: Resource Adequacy	3	4
2020 Total	**29**	**41**

Figure 10-7. Sample Risk Elements. NERC CMEP Implementation Plan (2020).

The CEA then:

- identifies information about the entity (from public sources and its own records), such as miles of transmission lines, number of substations, generation assets (if any), geographic location, etc.;
- reviews the NERC CMEP implementation plan's applicability to the entity (i.e., known risks to reliability based on the entity's registration); and
- leverages its existing understanding of the entity (based on prior audit results, compliance history, areas of concern, recommendations).

The IRA generates a list of relevant risk factors and criteria, specific standards and requirements to consider and a justification for risk factor criteria. Based on the results of the analysis, the CEA develops a preliminary entity-specific *Compliance Oversight Plan* (COP). One of the approaches to compliance monitoring the CEA may use is a compliance audit.

Internal Controls Assessment

As part of a compliance audit, the CEA assesses an entity's *internal controls* to see if they mitigate the entity-specific reliability risks and provide *reasonable assurance of compliance* with the Standards. Having documented controls and testing/monitoring the controls' effectiveness provide the CEA with an understanding of key reliability functions, an entity's resilience, and progress towards a high-reliability organization.

The internal controls assessment uses inputs from the IRA, including:

- The entity's inherent risks to reliability of the BES
- A list of prioritized requirements addressing those inherent risks
- Assumptions in the IRA
- Preliminary entity-specific COP
- Information gathered in response to IRA results

The internal controls assessment focuses on controls to manage reliability risks, provides reasonable assurance of compliance with applicable Standards, gives the CEA an understanding of entity risks and how to manage/mitigate them, and ensures controls are scaled to the entity's size and risk.

When performing an internal controls assessment, the CEA focuses on certain key issues/questions, such as:

- Does the entity have controls to mitigate the entity-specific reliability risks?
- What are the most important controls to mitigate those risks?
- Does each control produce its intended result?
- Does the entity have sufficient, credible evidence that each control produces its intended result?
- Does the entity monitor its controls?
- Can the CEA use the results of control monitoring to reduce compliance monitoring activities?

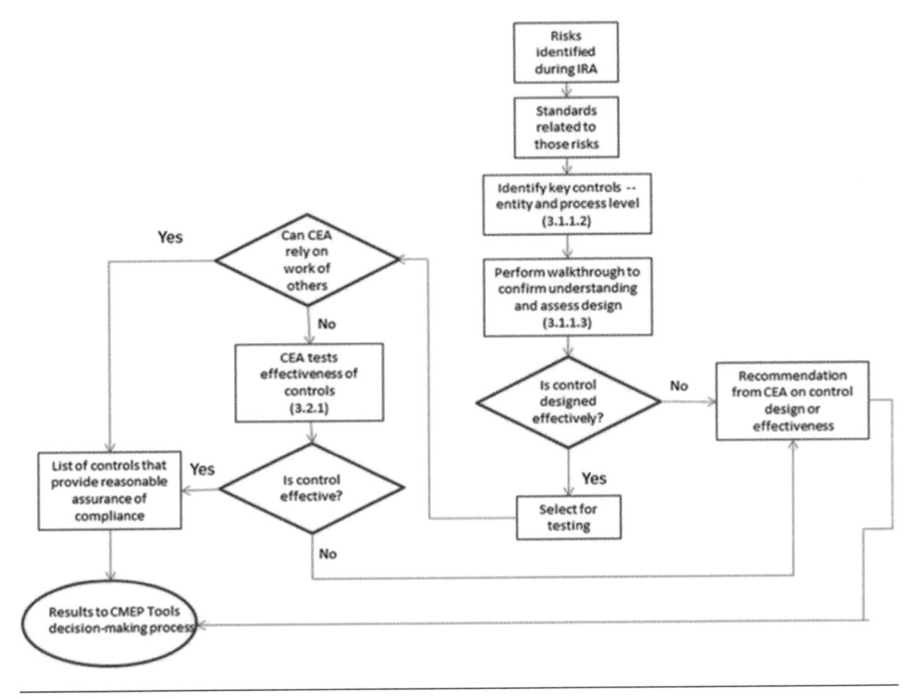

Figure 10-8. ICA Flow Chart.

Internal Controls

Companies employ three types of internal controls:

- Preventative
- Detective
- Corrective

Preventive controls are proactive and intended to encourage compliance and avoid unintended events/consequences. Detective controls identify errors and irregularities to support effective compliance. Corrective controls are used when an activity does not accomplish the intended result – they help return the activity to its desired state. Examples of internal controls include:

Preventative

- Documented processes
- Segregation of duties
- Preapproving actions/transactions
- Physical controls (i.e. card keys, locks)
- Access controls (e.g., computer passwords, two-factor authentication)
- Employee screening/training

Detective

- Spot checks
- Audits
- Reviewing and approving work products
- Inventory counts

Corrective

- Data backups
- Data validity tests
- Management reports for corrective actions
- Training
- Revising policies/procedures to prevent future errors/irregularities

Based on the results of the IRA and ICA, the CEA updates the preliminary COP – which *may* yield fewer requirements for audit, extend an entity's audit schedule, allow for less *deep* audits, allow self-certifications instead of audits, and/or allow the entity to enter the NERC self-logging process. The self-logging process allows Registered Entities to track their lower-impact *Potential Non-Compliance (PNC)* items and report them to the CEA on a periodic basis (typically quarterly).

Compliance Monitoring Tools

The NERC CMEP provides for audits, self-certifications, spot checks, investigations, self-reports, periodic data submittals, and complaint response. The full description of the NERC compliance monitoring and enforcement program appears in Appendix 4C to the NERC Rules of Procedure.

Audit

Compliance audits are in-depth reviews and analyses of a registered entity's compliance with the Reliability Standards based on *Generally Accepted Government Auditing Standards* (found in the government's *yellow book*). Annually, a CEA posts its audit schedule on its website (Figure 10-9) and, at least ninety days before the audit, notifies the registered entity of the Reliability Standards included in the audit and identifies the audit team. In response, the registered entity provides evidence of compliance to the CEA, which reviews the evidence (onsite, offsite, or both) to assess compliance. Upon the close of an audit, the CEA issues an audit report.

The registered entity must demonstrate compliance for the *audit period* (typically from the day after the end of the previous audit to the date of the current audit). If the CEA uncovers a PNC, the matter goes into the *enforcement* phase (described below).

WECC 2021 Audit Schedule—US					
NERC ID	Acronym	Entity	Audit Type		Date
NCR05015	AEPC	Arizona Electric Power Cooperative, Inc.	CIP	O&P	First Quarter
NCR05030	BHP	Black Hills Corporation	CIP	O&P	First Quarter
NCR05377	SDGE	San Diego Gas & Electric		O&P	First Quarter
NCR05032	BPA	Bonneville Power Administration	CIP	O&P	Second Quarter
NCR05106	CSU	Colorado Springs Utilities	CIP	O&P	Second Quarter
NCR05282	NWC	NorthWestern Corporation	CIP		Second Quarter
NCR10030	TSGT	Tri-State Generation and Transmission Association, Inc. - Reliability	CIP	O&P	Second Quarter
NCR05465	WASN	Western Area Power Administration - Sierra Nevada Region	CIP	O&P	Second Quarter
NCR05191	IPCO	Idaho Power Company	CIP	O&P	Third Quarter
NCR05377	SDGE	San Diego Gas & Electric	CIP		Third Quarter
NCR05398	SCEC	Southern California Edison Company	CIP	O&P	Third Quarter
NCR05464	WACM	Western Area Power Administration - Rocky Mountain Region	CIP	O&P	Third Quarter
NCR11226	IREA	Intermountain Rural Electric Association	CIP	O&P	Fourth Quarter
NCR05282	NWC	NorthWestern Corporation		O&P	Fourth Quarter
NCR05299	PGAE	Pacific Gas and Electric Company	CIP	O&P	Fourth Quarter
NCR05321	PRPA	Platte River Power Authority	CIP	O&P	Fourth Quarter
NCR05335	SNPD	Public Utility District No. 1 of Snohomish County	CIP	O&P	Fourth Quarter
NCR05372	SRP	Salt River Project Agricultural Improvement and Power District	CIP	O&P	Fourth Quarter
NCR03036	TBAY	Trans Bay Cable LLC	CIP	O&P	Fourth Quarter

Figure 10-9. Sample Audit Schedule.

Self-certification

The CEA may require registered entities to *self-certify* compliance with certain Reliability Standards (addressed in Section 3.2 of Appendix 4C to the NERC Rules of Procedure). Each year, the CEA posts its annual self-certification schedule on its website, and each registered entity must report to the CEA whether it is complying with the applicable Standard or is not in compliance. The CEA reviews the information from the registered entity and, if the CEA identifies a PNC, it follows up using one of the other CMEP tools (e.g., investigation, spot check, audit, etc.).

Spot Checks

CEAs also can use spot checks (addressed in Section 3.3 of Appendix 4C to the NERC Rules of Procedure) to monitor compliance. The CEA first notifies the entity of the spot check and provides the reason for it, its scope, the names of spot check participants, and a request for information. The entity then provides the information and the CEA reviews the evidence to verify compliance. In conclusion, the CEA issues a spot check report and sends the report to NERC (reports are *not* publicly posted). If the CEA identifies a PNC, it sends the item into the enforcement process (described later in this chapter).

Compliance Investigations

The CEA can use a compliance investigation to monitor compliance (addressed in Section 3.4 of Appendix 4C to the NERC Rules of Procedure). The CEA requests information/data from the registered entity and, in response, the entity provides the requested information. The CEA reviews that evidence to verify compliance. If the CEA identifies a PNC, it follows up (perhaps seeking additional information/data) and, if necessary, refers the matter to the enforcement process (described below). If the CEA does not identify a PNC, it informs NERC of that fact and NERC informs FERC.

Self-reports/Periodic Data Submittals

A CEA may use self-reports and periodic data submittals to monitor entity compliance (addressed in Sections 3.5 and 3.6 of Appendix 4C to the NERC Rules of Procedure, respectively). In a self-report, the registered entity submits a suspected compliance deviation to the CEA, which reviews information associated with the self-report and may request additional information. If a PNC exists, the CEA follows up by either investigating further or referring the item to the enforcement process (described below).

For periodic data submittals, the CEA posts a reporting schedule on its website describing the required data, and the registered entity provides that data (e.g.

vegetation outages, equipment misoperation reports, etc.). The CEA reviews the information and, if needed, requests additional data. If a PNC exists, the CEA follows up by either invoking another CMEP tool or referring the item to the enforcement process (described below).

Complaints

If the CEA receives a complaint about a registered entity, it follows the process described in Section 3.7 of Appendix 4C to the NERC Rules of Procedure and considers whether to initiate another CMEP process to address the complaint. If the CEA opts to use another CMEP tool, it initiates that process and follows the appropriate steps. If the CEA finds the complaint groundless, it closes its file on the matter. A CEA may transfer a compliant directly to NERC if warranted.

Enforcement

Enforcement begins when the CEA determines a *possible violation* exists.

Notice of Possible Violation

When a CEA discovers a possible violation, it issues a *Notice of Possible Violation* (NPV) to the registered entity. The NPV contains:

- a statement that the CEA identified a possible violation;
- a brief description of the possible violation, including the Reliability Standard requirement(s) and, if known, the date(s) involved; and
- an instruction for the registered entity to retain/preserve all data and records relating to the possible violation.

The CEA also reports the NPV to NERC, who, in turn, reports it to FERC.

Violation Investigation by CEA

After issuing an NPV, the CEA assesses the facts and circumstances surrounding the event to determine if the registered entity violated the requirement(s) identified in the NPV or whether to dismiss the possible violation. In making this determination, the CEA may consider any additional information to determine if it should dismiss or modify the possible violation or resolve it through one of the two expedited processes discussed below.

Notice of Alleged Violation and Proposed Penalty or Sanction

If, after investigation, the CEA believes the registered entity violated a Reliability Standard, it issues to the entity a *Notice of Alleged Violation and Proposed Penalty or Sanction* (NOAV) and reports its findings to NERC.

If the registered entity agrees with, does not contest, or does not timely respond to the NOAV, the CEA deems the entity to have accepted the violation and penalty or sanction, issues a *Notice of Confirmed Violation,* and reports the outcome to NERC. In response, the registered entity may provide a written explanatory statement to accompany the filing to FERC and NERC's posting of the confirmed violation on its website.

Violation Contest by Entity

If the registered entity contests the alleged violation or proposed penalty/sanction, the entity submits a response explaining its position, with supporting information and documents. The CEA then schedules a conference with the registered entity to try to resolve the matter. If the CEA and registered entity cannot resolve all issues, the registered entity may request a hearing. If the entity does not request a hearing, the violation becomes a Confirmed Violation. The NERC Rules of Procedure contain a hearing process (described in Attachment 2 to Appendix 4C of the NERC Rules of Procedure) too cumbersome to describe in this text. The procedure resembles the process of a traditional lawsuit. If the registered entity or CEA does not agree with the hearing's outcome, it may appeal the finding to NERC.

Appeal Decision

If the entity requests a hearing, the Compliance Committee of NERC's Board of Trustees decides the appeal based on the notice of appeal, the record of the hearing body's proceeding, responses, and any reply filed with NERC. The Compliance Committee may invite representatives of the entity making the appeal and the other participants in the proceeding to appear before the Committee. Decisions of the Compliance Committee are final, but the aggrieved party can appeal to the applicable governmental authority (FERC in the U.S. or a provincial body in Canada). Finally, if the aggrieved party wishes to appeal the governmental authority's decision, it may do so through the applicable court system.

Expedited Processes

NERC also applies two *expedited* approaches to compliance enforcement. The first – *Find, Fix, Track & Report* (FFT) – applies to violations posing a *minimal* or *moderate* risk to the grid reliability and requires the registered entity to mitigate the noncompliance before applying.

The second – *Compliance Exceptions* – applies to violations posing only a *minimal* risk to grid reliability. NERC does not include Compliance Exceptions in a registered entity's compliance history for penalty purposes. A CEA, however, must:

- consider a history of Compliance Exceptions where the failure to fully remediate the underlying compliance matter contributes to a *subsequent* serious or substantial noncompliance; and

- assess subsequent noncompliance to determine whether a registered entity should continue to qualify for Compliance Exception treatment. The Compliance Exception process requires the registered entity to have mitigated the noncompliance.

Penalties and/or Sanctions

Financial penalties are assessed based on two factors: significance and severity.

Violation Significance

As discussed earlier in this chapter, every Reliability Standard has a VRF adopted when that standard was approved. VRFs are classified as Lower, Medium, or High, and identify the *significance* of noncompliance with the Standard. Paperwork violations, for example, create less reliability risk than violations involving tree trimming, equipment testing, and facility design. For clarity, the significance of the violation is predetermined when the standard is written – each standard has one VRF associated with it.

Violation Severity

After determining an entity violated a Reliability Standard, NERC establishes the violation severity using the Reliability Standard requirement's VSL – Lower, Moderate, High, and Severe.

The VSL involves how severely (badly) the entity violated the regulatory requirement. The CEA assesses the VSL *post*-violation based on the extent to which the registered entity violated the Standard. If an entity, for example, has 100 protection systems on its portion of the interconnected grid and fails to test or maintain only one device, the CEA will likely consider that a *lower* VSL. On the other hand, if that entity failed to test or maintain dozens of its protection systems, NERC would probably consider that a *moderate* or *high* severity. The amount of the financial penalty assessed by NERC depends on VRF and VSL, as shown in Figure 10-10.

	Violation Severity Level							
Violation Risk Factor	Lower		Moderate		High		Severe	
	Range Limits		Range Limits		Range Limits		Range Limits	
	Low	High	Low	High	Low	High	Low	High
Lower	$1,000	$3,000	$2,000	$7,500	$3,000	$15,000	$5,000	$25,000
Medium	$2,000	$30,000	$4,000	$100,000	$6,000	$200,000	$10,000	$335,000
High	$4,000	$125,000	$8,000	$300,000	$12,000	$625,000	$20,000	$1,000,000

Figure 10-10. NERC Penalty Matrix.

Monetary Sanction

When determining monetary sanction, the CEA uses the following factors:

- VRF and VSL Table (Figure 10-10)
- Entity size
- Assessed risk to the reliability of the bulk electrical system
- Violation duration
- Violation time horizon

The CEA then adjusts the base monetary penalty depending on relevant aggravating or mitigating factors. The result is an *adjusted monetary penalty amount.* The aggravating/ mitigating factors include the following:

Aggravating

- Repetitive violations and compliance history
- Failure to comply with a remedial action directive
- Intentional violations
- Attempts to conceal the violation, or resist, impede, be nonresponsive, or otherwise exhibit a lack of cooperation
- Management involvement in any intentional violation or attempt to conceal the violation

Mitigating

- Presence and quality of compliance program
- Degree and quality of cooperation in the violation investigation
- Any mitigating activities directed to the violation
- Self-reporting of the event and/or undertaking voluntary mitigating activities

Final Adjustments

Finally, the CEA adjusts the amount for other circumstances such as the entity's agreeing to settlement, extenuating circumstances, disgorging unjust profits or economic benefits associated with an economic choice to violate, and the entity's ability to pay the monetary penalty. Ultimately, the CEA may reduce a monetary penalty to an amount the registered entity has the financial ability to pay, extend the payment period using an installment schedule, or excuse the monetary penalty.

Nonmonetary Penalties

In addition to monetary sanctions, the CEA may consider nonmonetary penalties to promote reliability, address risks to reliability, and ensure compliance with the Reliability Standards. The CEA considers the factors described above when evaluating whether to impose nonmonetary penalties and whether the nonmonetary penalties bear a reasonable relationship to the seriousness of the violation.

Nonmonetary penalties may include:

- requiring the CEO or equivalent to sign a settlement agreement;
- requiring periodic reporting on reliability;
- adding security or compliance-related efforts to the entity's board of directors;
- issuing a nonpublic or public letter of reprimand;
- conducting additional compliance monitoring;
- placing the entity on a reliability watch list;
- setting conditions for carrying on certain activities, functions, or operations;
- imposing other nonmonetary penalties, using professional judgment, that bear a reasonable relationship to the seriousness of the violation.

When the CEA imposes a nonmonetary penalty impacting the final monetary penalty, it explains in the notice of penalty how the nonmonetary penalty impacted the final monetary penalty amount.

United States – State Regulation

Each state has a *public utilities commission* (PUC), *public service commission* (PSC), or similar agency to regulate the electric industry at the state level. PUCs typically oversee matters like consumer protection, facility siting, and a variety of other issues. Exploring regulations on a state-by-state basis is beyond the scope of this text.

Canada – Non-NERC

Other than the NERC Reliability Standards adopted in Canada, the Canadian regulatory framework depends on the province or territory. Provincial regulators preside over their province's electric generation, intra-provincial transmission, distribution, retail pricing, and wholesale markets (if one exists). *Unbundling* and *separation of function* also change from province to province, with Alberta and Ontario having stringent requirements (regarding generation and transmission separation), while the provinces in which *Crown corporations*[75] dominate generally have fewer regulations.

Due to the large involvement of the individual Canadian provinces in the electric industry, the federal government has a much smaller role in Canada than in the U.S.

Provincial regulators largely regulate on a *public utility* model, requiring "certificates of public convenience and necessity" (or similar) to expand facilities and dictating the service terms and conditions between a regulated utility and its customers (using tariffs and rate cases). Provinces with competitive markets use market monitors to oversee competition (like Alberta's Market Surveillance Administrator).

Important regulatory revisions have occurred in connection with Bill C-69 enacted in 2019. Through Bill C-69, the federal government supplanted the National Energy Board (NEB) with a federal regulatory agency (Canada Energy Regulator or CER). The new agency serves as part of a group of reforms overhauling the regulatory makeup of Canada's electric sector.

Other Parts of the World

Asia and Africa have larger landmass than North America, but the electrical systems on those continents did not start as early, grow as large, or become as complex as the North American electric grid. The complexity and age of the North American electric grid meant it had to address large-scale reliability. The balance of this chapter contains a brief review of how other countries and regions regulate their grids.

United Kingdom

Because the United Kingdom is on an island, it is no surprise its electric grid differs from the grid on the European continent. The regulatory regime in the U.K. involves legislation, licenses and industry codes. An independent regulator oversees the sector and enforces rules and regulations. The primary legislative framework appears in the Electricity Act 1989, which created a licensing regime and established statutory duties for the regulator (the Gas and Electricity Markets Authority or GEMA). Unless exempted for some reason, a company must obtain a license for the following activities:

- Generation
- Transmission
- Distribution
- Supply
- Participation in operating an electricity interconnector
- Providing smart meter services

The Act requires separated activities by prohibiting an entity from performing other licensed activities. For example, the transmission license of the operator of the Great Britain transmission system prohibits it (and all affiliated and related entities) from owning electricity supply or generation interests. Additionally, an interconnector licensee cannot hold a license for generation, transmission, distribution, or supply.

In addition to licenses, industry codes create rules governing the industry. The transmission operator (NGET) establishes and maintains most of the transmission system pursuant to industry codes. Those codes include:

- Connection and Use of System Code (CUSC), which is a contractual framework for connecting to and using Great Britain's high-voltage transmission system;
- Balancing and Settlement Code (BSC), which contains provisions for the wholesale market (particularly electricity balancing and settlement);
- Grid Code, which establishes the operating procedures and principles governing the transmission operator's relationship with its users. It includes generating companies, suppliers or suppliers' customers, externally interconnected parties or users with systems directly connected to the transmission system;
- System Operator–Transmission Owner Code (STC), which defines the basic association between the system operator and transmission system owners;
- Master Registration Agreement (MRA), which provides a way to manage the processes between electricity suppliers and distribution companies, allowing electricity suppliers to transfer customers;
- Distribution Connection and Use of System Agreement (DCUSA), which provides a central document dealing with the connection to and use of distribution networks;
- Distribution Code, which creates the day-to-day procedures for the distribution licensee and users of its system for planning and operations purposes. It also ensures the distribution licensee can meet its Grid Code regulatory requirements.

European Continent

The European Union has unique challenges due to the patchwork nature of national regimes. Electricity regulation in Europe is highly complex because no two countries regulate the industry in the same way. Europe is not likely to have a single regulatory system in the short term, and different models will likely coexist. European regulators lean heavily on industry associations such as the Association for Electrical, Electronic & Information Technologies (VDE) for their standards. VDE is headquartered in Frankfurt am Main and has branches in Brussels and Berlin.

The European Commission will likely continue driving toward greater consistency in regulatory approaches. Rising energy costs, however, create political controversy, and the desire for regulatory convergence could be countered by government interference with policy decisions. This situation could inject instability and risk for companies operating in the international arena.

In continental Europe, Transmission operators must balance electricity generation and consumption in real time to sustain system frequency of 50hz. Every generation and consumption site connected to the grid is assigned to a balancing group that accounts for electricity injections and withdrawals from the system and virtual injections/withdrawals from other balancing groups. Balancing group operators manage their assets to comply with applicable balancing provisions. Each balancing group operator must keep its system in balance. To meet their balancing obligations, balancing group operators can:

- perform network-related measures (like modifying system parameters using ancillary services);
- request emergency assistance from other countries/operators; and
- activate additional measures like interrupting loads, or purchasing/selling electricity on the intraday market.

In extreme circumstances, Balancing group operators can access additional reserves like network reserves or capacity reserves.

Germany

Operating transmission lines in Germany requires authorization by the state regulators. Also, the Federal Network Agency for Electricity, Gas, Telecommunications, Post, and Railway certifies Transmission operators. Transmission operators must operate and maintain secure, reliable, and efficient networks so long as economically reasonable and must especially ensure network stability when trading electricity with other networks. Finally, they must comply with technical standards established by VDE.

Generation plants must comply with technical standards, including rules established by VDE and must also comply with general building safety and environment requirements.

Russian Federation

Russia is a federated state with eighty-five *subjects* – or regions. Some powers rest exclusively with federal regulators, while some come under joint jurisdiction by federal and regional authorities, and still others only at the regional level. Local (municipal) governments have some authority – at the lowest level. The following laws apply to the electric sector:

- Law on Use of Nuclear Power No. 170-FZ dated 21 November 1995
- Law on Electricity No. 35-FZ (March 26, 2003)

The electric power industry includes generating facilities, transmission grids, operational dispatch management, and sale of electricity to customers. Except for nuclear generation, no activities require a license or other permit.[76]

In 1998, Russia's economy began growing dramatically (after the USSR's collapse) and, over the next decade, soaring electricity demand revealed the dilapidated state of the nation's electric infrastructure. Realizing the state could not cover the costs of maintaining and upgrading the infrastructure, and after considering dozens of models, the Russian Duma approved a reform plan that involved unbundling the incumbent monopoly, creating an independent regulator, privatizing generation, and liberalizing electricity prices.

In 2008, the state electric monopoly – Unified Energy System (UES) – saw its holdings unbundled, with generation, transmission, and distribution structurally divided and managed by various companies. Fourteen territorial power and heating companies (referred to as TGK) and seven wholesale power-generating companies (OGK) generate electricity. An antimonopoly organization prevents one private owner from controlling more than 20% of generating capacity in any one of the eight regional zones. The state continues to own 100% of nuclear generation and most hydropower and major transmission facilities.

The Ministry of Industry and Energy oversees the power sector, and the System Operator (or Centralized Dispatching Administration), a state-owned joint-stock company, ensures the dispatch of electricity and stability of the unified grid. Due to the unified electric grid in Russia, the state-owned Federal Grid Company (FGC) operates and develops the electric transmission grid.

Australia

To construct electricity generation plants, operators must meet requirements that vary among the states and generation type. The primary requirements include getting planning consent for construction and meeting relevant environmental requirements.

Generally, states require a development consent or planning permit that establishes conditions on construction. In some cases, the applicant must obtain further consent or licenses before beginning construction (typically listed in the planning permit and including permits for performing building works, operational works (like road widening), drainage works and plumbing works).

To operate a generation plant connected to the transmission grid, the grid operator must grant a generation authorization for the plant. The generation owner must classify its plant as a scheduled, semi-scheduled, or non-scheduled generating plant.

Victoria, Queensland, South Australia, the Australian Capital Territory, and Tasmania require generators to obtain a generation authorization in addition to the National Electricity Market authorization. New South Wales does not require separate approval from the state. Authorization requirements are based on several factors, including generation size and type, and impact on the surrounding environment and the jurisdiction.

To connect to the transmission grid, a generator makes a connection request to the transmission operator. The transmission operator informs the generator of the required connection data and application fee. After the generation operator pays the fee and the transmission operator makes a *connection offer*, the parties negotiate a final connection agreement containing the technical and commercial connection requirements. While operating, the generator must comply with generator performance standards (GPS) of the transmission provider and/or the Australian Energy Market Operator (AEMO). The GPS create technical specifications and requirements to ensure grid stability.

On the transmission side, transmission providers (TPs) generally must obtain development approval or a planning permit (although state-owned TPs are exempt under certain circumstances). The TPs must obey legislation in their jurisdictions and all other applicable laws, rules, and regulations. States provide separate authority (a license) allowing TPs to operate their networks, and a separate entity ensures compliance with those licenses (e.g., New South Wales has the *Independent Pricing and Regulatory Tribunal*). TPs also must meet standards, usually implemented by the state government, addressing designing, building, and operating facilities to avoid or manage electricity outages.

Distribution network providers (DPs) must follow the same state-specific requirements applicable to TPs. DPs receive state licenses to distribute electricity administered by a separate body than the one overseeing TPs. DPs also must meet individual standards regarding design, construction, and maintenance of network equipment, including minimum equipment specifications, environmental, property, protection, testing, and metering.

Summary

- North America (especially the U.S.) has stringent regulatory requirements regarding the reliability and security of the bulk power system.
- In addition to a federal regulatory regime, U.S. states have public utility commissions (or similar) overseeing certain aspects of the electric industry.
- Canada uses primarily a province-focused approach to regulating the reliability of the electric grid.
- The U.K. has a more centralized approach to electric industry regulation.
- The E.U. has a patchwork quilt of regulatory regimes of which Germany is one example (with shared responsibility between the federal and state governments).
- Russia shares responsibility between the federal government and its subjects.
- Australia, like Canada, uses an approach focused primarily on state requirements.

11

Electricity Markets

This above all: to thine own self be true,
And it must follow, as the night the day,
Thou canst not then be false to any man.

—Polonius in *Hamlet*

"Grandpa Tom, I'm bored; can you take me to Montie Beach Park?" asked Luke. "Mom and Dad say our electricity contract is coming to the end of its term, and they have to decide whether to choose a new electric provider."

"Sure, let's go!" the grandfather exclaimed.

On the way to the park, Luke asked, "Who do you get your electricity from, Grandpa Tom?"

The grandfather replied, "You have retail competition in Houston, so your Mom and Dad can choose your electric provider. In Austin, we get our electricity from the electric utility department of the city government called Austin Energy. We don't get to choose our electric provider."

"That's crazy," replied Luke. "Two cities in the same state treat electricity that differently." He paused. "Wait, if we change who we get electric from, will we get new electric lines to the house?" asked Luke.

"No," replied Grandpa Tom. "The same wires will still deliver electricity to your house from the same substation connected to the same grid as now," the grandfather added.

"Sounds complicated," concluded Luke.

Introduction

On a broad basis, economists say markets are either free, monopolistic, or somewhere in between. Economics teach that facility ownership in industries requiring large capital expenditures eventually consolidate, reducing competition among the players and raising prices to consumers. This trend gave rise to government regulations aimed at fostering competition and controlling prices to protect consumers.

Free Markets

> **Market Power**
> The ability of an entity to set or influence prices.

In truly free markets, no single competitor has *market power*. Regulators generally agree free market prices do not need government regulations, because competitors effectively regulate pricing as they consider whether to raise or lower prices in the quest for maximum long-term profitability.

Monopolistic Markets

By definition, monopolistic markets have only one market participant that can set the price and, if customers want the item or service, they must pay the established price. Regulators say these markets require consumer protection, and have devised ways to calculate *fair price* and *competitive markets.*

In Between Markets

Few, if any, markets are completely free or monopolistic, leaving regulators to devise various regulatory schemes in their quest to protect consumers.

Fair and Reasonable Price

In the late-1800s, U.S. railroads used their market power to limit competition and charge high prices. As a result, the United States, Congress passed the Interstate Commerce Act of 1887 to regulate the railroad industry's monopolistic practices. The Act required railroad rates to be "reasonable and just." The electric industry originally followed the same approach.

Economists calculate the reasonable and just price using some form of Cost of Service (COS) method. Theoretically, COS allows the seller to recover its cash costs incurred in providing the goods or services plus a return on the capital invested to provide that good or service. COS calculations sound simple and straightforward, but, as always, "the devil is in the details" and those calculations often become convoluted. COS calculation details are left to the economists.

Market Access

In 1903, the U.S. Congress passed the Elkins Act, amending the Interstate Commerce Act, and authorized the Interstate Commerce Commission (ICC) to impose heavy fines on railroads that offered rebates to certain shippers and upon those who accepted the rebates. The twentieth century saw growing price and market regulations around the globe.

Electrical Power Market Segments

Currently, the electrical power value chain has these main functions: generation, transmission, distribution, and sale (wholesale and retail), as shown in Figure 11-1.

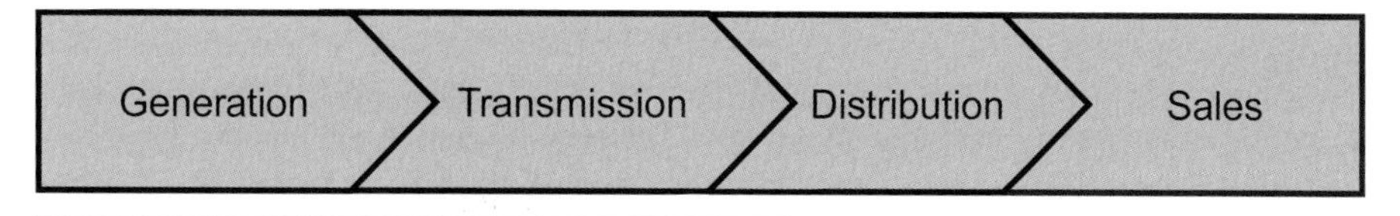

Figure 11-1. Electrical Power Value Chain.

Moving to intermittent generators (primarily wind and solar), a fifth electrical power function – storage – is emerging. A brief review of these five functions or markets, their assets, and how those assets compete with each is next.

Generation

Electricity is a commodity, and generators cannot differentiate the electricity they generate from that generated by others. They can supply any customer connected to the same grid to which they are connected. Like other commodities produced by multiple producers and serving multiple customers, the bulk electric market is competitive – absent transmission and distribution constraints.

Transmission

Transmission lines transmit electricity from generator switchyards to distribution step-down substations. Once a transmission line is built between point A and point B, building a competing line serving the same route is not economically feasible – as long as the first line still has available capacity. Most companies owning transmission lines operate as regulated monopolies.

Distribution

Distribution lines are a lot like transmission lines because it would be inefficient to build two distribution systems from the same substation to the same customers. Owners of distribution lines often operate under a franchise from a governmental entity giving them an exclusive distribution right in return for agreeing to rate regulation. Transmission and distribution are so similar (except for voltages) they are sometimes grouped together and abbreviated T&D, even though different, and potentially competing, entities may own them.

Wholesale and Retail Sales

The sales function requires low-capital investment, meaning it has a low barrier to entry. Sales competition ebbs and flow and is generally considered competitive in deregulated markets.

Storage

Electric storage is similar to generation because it can supply electricity to any customer connected to the same grid. Like generation, it is a competitive market so long as enough storage exists to meet customer needs.

Electricity Market Regulation

Over the years, market regulators across the globe experimented as they first regulated, then deregulated, wholesale and retail electricity prices. The United States Court of Appeals for the D.C. Circuit stated, "Of the [Federal Energy Regulatory Commission's] primary task there is no doubt, however, and that is to guard the consumer from exploitation by non-competitive electric power companies."[77] The Commission has always used the following two general approaches to meet this responsibility:

1. Regulation – The primary approach for most of the twentieth century and remains the primary approach for wholesale transmission service.
2. Competition – The primary approach in recent years for wholesale generation service.

Advances in technology, exhaustion of economies of scale in most electric generation, and new federal and state laws have changed the Commission's views of the right mix of these two approaches. The Commission's goal has always been to find the best possible mix of regulation and competition to protect consumers from the exercise of monopoly power.[78]

Regulated Market Structure

Historically, the electrical power industry was vertically integrated with an electrical utility owning all the assets and performing the functions shown in Figure 11-1. In the case of vertical integration, the prices of the functions are bundled (aggregated) and the customer pays one price. Vertical integration gives entities market power over the geographic areas they serve via the access

T&D

Vertical integration combined the transmission and distribution functions and abbreviated them T&D. In reality, they are two different but related functions. T&D is still a common electrical industry term.

control provided by T&D assets. Because of this market power, pricing and market access is typically regulated for T&D assets.

In *regulated* markets, either:

- the government owns and operates electric generation, T&D, and retail sales; or
- the government regulates non-governmental entities owning the assets and performing those services.

Because government motives differ from those of private companies, governmental ownership sometimes results in decisions motivated by politics, not economics, leaving customers no choice but to pay for the cost of those activities.[79]

In areas where the government does not own and operate the electric utility but, instead, regulates the prices charged by nongovernmental entities, the government usually sets prices based on the COS approach introduced earlier in this chapter.[80] When jurisdictions move away from the traditional, regulated, vertically integrated utility model to open markets, they generally break up the business into *wholesale* and *retail* markets.

Deregulated Market Structure

Increasingly, regulatory jurisdictions have *deregulated* (or *re-regulated*) markets into generation, wholesale sales, and retail sales and kept the middle parts – T&D – regulated. Deregulating generation and sales requires *unbundling* vertically integrated utilities into the separate segments.

Deregulating markets does not mean other areas like safety, reliability, environmental performance, or consumer protection are deregulated. For example, the Texas Public Utility Commission has *Customer Service and Protection* rules codified in its Electric Substantive Rules, Texas Administrative Code, Title 16, Part 2, Chapter 25, Sub-Chapters B and R.

Deregulated Wholesale Energy Markets

Deregulating markets requires one or more reforms, including privatization, vertical unbundling, or horizontal restructuring. The required reform depends on the market structure before deregulation. Consequently, postderegulated wholesale and retail market practices vary by country and even within regions of the same country. Electricity market deregulation seeks to establish competitive markets to improve efficiency and provide long-term consumer benefits.[81] Deregulation also requires *nondiscriminatory access* to the T&D network so resellers can match geographic supply with geographic demand. Following are examples of how countries have deregulated electricity markets.

Australia

Australia deregulated its electric market in 1998.[82] The Australian electricity market involves energy generators, a transmission network, distribution networks, and retailers. The Australian Energy Regulator (AER) oversees the market. In each state and territory, electricity distributors sell wholesale power to retailers who, in turn, enter contracts with end-use customers. As part of deregulation, Australia established a *market operator* (National Electricity Market Management Company) to dispatch generation across the market, register participants, perform settlements (charges and payments), and undertake network planning to ensure system reliability. The NEMMC works much like an ISO or RTO in the U.S.

The AER regulates the National Electricity Market (NEM) networks by monitoring and regulating wholesale prices. The number of customers supplied by each network varies (with the New South Wales and Queensland networks having the most consumers). The AER does not *establish* retail prices for electricity but does regulate and approve the prices generators charge to supply wholesale electricity. Individual retailers set the price for the electricity they sell to end-use consumers.

> **Market Clearing**
> When prices and MWs for sale exactly match prices and MWs needed to serve load, the auction stops, orders are matched against each other, and the market *clears.*

The Australia wholesale market uses an auction process through which generators offer to supply electricity at specified prices and retailers provide bids to buy at specified prices. Buyers and sellers adjust prices until the market *clears* to meet electricity demand (or *load*). Because generators cannot easily store electricity, providing it to consumers takes place through a *spot market.* The market operator compiles generation needs every five minutes and tells generators how much energy to produce for each interval to exactly match generation and load. Generators can resubmit offers every five minutes. The market operator chooses bids to meet market demand starting with the cheapest generation. This market design satisfies electricity demand in the least expensive way.

> **Market Settlement**
> "Settling" the wholesale market means collecting money from retailers and paying money to the generators based on the clearing price.

Once the market operator matches supply and demand, it calculates the market price using the final (most expensive) bid required to clear the market. The market operator clears the market every five minutes (creating a clearing price each time) and averages those six prices every half-hour to set the "spot price" and pays all generators the average thirty-minute clearing price until the next auction. Paying all generators based on the clearing price provides an incentive for them to offer cheaper generation because their bid will more likely be chosen and provide them the most profit. The market operator uses that *spot price* to *settle* the financial transactions for supplied electricity. Australia limits the spot price to a maximum of $15,000 per MWh (adjusted for inflation). The government reviews the price cap every four years. The government also attempts to ensure dependable electric supply by licensing and registering generators.

End-use customers enter contracts with retailers who buy power from generators. Retailers manage the risk of wholesale price variations by contracting with generators for *firm* prices (i.e., prices that do not fluctuate with the spot market).

Generators can either participate in the wholesale market or in a *nonmarket* capacity, meaning they agree to provide *ancillary services* or network support for transmission reliability but not sell directly into the wholesale market. *Market* capacity is when the generator participates in the wholesale spot price market while *nonmarket* capacity means the generator has contracted its generation capacity to a customer and does not participate in the spot price market.

Ancillary Services

Services to support electric power transmission from generators to end users to maintain transmission system reliability. They are specialty services and functions to facilitate and support the flow of electricity to meet demand in real time.

Japan

In the year 2000, Japan allowed *partial* retail competition (for electricity consumers using >2,000 kW). In 2004, it expanded retail competition to users of >500 kW and in 2005 to users of >50 kW. Also in 2005, Japan established the Japan Electric Power Exchange (JEPX) to facilitate purchases and sales of wholesale electricity (forward, day-ahead, and intraday markets). In April 2016, Japan fully deregulated its retail market and, in 2020, abolished its regulated retail tariff and required unbundling of vertically integrated utilities.[83]

In 2015, Japan established the Organization of Cross-regional Coordination of Transmission Operators to promote development of T&D networks and enhance supply-demand balance in normal and emergency situations (somewhat like an ISO or RTO in the U.S.) and the Electricity and Gas Market Surveillance Commission to support competition and provide consumer protections.

European Union

Deregulation of electric markets in the EU began in earnest in 1996 when it opened electric markets. All EU countries had to open their markets by mid-2007. The wholesale electricity market in the EU consists of:

- electricity suppliers (who buy electricity from generators and sell it to consumers);
- transmission system operators (TSOs) (providing long-distance transport of electricity and ensuring system stability);
- distribution network operators (DSOs) (delivering electricity to consumers);
- regulators (setting rules and overseeing the market).

The European transmission grid contains more than 186,000 miles of power lines (with 355 cross-border lines).[84]

The retail market involves suppliers (who offer electricity to end-use consumers) and consumers (who choose their supplier). Suppliers buy electricity from generators and invoice consumers based on the price for electrical energy plus T&D and taxes/levies. Participants in the wholesale market include generators, electricity suppliers, and large industrial consumers.

Depending on the contract or market, transactions include:

- long-term contracts (for many years);
- forward and future markets (weeks to years ahead);
- day-ahead market
- intra-day market (within a specified time period, e.g., an hour);
- balancing market (real-time balancing of supply and demand).

Marginal Price

The price paid for the last MW of generation needed to serve load at the time the market clears.

Private parties may trade electricity in bilateral contracts or through an energy exchange.

Generators provide electricity based on price – from lowest to highest until a *clearing* price is reached (i.e., the *marginal* price). Europe uses day-ahead and intraday markets for physical trades in which participants set their bids. The financial markets (described more fully below) usually include future, forward, and option markets. Some financial markets are physical (i.e., settlement does not occur via cash but by physical delivery of electricity) while others do not require physical delivery of electricity and simply provide a mechanism to speculate on market prices. TSOs work with the power exchanges to provide transmission capacity and inform the exchanges regarding how much daily transmission capacity exists.

Europe uses a *zonal pricing* method. If insufficient transmission capacity exists between zones, they split the area into a number of zones each with its own price. On the other hand, if several exchanges calculate their own system price and enough transmission capacity exists between the areas, they combine the zones into a single pricing area.

Russian Federation[85]

The Russian Federation began reforming its electricity sector in 2006. Previously, all facilities were part of a state monopoly, Unified Energy System of Russia (UES), and the state regulated prices. In 2008, the government unbundled UES into more than twenty independent power companies. The government gradually relaxed prices and, currently, about 80% of electric power trades at nonregulated prices. Although planned reforms are for all electric power to be traded at market prices, the state will probably regulate prices for the foreseeable future. In addition, the reforms are not expected to reach all of Russia's geographic area (e.g., the Russian Far East, Kaliningrad, and Arkhangelsk regions).

Reconsolidation: More than ten years after reforms began, reconsolidation of power assets is occurring. The T&D facilities were reconsolidated in a public joint stock company, Russian Grids, which acquired the assets of the Federal Grid Company and Holding IDGC, and the state now controls them. Several power generation facilities were reconsolidated under *Inter RAO PJSC*, the successor of RAO Unified Energy System of Russia (RAO UES). The state controls Inter RAO PJSC and, between 2008 and 2012, repurchased power generation assets previously privatized.

Electricity Market: Russia divides its power market into two parts: wholesale and retail. The wholesale market trades power between generating companies and suppliers and some large end users and includes trading *capacity*. The basic regulation governing the wholesale power market went into effect in 2010. The retail market involves selling power from suppliers to consumers (industrial, commercial, and residential).

Transmission Pricing: A regulation instituted in 2004 governs transmission operations and provides the procedure to obtain access to transmission facilities and services. Pricing is covered by a government resolution promulgated in 2011. The Ministry of Energy oversees the Russian electric power sector by implementing state policy and issuing regulatory acts to do so. The Federal Antimonopoly Service establishes the remaining state-controlled prices (e.g., transmission services or residential tariffs) and supervises compliance with anti-monopoly regulations (e.g., nondiscriminatory access to transmission services, avoiding monopolistic prices). The Service can issue orders or penalize companies for violating anti-monopoly regulations.

United States

Most deregulated electric markets in the U.S. are designed like the Australian market. For example, the market in most of Texas, administered by Electric Reliability Council of Texas, Inc. (ERCOT), relies on unregulated power generators, regulated T&D entities and unregulated retail electric providers (REPs). End-use customers enter a contract with a REP who can enter bilateral contracts and/or obtain power from the spot market administered by ERCOT. ERCOT ensures grid stability through, among other tools, ancillary services. One unique aspect of the ERCOT market is its *energy only* design. ERCOT does not have a *capacity market,* which means it does not compensate generators for installing additional capacity to enhance reliability.[86]

Unlike the ERCOT market, many deregulated electricity markets provide power generators payments for capacity ready to produce electricity, not just electricity actually produced.[87] For example, if a generator has a 100 MW power plant and the capacity payment is $1,000.00 per MW per month, the generator would earn $100,000 per month ($1.2 million per year) regardless of how much electricity it actually produced (if any). Capacity payments have become a large part of generators' revenue streams in those markets.

Most deregulated energy markets in the U.S. have four main pricing components:

1. Energy – the electricity consumed by end users
2. Capacity– generation resources available for dispatch
3. Ancillary Services – services to ensure grid reliability and stability (discussed in more detail below)
4. Transmission congestion and losses – costs associated with (1) getting power over transmission lines not rated to handle the amount needed, and (2) line losses from moving power across a long distance

In the United States, different market operators use different terms and pricing mechanisms for the components, but all four components factor into the ultimate price paid for electricity. The energy component usually makes up the largest part of the price, followed by capacity. Other components add cost depending on the circumstances.

Final Price

Conceptually, the price paid by customers for electricity should follow Figure 11-1 and equal:

Price to Generator + Price to T&D + Taxes and Other Costs

In the electrical power industry, this concept holds true. One of the most common terms applied to this price in electrical markets is *locational marginal price* (LMP) – marginal because it is based on marginal price and locational because it is based on the cost of transmitting the electricity from the generation plant to the node (connection point) from which it is moved to the consumer.

The term *marginal* comes from the fact the most expensive (i.e., marginal) generation resource defines the price for *all* electricity

> **Nodes**
>
> Locations where electricity leaves a transmission line and enters another transmission line or a distribution substation.

used at a specific place and point in time. With no transmission line constraints or line losses, the LMP at every location will be the same.

Rarely, though, does a system not have transmission line constraints or line losses. Consequently, LMPs tend to vary by node. The LMP incorporates:

- marginal energy cost (MEC) across the system;
- marginal congestion cost (MCC) by location;
- marginal loss cost (MLC) by node.

Marginal congestion costs and marginal losses vary by location and can be positive or negative. The LMP at any location is the sum of the marginal energy, congestion, and loss components, shown as:

$$LMP_i = MEC + MCC_i + MLC_i \text{ (where }_i\text{ is the node's location on the grid)}$$

In deregulated markets, wholesale prices are determined by:

- Auctions – forward prices established in a bidding process with a clearing price; forward price auctions take place ahead of *real time*.
- Real-time price – based on a bidding process with a clearing price based on market conditions at the time power is consumed.
- Bilateral contracts – contracts between electric providers and electric users; the supplier is usually a generation plant and the buyer is usually a retail electric provider, electric utility, or end user (e.g., an industrial facility like a refinery or factory).

Auctions

Auctions match electricity supply to demand at the lowest possible price. The market operator estimates the hourly demand for the day, and each generator offers a specific amount of generation supply into the market at a specific price. In theory, generators base their offer prices on the cost to operate the facility plus a return on investment.

Once generators make their offers, the market operator sorts them in ascending order to determine how much supply is available at different price points. It then selects the "winning" bids – the lowest-priced combination of offers to meet demand – to dispatch at the hour dictated by the auction. As in Australia, the market operator sets the clearing price based on the marginal (most expensive) unit of generation to meet demand.

Infrastructure

In deregulated markets, generation, T&D, and retail sales are *unbundled*, as discussed earlier in this chapter.[88] The physical infrastructure, though, remains the same as in regulated markets. In other words, generation facilities (nuclear, coal, natural gas, wind, water, and others) continue operating, but the same company cannot own generation facilities *and* T&D facilities and cannot be a retail electric provider.

T&D lines, substations, switches, breakers, protection systems, etc., reside within companies providing only T&D services. Finally, the role of REP (defined earlier) must be separate from the generation companies and T&D companies. REPs generally do not own any part of the bulk electric system. The T&D company owns and operates the wires and associated equipment (including meters) to get the electricity from the generation facilities to the end users. REPs merely provide electricity to the customers and bill them based on usage.

The *players* in the electric market, as mentioned above, are:

- REPs – Supply energy to consumers who can choose their REP
- T&D Company – Delivers electricity and maintains poles, wires, and meters
- Power Generation – Generates electricity
- Public Utility Commission (or other regulatory body) – Maintains and enforces a regulatory regime
- Market Operator – Manages and maintains the electric market
- Utilities – Primarily responsible for generating and distributing electricity to all consumers within a designated geographic area
- Electric Cooperatives (Co-ops) or public utility districts – Provide electricity to customers in a defined geographic area

Financial Markets

> **Elasticity of Demand**
>
> An economic term describing how prices and demand interact with each other. "Inelastic" means consumers buy the same amount despite changing prices.

Electricity demand is essentially *inelastic* because people need electricity for day-to-day life activities and do not use more if prices get cheaper or less as prices get higher – at least not in the short or mid-term. Because recessions tend to have less effect on electricity usage than on other commodities, investors use the electricity market for investing.

Electrical supply must always equal electrical load on a real-time basis across the grid. This instant nature of electricity delivery means prices can be volatile especially during power outages and seasonal fluctuations.

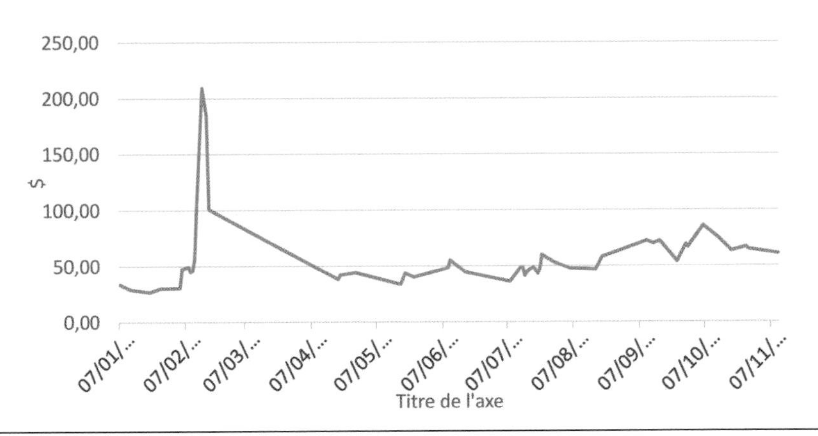

Figure 11-2. Average Wholesale Price Paid for Electricity at Indiana Hub.

Hedges

Buyers and sellers use a variety of financial instruments to control risk. A *hedge* reduces investment risk using *call options, put options,* or *futures contracts* to lock in profits and reduce the risk of loss.

In a *call option,* the holder has the right (but not the obligation) to buy the electricity at an agreed-upon price on or before the contract expiration date, regardless of the market price at that time. Someone buys a call option if he believes the price of electricity will rise by the end of the contract. If the price does rise, the holder may buy the electricity for a profit. If the price does not rise, the option expires and the holder limits his loss to the price of the option contract.

In a *put option,* the holder has the right (but not the obligation) to sell electricity at an agreed-upon price on or before the contract expiration date, regardless of the market price at that time. Someone buys a put option if he believes the price of electricity will fall by the end of the contract. If the price does fall, the holder may buy the electricity for a profit. If the price does not fall, the option expires and the holder limits his loss to the price of the option contract.

A *futures contract* provides for the delivery of a specified asset in exchange for the selling price at a specified future date. In a futures contract, both parties are obligated to honor their commitment (the buyer to purchase the electricity and the seller to deliver it). Electricity traders generally fall into two categories: *hedgers* and *speculators.*

Hedgers

Generators and retail electric providers typically use futures to *hedge risk.* Local and regional electricity producers and suppliers use a *short hedge* to obtain a set price for selling electricity they will produce in the future. It ensures the producer a known amount even if prices fall or rise. Consumers, in turn, can use a *long hedge* to establish a set purchase price for a quantity of electricity in the future.

Speculators

Speculators use the NYMEX (and other exchanges) to trade electricity futures. Speculators generally do not intend to deliver (nor receive delivery of) electricity. Instead, speculators look to profit by accepting the price risk hedgers shift to them while hedgers insulate themselves from price fluctuations. Speculators try to gain from equipment outages, network congestion, weather, fuel source price volatility, mistakes, or forecast errors by electric market participants. They buy electricity futures if they expect prices to increase and sell electricity futures if they think prices will decrease.

NYMEX

The New York Mercantile Exchange was the first to trade electricity futures.

Ancillary Services

In addition to generating electricity, generators provide *ancillary services* to maintain grid stability. Grid operators primarily use two types of ancillary services – *regulation* and *reserves*. Historically, generators provided ancillary services, but with the integration of intermittent generation and development of "smart grid" technologies, smaller distributed generators and even end-use consumers can provide *some* ancillary services.

Regulation vs. Reserves

Grid operators manage small deviations between load (electricity consumed) and generation (electricity produced) with regulation service.

They manage large deviations caused by large generators going offline unexpectedly by using reserves.

Regulation Service

Generators providing regulation service adjust electricity output in small increments after receiving an automated signal to do so from the grid operator and receive compensation for providing the service. Regulation service matches generation and load to keep the grid stable by:

- maintaining system frequency of 60 Hertz;
- tracking instantaneous fluctuations in load;
- correcting for unintended generation changes;
- controlling differences between forecast/scheduled power flow and actual power flow.

Area Control Error

Area Control Error measures the grid frequency difference between actual grid frequency and desired grid frequency (60 hertz or 50 hertz). Grid operators use regulation service to keep ACE within acceptable parameters.

When ACE is greater than zero, the grid operator issues Regulation *Down* instructions (telling generators to decrease output). When ACE is less than zero, the grid operator issues Regulation *Up* instructions (telling generators to increase output). Generators providing regulation service respond to Regulation Up and Regulation Down dispatch instructions from the grid operator through Automatic Generation Control (AGC).

The grid operator controls loads through Load Frequency Control (LFC) to respond to minor frequency deviations. The grid operator uses ACE to determine the amount of correction necessary to control system frequency.

As technology evolves, more types of market participants can provide Regulation Service. For example, distributed energy resources, like wind turbines, photovoltaic arrays, and fuel cells, can work with storage devices to provide instantaneous grid support

Automatic Generation Control (AGC)

A system that automatically moves a generator's output up or down every few seconds to adjust for changes in frequency (to keep frequency at 50 Hz or 60 Hz, as appropriate).

Reserves

Generation *reserves* are electricity resources (generating units or storage) not currently in use but the generation or storage owner can quickly make them available in response to an unexpected loss of generation.

Primary Reserve Service (PRS)

The grid operator uses PRS to restore transmission system frequency within the first few minutes after a *significant* deviation from the standard frequency of 50 Hz or 60 Hz. This category of service includes "quick start" or "peaker" generation units that can start and synchronize quickly (usually within ten minutes). Historically, quick start units were natural gas turbine generators discussed in Chapter 3. As the amount of grid storage increases, so does its availability as PRS.

Nonspinning Reserve Service

NonSpinning Reserve Service (NSRS), also known as supplemental reserves, provides additional capacity from offline generation resources, online resources not producing at full power, or grid-scale storage resources available within a short period. NSRS is extra generating capacity not currently supplying the grid that can be brought on line relatively quickly.

To provide NSRS, the generation resource must be *dispatchable,* meaning wind and solar resources do not qualify (unless coupled with energy storage facilities). Controllable load resources can also provide NSRS by decreasing load in response to a dispatch instruction, as discussed in Demand Response section. The grid operator deploys NSRS in economic order, working from least to most expensive, as needed to support grid stability.

Demand Response (DR)

Demand response means reducing demand (load) in response to one of three situations:

- Emergency response – to avoid rolling blackouts during generation shortages
- Economic response – electricity customers lower usage when electricity prices are high
- Ancillary services response specialty services – promote reliable operation of the transmission grid as discussed in the previous section

DR allows electricity end users (customers) to help preserve transmission system reliability, enhance competition, and mitigate high prices. The electricity users participating in DR programs are called *Load Resources.*

Load Resources willing to reduce or modify electricity usage in response to signals from the grid operator can participate in the wholesale market through demand response service.

Emergency Demand Response: Emergency demand response service decreases the need for load shedding (i.e., rolling blackouts). Customers capable of reducing load in response to a dispatch instruction from the grid operator and who meet predetermined performance criteria agree to act as Load Resources. Grid operators choose which of these Load Resources to shed in an electric grid emergency based on price. Load Resources can participate in the wholesale market *or* provide ancillary services.

Grid operators may obtain emergency response service through pre-established response times – for example, thirty minutes or ten minutes. Customers must respond by decreasing electricity consumption within the specified time period or forfeit compensation. Typically, the grid operator pays the Load Resource (for decreasing usage) the same price it would pay a generator for providing electricity. Load Resources submit bids (like generators), and the grid operator instructs them to reduce load if market prices equal or exceed the bid level. Load Resources scheduled or selected by the grid operator can receive capacity payments regardless of whether they actually receive a dispatch instruction to decrease load.

Customers not participating in demand response programs sometimes choose to reduce consumption in response to price spikes or grid reliability issues. By doing so, those customers may benefit financially when wholesale market prices are high.

Deregulated Retail Energy Market

In theory, the retail electric market is as "simple" as end users choosing providers (REP) just as they choose cellular phone providers or insurance companies. In practice, however, the process can appear more complex to retail customers who know less about electricity than industrial or commercial customers and may not understand the terms and conditions of the product they agree to purchase.[89] The authors discuss some of those factors under Rates and Contracts below.

Regulated Utilities

The traditional retail electricity sales model was the vertically integrated utility discussed earlier in this chapter. Markets served by vertically integrated utilities are typically regulated. Customers served by vertically integrated utilities pay the regulated prices established by their supplier and depend on regulations for protection. Many parts of the world still employ this approach to selling electricity.

Deregulated Utilities

As mentioned previously, many jurisdictions have *deregulated* (or *re-regulated*) their electric markets (at the wholesale level, retail level, or both). Those jurisdictions typically require *unbundling* of vertically integrated utilities into three segments: generation, T&D, and retail sales.

Unbundled Utilities

When required to unbundle, a utility may create separate subsidiaries to own the generation, T&D, and retail portions of the business. To prevent the subsidiaries from sharing sensitive market information that might give one of the subsidiaries a competitive advantage, regulations typically contain requirements to shield various types of information from passing between the subsidiaries.[90] Consumer protection rules also cover the relationship between retail sellers and buyers of electricity.

Ownership Forms

Electricity providers have taken on various business structures over the years, driven by different business factors. These business forms are discussed in the next section.

Cooperatives

Cooperative ownership started in areas where the cost to provide services was not worth the investment required and, therefore, public and private utility companies were not willing to make an investment. Most co-ops received financial support incentives from governments to get them started and are owned by the residents, businesses, and industries they serve.

Distribution electric co-ops distribute electricity to their owners (as the name implies) and are typically managed by a board of directors elected by the membership. Generation and transmission cooperatives typically sell wholesale power to member distribution cooperatives.

United States

Per United States federal law, electric cooperatives are not-for-profit companies, financed in whole or in part under the Rural Electrification Act of 1936, that own or operate facilities to generate, transmit, or distribute electricity, or a not-for-profit successor of such company. According to the National Rural Electric Cooperative Association (NRECA):

- Co-ops own and maintain 42% of U.S. electric distribution lines.
- Co-ops serve 42 million people.
- Co-ops power over 20 million businesses, homes, schools, and farms.
- In 2019, U.S. co-ops returned more than $1.3 billion in capital credits to their members.
- 832 distribution co-ops were built by and serve co-op members.
- 63 generation and transmission co-ops provide wholesale power to distribution co-ops, electric generation facilities, or by purchasing power on behalf of the members.[91]

Continental Europe

In Continental Europe, a federation of energy cooperatives began in 2013 (called REScoop) and currently has a network of 1,900 co-ops representing more than 1.25 million customers.[92] REScoop has staff based in Belgium – with different nationalities and backgrounds. The REScoop Coordinator oversees daily operations and reports to a Board of Directors elected by a General Assembly that includes all members. Full members (individual energy co-ops or national/regional federations) can vote (associate members cannot vote). The General Assembly follows the "one member, one vote" philosophy, and a majority makes all decisions. The board makes decisions related to policy, strategy, organizational planning, and budget. It consists of eight members.[93]

United Kingdom

In 2010, the United Kingdom established Co-op Energy, an energy supply company selling renewable electricity (and natural gas) to its members. Its parent is

Midcounties Co-operative, the only cooperative supplier in the British market. Since 2016, Octopus Energy has served as the supplier for the co-op.

Philippines

In the Philippines, distribution co-ops serve most of the provinces. The national legislature created the National Electrification Administration (NEA) in 1973 to administer the country's electrification by organizing, financing, and regulating electric cooperatives. In 1979, Presidential Decree (P.D.) No. 1645 broadened the NEA's regulatory powers by giving it the power to conduct investigations and impose sanctions on the boards of regulated co-ops.

In 2001, the Electric Power Industry Reform Act of 2001 (EPIRA) instituted institutional reforms in the electric power industry. EPIRA created the Energy Regulatory Commission (ERC) to regulate the restructured electric power industry, promote competition, encourage market development and choice, and apply sanctions for abuse of market power. The EPIRA considered electric cooperatives as distribution utilities under ERC jurisdiction. Under the EPIRA, the NEA and ERC both have jurisdiction over electric co-ops.

In 2013, the national legislature enacted the National Electrification Administration Reform Act based on three policies: empower and strengthen the NEA, empower and enable electric cooperatives, and promote sustainable development in rural areas by rural electrification.[94] In that law, the Philippines Congress confirmed the NEA's authority to supervise the management and operations of all electric cooperatives.

Electric cooperatives play an important role in the delivery of electricity to end-use consumers – especially in the rural areas investor-owned utilities are reluctant to serve due to low profitability.

Municipal Utilities

In the early days of electricity, the local governments in small cities and towns began establishing their own electric utilities. By the late-1800s and early 1900s, municipalities ran most utilities (>3,000 existed by 1923). Over time, technological advances in generation and transmission made small plants uneconomical, so many municipal utilities sold their facilities and transferred their customers to Investor-Owned Utilities (IOUs).

The American Public Power Association (APPA), the trade association for community-owned electric utilities, explains how "not for profit, community-owned, locally controlled" utilities provide better service at lower rates than investor-owned utilities and also provide revenue for residents. Generally speaking, meetings of governing bodies (often the City Council) are open to the public and televised, and a portion of profits go to the city to be spent as determined by the city council.

Some cities use a progressive rate structure that charges more per kilowatt-hour to customers using larger amounts of electricity. The theory behind this approach is lower-income customers save money while rich(er) customers (with larger houses and more items to power) pay more. Additionally, according to the APPA, publicly owned electric providers are switching to carbon-free sources faster than privately owned utilities.

Investor-Owned Utilities (IOUs)

According to the U.S. Energy Information Administration (EIA), IOUs are large electric distributors that issue stock owned by shareholders. IOUs serve approximately three-fourths of utility customers in the U.S.— mostly in densely populated areas on the East and West coasts. Per the EIA, in 2017, 168 IOUs served an average of 654,600 electric customers. In some countries and Canadian provinces, the government holds much of the stocks in IOUs.

IOUs are typically regulated to protect the interests of consumers. IOUs have a specific service territory granted by a regulator and must serve all customers in that area. No other utility can provide service in the IOU's service territory.[95] The regulatory agency determines the rates an IOU can charge and establishes the rate of return for the utility.

In the U.S., the Public Utility Holding Company Act (PUHCA) prohibited IOUs from crossing state lines, leading to many IOUs throughout the U.S.[96] Congress repealed the PUHCA in 2005. As mentioned previously, in a deregulated market, IOUs typically unbundled their operations so they were no longer *vertically integrated.*

Figure 11-3 shows electric utility types according to the EIA. Figure 11-4 shows the distribution of IOUs, Co-ops, and Municipal Utilities.

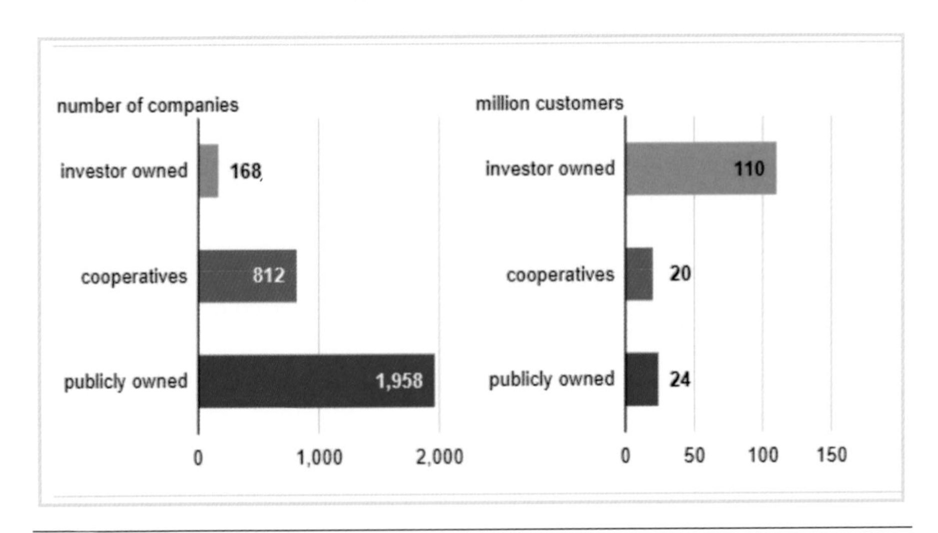

Figure 11-3. Electric Utilities by Type.

Figure 11-4. Counties Served by Utility Type.

Rates and Charges

Finally, this section deals with how the concepts already covered translate into rates and charges. Utilities use several different approaches for setting rates and billing customers. The list below is not exhaustive, but includes the common rate and charge categories:

- Administrative Charge
- Energy (Usage) Charges
  - Electricity generation costs, line losses, other variable costs
  - May be market-based (supply & demand) or regulated (cost plus allowable profit)
  - Typically billed as $/kWh
- Demand Charges
  - Fixed infrastructure cost ("wires charges") (paid even when electricity is not used)
  - May include generation capital costs in regulated areas
  - Typically billed as $/kW based on peak metered load
  - Some states may allow several demand charges (on and off peak, utility and transmission)
- Reliability or Ancillary Charges
  - Cost of various programs to ensure grid reliability
- Taxes
- Miscellaneous and Other

Figure 11-5 is an example of an electric service invoice for an oil pipeline pump station showing various rates and charges.

Sample Invoice 1			
Pipeline Company, LLC		Account #:	1200-001
Bill Start Date: 3/1/2020		Bill End Date:	4/1/2020
Customer Charge:			$50.00
Meter Charge:			$100.00
Billing Demand:			
Distribution Demand Charge:	10,000 kW x	$2.9450 =	$29,450.00
Transmission Demand Charge:	9,500 kW x	$1.9250 =	$18,287.50
Transmission Cost Recovery:	9,500 kW x	$3.8680 =	$36,745.87
Subtotal:			$84,483.37
Energy Charges:			
Retail Adder	5,000,000 kWh x	$0.01000 =	$50,000.00
Fixed Price Retail Supply	4,000,000 kWh x	$0.02975 =	$119,000.00
RTSPP Retail Supply	1,000,000 kWh x	$0.02432 =	$24,320.00
Subtotal:			$193,320.00
Sales Taxes:		0.1500%	$416.71
Previous Balance:			$ -
Total Amount Due:			**$278,370.07**

Figure 11-5. Sample Invoice for an Oil Pipeline Pump Station Showing Various Rates and Charges.

Administrative Charge

Typically, this charge is a flat rate per month and is the same for all customers. It recovers the cost of providing administrative functions including, among other things, billing-related costs, servicing and reading meters, mailing bills, and maintaining customer records. Two administrative charges appear in Figure 11-5: a customer charge of $50.00 and a meter charge of $100.00.

Usage Charge

As the name implies, this charge covers the *amount* (usually in kWh) of electricity consumed during the billing period. Usage rates come in a variety of forms. Some entities apply decreasing cost per kWh to encourage usage; others use increasing cost per kWh to *discourage* usage, while others have varying rates by season or time of day. More about usage rates later in this text. Whatever the rate, the usage charge is simply:

$$\text{kWh Used} \times \text{Applicable Rate} = \text{Usage Charge}$$

The total usage in Figure 11-5 is 5,000,000 kWh, and the total usage charged is $193,320.00, yielding an average rate of $0.0387/kWh for the energy used.

Demand Charges

In addition to the usage charge shown in Figure 11-5, there are three demand charges totaling $84,483.37. Demand charges recover the cost of building generation and T&D resources. Every electrical device in a house or business requires a certain amount of electricity when it operates. Operating them all at once requires constructing more resources than operating them one (or several) at a time. Instead of total usage, demand charges focus on the *maximum* amount of energy required over a time period (usually fifteen minutes) in a month.

Because the demand charge depends on the maximum energy requirement, customers have an incentive to reduce how much electricity they require *at once* in addition to overall usage levels. Demand charges were traditionally used for industrial customers but are starting to apply to residential customers.

Utilities bill customers with a demand charge for monthly usage *plus* demand. However, in comparison to a purely usage-based bill, the focus of the bill shifts from solely *overall consumption* to consumption *and* peak demand. Customers using electricity at higher rates pay higher demand charges to cover capital costs required to build the infrastructure to generate and transmit the electricity.

Figure 11-6 shows an electrical demand profile for a pipeline pump station over a one-month period. In this example, the pump station has two pumps – one runs continuously and the other runs intermittently. Running the second pump eight days during the month doubles the demand charge.

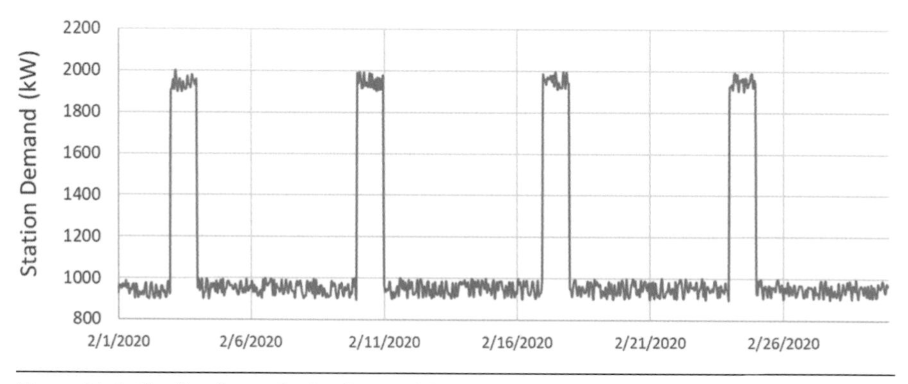

Figure 11-6. Pipeline Pump Station Demand Profile.

Utilities may combine generation, transmission, and distribution demand charges on the invoice or charge them separately, as shown by the three demand charges in Figure 11-7.

Using the physical size of a water pipe as an analogy, a one-inch-pipe provides water to a typical home and delivers sufficient flow and pressure to satisfy normal water demand. A larger pipe (like a three inch) costs more to install but delivers more water at any given moment. When designing a facility's electrical system, the

Sample Invoice #2			
Pipeline Company, LLC		Account #:	1200-001
Bill Start Date: 3/1/2020		Bill End Date:	4/1/2020
Meter Charge:			$1,000.00
Facility Charge:			$15,000.00
Billing Demand:			
Distribution Demand Charge:	10,000 kW x	$6.50 =	$65,000.00
Generation Demand Charge:	350 kW x	$10.00 =	$3,500.00
Substation Demand Charge:	10,000 kW x	$2.00 =	$20,000.00
Subtotal Demand			$88,500.00
Energy Charges:			
Co-op Energy Charges:	6,000,000 kWh x	$0.01000 =	$60,000.00
Pass Through Energy Charges:	6,000,000 kWh x	$0.03000 =	$180,000.00
Subtotal Energy			$240,000.00
Sales Taxes:		0.1500%	$492.75
Previous Balance:			$ -
Total Amount Due:			**$344,992.75**

Figure 11-7. Sample Invoice for an Oil Pipeline Pump Station Showing Various Demand Charges.

designer estimates the peak energy consumption and charges for those periods because the customer *demands* more from the electric grid.

When a customer uses a lot of power *briefly*, its monthly bill contains demand charges based on the peak demand. If the customer uses power consistently through the billing period, its demand charge will make up a smaller part of the bill. Within a particular customer class, if two customers use exactly the same *amount* of electricity but one has a higher demand, the customer with higher demand has higher bills.

Load Management

Large users may have energy management professionals on staff or may hire consultants to help them manage energy loads and costs, including demand costs. Figure 11-8 is an example of an invoice for an oil pipeline pump station where demand charges are $175,000 and energy charges are only $50,360. This is an extreme hypothetical example demonstrating the importance of managing demand charges.

Sample Invoice #3			
Pipeline Company, LLC		Account #:	1200-001
Bill Start Date: 3/1/2020		Bill End Date:	4/1/2020
Customer Charge:			$50.00
Meter Charge:			$100.00
Billing Demand:			
Winter On-Peak Demand	10,000 kW x	$15.00 =	$150,000.00
Winter Off-Peak Demand	5,000 kW x	$5.00 =	$25,000.00
Subtotal Demand			$175,000.00
Energy Charges:			
Winter On-Peak Energy	440,000 kWh x	$0.06500 =	$28,600.00
Winter Off-Peak Energy	1,088,000 kWh x	$0.02000 =	$21,760.00
Subtotal Energy			$50,360.00
Sales Taxes:		6.2500%	$14,085.00
Previous Balance:			$ -
Total Amount Due:			**$239,595.00**

Figure 11-8. Sample Invoice for an Oil Pipeline Pump Station where Demand Charges Exceed Energy Charges.

Coincident Peaks

Electricity demand has peaks and valleys, as shown in Figure 11-9.

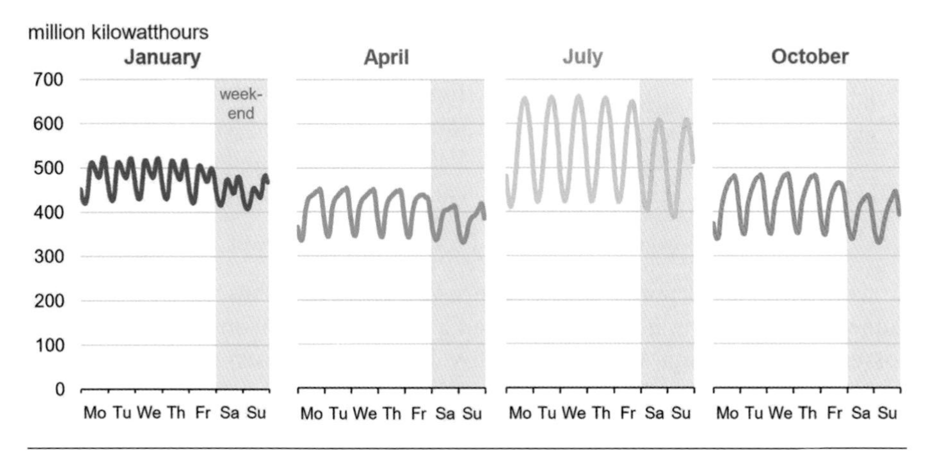

Figure 11-9. Average Weekly Electrical Usage Peaks and Valleys.

As discussed early in this chapter, grid operators select generation resources using a bid process that selects the lowest-price electricity first. Figure 11-9 shows the lowest average cost occurs in April and the highest average cost and largest swings occur in July.

Coincident peaks occur when a user's peak electricity demand coincides with peak demand on the electric grid. Utilities charge for coincident peaks in a variety of ways. Figure 11-8 shows different demand charges for winter – an *off-peak* demand charge of $5.00 per kW and *on-peak* demand charge of $15.00 per kW (three times higher). Rather than charge more during peak times, electrical suppliers sometimes offer financial rewards to users who reduce or even shut down demand during peak times. Some electric providers define four coincident peak periods per year (one for each season) to set the demand peak for the following year, while others use real-time tracking and alert users when the supplier thinks the monthly peak will occur.

Reducing the coincident peak demand charge involves using external information such as weather and the current state of the T&D system to predict the hour the peak will occur and reducing demand during that hour. Increasingly, consumers use software to analyze the internal and external inputs to predict the peak.

Power Factor Charge

Grid Efficiency

Power delivered to loads as related to power delivered to the grid by all generation resources. T&D grid efficiency in the U.S. is typically between 92% and 94% according to the U.S. EIA. Between 6% and 8% of the power delivered to the grid turns into heat caused by impedance and equipment losses.

A motor produces alternating magnetic fields making its rotor turn. In the process, it causes current to lag voltage and generate reactive power, measured in Voltage-Amperage-Reactive (VAR). VARs are needed for grid stability but push against real power, leaving only apparent power and causing grid inefficiencies. Power factors express the relationship between the apparent power (amount of electricity transmitted to the load) and the real power (amount of work performed by the load). Based on apparent and real power, electrical suppliers calculate power factors:

Power Factor = Real Power/Apparent Power

Loads with lower power factors reduce grid efficiency. Thus, based on the customer's power factor, it may be charged a *penalty* to compensate for efficiency losses it caused. Large users of inductive loads (like motors) sometimes install capacitors to help manage VARs and avoid a power factor penalty.

Seasonal Charges

As shown in Figure 11-9, in summer-peaking areas, average electrical usage in July is higher than in January, April, or October. Because average power costs increase as usage increases, many utilities charge seasonal rates to address large differences in load during different times of year. The utility passes on to customers the higher generation costs for the high-use period. Figure 11-8, for example, shows the winter rate.

T&D Charges

In deregulated markets, a retail electric provider (REP) typically splits a customer's bill into *usage* (in kilowatt-hours) and *T&D charges* – the cost of transporting the electricity from point of generation to point of use. The REP calculates bills this way because it must pay an electricity supplier for the actual electricity *and* pay a separate T&D company for transporting the electricity from the generator to the end-use customer.[97] A typical customer's bill may look something like this:

1,000 kWh	
Usage:	1,000 kWh × 5¢/kWh = $50.00
T&D Charge:	1,000 kWh × 3¢/kWh = $30.00
Customer Charge:	$10.00 / billing cycle
Admin Charge:	$3.00 / billing cycle
Total Due:	$93.00

Tiered (or Step) Rates

This type of rate is common for residential service to either encourage or discourage consumption, depending on whether successive tiers have higher or lower rates per kWh. For example, if the utility uses a two-tiered rate structure in which the rate for 500 kWh is 8¢ per kWh and 10¢ per kWh for any amount over 500 kWh and the customer uses 2,300 kWh in a month, the invoice totals $220.00, as calculated below:

500 kWh at 8¢/kWh	$ 40.00
1800 kWh at 10¢/kWh	$180.00
Total Due:	$220.00

Reversing the tier structure costs per kWh results in a total cost of $194.00, as calculated below:

500 kWh at 10¢/kWh	$ 50.00
1800 kWh at 8¢/kWh	$144.00
Total Due:	$194.00

Time of Use (TOU) Rates

Many utilities use TOU rates to maintain or control peak demand by charging customers higher rates during high-use periods to encourage conservation. Utilities generally have peak periods in the morning and evening – for example, from 8:00 a.m. to 10:00 a.m. and between 5:00 p.m. and 8:00 p.m. Some utilities may have several TOU periods during a day based on the capacity to provide electricity during those times. TOU rates are typically described as "off-peak," "mid-peak," "on-peak" and "critical." Pricing for each period is predictable – lowest for off-peak and highest for critical. TOU rates incentivize users of home storage. As an example, a personal acquaintance living near San Diego, CA, installed photovoltaic (PV) cells and a wall of batteries. During the night, when rates are low, he charges the batteries. During the day, he uses electricity generated by his PV cells and discharges his battery wall when rates are high.

TOU rates require special electric meters to accurately track usage during specific times and periodically provide usage data to the utility to allow for accurate billing. Utilities usually allow customers to track usage through a computer or smart phone application providing real-time usage data so customers can adjust usage accordingly.

Fuel Pass-Through Costs

Some utilities include a separate *fuel pass-through* charge based on their actual fuel cost compared to the fuel cost included in the approved rate. The prices of fuels to generate electricity can fluctuate dramatically, moving up and down based on market conditions. Utilities adjust the fuel pass-through charge periodically based on the utility's cost to procure the fuel during each time period.

While many consumers think of their REP as their "power company," the REP often buys electricity from utility-scale generators in the wholesale market because producing electricity is much more efficient and affordable when done on a large scale. Therefore, the electricity many end users consume is generated by a large generation company (separate from the REP). After purchasing electricity from a wholesaler, the REP either transmits the electricity across its own distribution system to deliver it to its customers or pays a T&D utility to transmit and distribute the electricity. Most wholesale contracts are cost-based to guarantee a reliable source of electricity and include a fuel cost adjustment that the REP passes on to its customers without markup.[98]

Weekend/Holiday Rates

Electricity usage for some customers (especially commercial and industrial customers) changes significantly on weekends and holidays. Some REPs charge lower weekend rates because their cost to purchase electricity is lower. Other REPS may offer "free" weekend electricity to attract customers. They generally offset the "free" electricity with higher rates during the week.

Miscellaneous Rates

Rate categories are limited only by the creativity of those creating them, meaning a variety of other rates and charges exist, including:

- Customer connection charges
- Rate riders for energy efficiency programs
- Administration fees
- Debt retirements charges
- Economic development (typically expire after a certain time to promote new development)
- Dual feed (charges associated with providing electricity from two service lines to help ensure continued service; often used for hospitals and other critical facilities)

Summary

- Energy markets can be regulated or deregulated.
- Deregulated markets typically involve unbundling vertically integrated utilities.
- Most deregulated electric markets use an auction process to determine the clearing price for electricity, often using locational marginal pricing (LMP).
- Market operators use ancillary services to support the reliability of the power grid.
- Most deregulated markets include a capacity market.
- Most deregulated electric markets use four main pricing components: energy, capacity, ancillary services, and transmission congestion/losses.
- Deregulated electric markets typically use auctions for day-ahead pricing and real-time pricing and bilateral contracts for direct transactions between generators and customers.
- The participants in electric markets include: retail electric providers; T&D companies; electricity generators; governmental regulatory bodies; grid/market operator; vertically integrated utilities; and co-ops/public utility districts.
- In most deregulated electric markets, grid operators pay market participants based on locational marginal pricing.
- Electric financial markets allow wholesale market participants to hedge their electric price risk by shifting it to other market participants, one of which is speculators who accept that price risk for financial profit and put themselves at risk for financial loss.

- Ancillary Services are products provided by generators or Load Resources to support the reliability and security of the transmission system and include products such as Regulation Service, Responsive Reserve Service, NonSpinning Reserve Service, and Demand Response.
- Electric markets consist of wholesale and retail markets.
- Electric providers can be vertically integrated or unbundled utilities, which can include co-ops, municipal utilities, and investor-owned utilities.
- Consumers can purchase electricity through many different types of contracts, using one or more of the following rates: general service; demand charge; large general service; seasonal charges; T&D charges; tiered (or step) rates; time of use rates; fuel pass-through; weekend/holiday rates; and other miscellaneous rates.

12

Challenges for the Future

We have met the enemy and he is us.

—*Pogo* by Walt Kelly (1913–1973)

"Grandpa Tom, today our teacher told us about clean energy, and I wonder why the whole world does not just switch to clean energy," Luke said to his grandfather.

"Well, Luke, it is easy to say 'clean energy,' but people should really say, 'cleaner energy' instead," replied the grandfather.

"What do you mean? Electricity is clean!" exclaimed the boy.

Grandpa Tom responded, "Do you remember when we drove to Marfa to see the lights and we saw lots of wind turbines? When we went to see one of the wind turbines up close, it was loud, wasn't it? Can you imagine what it would be like to live near one of those and hear that noise day in and day out? And, do you remember that rancher we talked with who didn't like the transmission towers ruining his view of sunset on the prairie? Finally, do you remember those dead birds we saw on the ground around the wind turbine?"

"That's right, and I remember when you had difficulty finding a place to recycle your batteries when you had to replace them in the Escape®," Luke responded. "I also remember my teacher telling me about mining stuff for batteries and how it destroys the land and pollutes the water," Luke added.

"We all want cleaner energy and lots of it," mused the grandfather. "But until we figure out how to produce, transport, and use truly clean energy, we need to make a lot of choices involving tradeoffs," Grandpa Tom concluded.

Introduction

Public policy, economics, and technological advances drive rapid changes to the bulk power system (BPS), comprised of generation resources and transmission facilities. While demand forecasts vary, they all show electrical demand rising into the foreseeable future, requiring new generation, transmission, distribution, and storage resources. Two forces drive the rising electricity demand are:

- increasing electrical usage from traditional electrical loads – residential, commercial, and industrial customers; and
- new demand for "cleaner" electricity to replace carbon-based fuels such as gasoline and diesel fuel.

Supplying traditional loads with increased electricity and supplying the electrical power required to replace carbon-based-fueled vehicles with electrical vehicles (EVs) while, at the same time, replacing higher-capacity-factor carbon- and nuclear-based generation resources with lower-capacity-factor generation resources (wind and solar) that require storage, and building reliable and resilient transmission lines and distribution systems has created a "perfect storm" for the global electrical power industry.

Satisfying Stakeholders

Against the backdrop of this "perfect storm" the challenge of the electrical power industry remains the same as that of any industry – balancing the disparate and often conflicting needs and wants of its stakeholders. Figure 12-1 shows the electrical power industry surrounded by some of its many stakeholders.

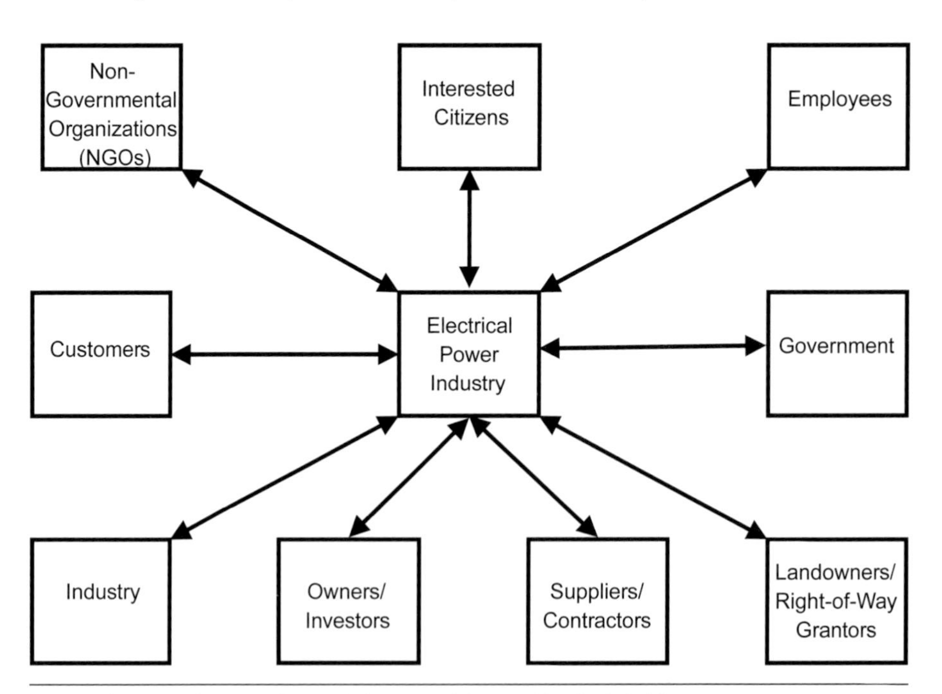

Figure 12-1. The Electrical Power Industry and Some of Its Stakeholders.

Growing Demand

The stakeholder group most people think of first is customers. Figure 12-2 shows forecasted electricity demand growth from 2019 to 2030. The trend lines on Figure 12-2, rise quickly, and the customers demanding this electricity expect it will be delivered to them. As an example of disparate expectations, some stakeholders want reliable, inexpensive, and responsibly generated, transmitted, and distributed electricity. Other stakeholders – like those who do not yet have electricity – just want it.

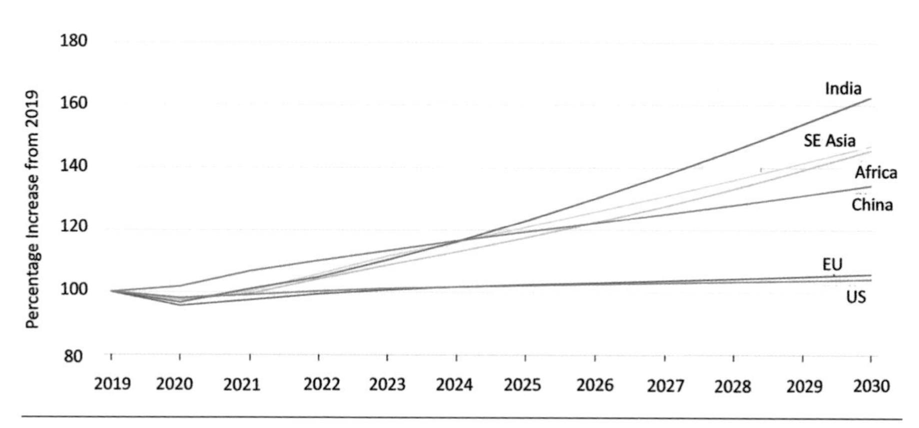

Figure 12-2. Electrical Energy Growth for Selected Countries and Regions.

New Infrastructure

The increasing demand for electricity to satisfy only existing customers will require massive infrastructure investments in generation, transmission, and distribution facilities. The switch from carbon-based energy to "cleaner" energy, much of which is forecast to be electricity, will increase the required investments even more. On top of those investments is the investment required for storing electricity due to the intermittent nature of renewable generation resources.

Generation

Figure 12-3 shows the percentage of primary sources currently producing electricity on a global basis. Adding the gray line (coal) to the red line (natural gas) totals 60%. Comparing that number to the orange line (renewables), currently 10%, means renewable generation capacity must expand by a factor of six to merely satisfy current (pun intended) electricity demand.

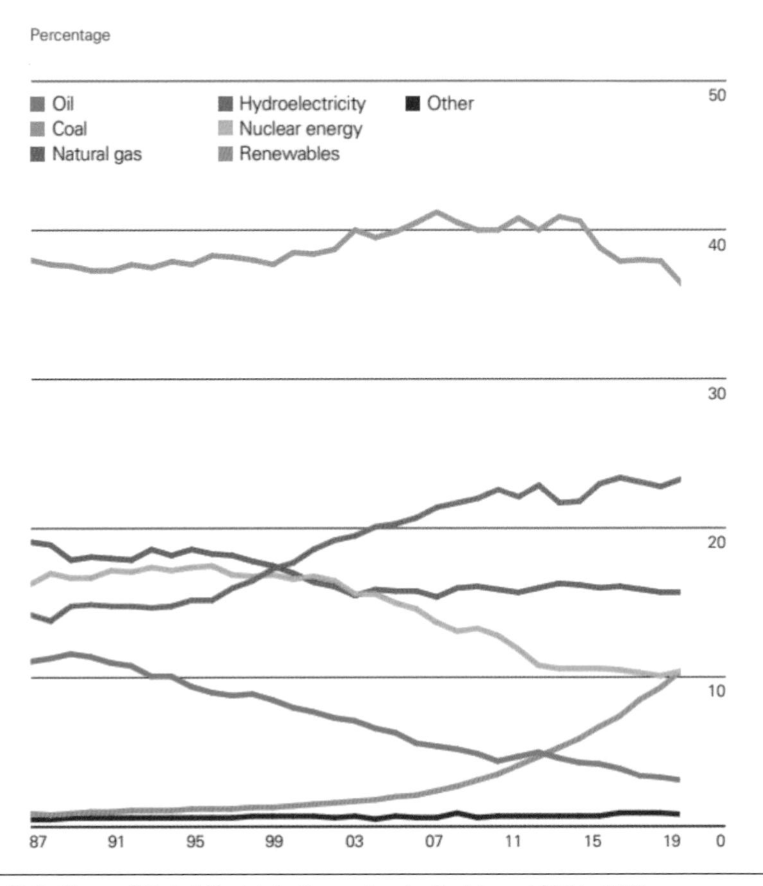

Figure 12-3. Share of Global Electricity Generation by Fuel from 1987 to 2019.

Transmission

Wind and solar farms occupy more area than equivalent utility-scale coal, natural gas, and nuclear plants, and are located in different parts of the country than current generation resources, requiring new transmission lines and substations, and expansions to existing transmission lines and substations. This text does not attempt to quantify the miles of new transmission that will be required across the globe – from a qualitative standpoint, that number is large.

Distribution

Electricity growth per capita will grow faster in less developed countries than in more developed countries, but both will require increased distribution systems to meet the needs of current customers. The switch from liquid fueled vehicles to EVs will require an extensive fleet of recharge stations, nearly all of which will be connected to distribution grids, necessitating new and expanded distribution lines.

Investment

Owners and investors, whether from the private or public sectors, are another key stakeholder. Private investors will expect reasonable returns and capital security, and governments investing in infrastructure will need to find ways to secure and repay that capital. This text does not provide solutions for the amounts of investment required but does raise it as an issue.

Environmental Performance

The environmental community, including interested citizens, landowners, and regulators, will expect due care is paid in constructing and operating this new infrastructure. The electrical power industry will need to dedicate resources to informing the public, including dialoging with them about this new infrastructure, and how it can be constructed and operated responsibly.

Aging Infrastructure

In addition to adding new infrastructure, the electrical power industry must maintain its existing, aging infrastructure. This section uses the United States as a case study of the aging infrastructure challenge.

The U.S. has more than 600,000 miles of electric transmission lines (240,000 of which consists of high-voltage lines) and an estimated 5.5 million miles of distribution lines.[99] According to the American Society of Civil Engineers, the U.S. faces an investment gap (in generation, transmission, and distribution) of $208 billion by 2029 and $338 billion by 2039.[100] That investment gap causes electricity interruptions due to equipment failures, capacity blackouts, and power quality irregularities of differing frequency and duration. Ultimately, system failures create an unreliable electricity supply, affecting homes and businesses.

BPS Reliability

The reliability of the legacy BPS is declining and will continue to do so. Electric grid reliability consists of two parts: *adequacy* and *security*. *Adequacy* (or *reliability*) means the system can, at all times, supply total demand (taking into account planned and unplanned facility outages).

Security (or *resiliency*) means the grid's ability to withstand *unexpected* disturbances like short circuits or unforeseen facility loss. The electric industry tracks reliability by the frequency, duration, and magnitude of adverse effects

SAIDI and SAIFI

SAIDI describes the total *duration* of interruptions for the average customer across the system over a time period (*e.g.,* a month or year), commonly measured in minutes. Mathematically, it is the number of customer-minutes of interruption divided by total customers on the system.

SAIFI describes how *often* the average customer experiences a sustained interruption over a time period (typically a year), by dividing the total number of customers interrupted by the total number of customers served.

on customers, relying primarily on two measures of reliability: System Average Interruption Duration Index (SAIDI) and System Average Interruption Frequency Index (SAIFI).

A significant portion of the U.S. electric grid was built in the 1950s and 1960s, with a life expectancy of fifty years.[101] In the Executive Summary of a report, *Keeping the Lights On in a New World,* the U.S. Department of Energy (DOE) stated, "the current electric power delivery system infrastructure...will be unable to ensure a reliable, cost-effective, secure, and environmentally sustainable supply of energy for the next two decades."[102] It continued, "much of the electricity supply and delivery infrastructure is nearing the end of its useful life."[103]

Utilities in the U.S. are, for the most part, not investing in new equipment sufficiently to replace deteriorating infrastructure because:

- growth of electricity consumption has slowed (mostly due to demand-side management);
- regulatory barriers due to environmental and sustainability concerns make siting and building infrastructure more difficult;
- the increasing cost of equipment causes higher rates in a competitive market; and
- significant risk of not recovering costs plus a return on investment exists.[104]

Permitting

Interested citizens, landowners, and regulators are other stakeholders in the electrical power industry. Difficulties in siting new infrastructure make updating the electric grid more challenging due to property owners' attitude of "not in my backyard" (NIMBY), which has become more prevalent in recent years. Many special interest groups or local committees oppose construction of electric infrastructure projects near their homes or businesses, even though the improvements typically benefit the public. One study found a significant increase in *NIMBYism* between 2007 and 2017.[105]

The Federal Energy Regulatory Commission (FERC) regulates transmission systems at the federal level for the U.S. However, oversight and approvals for specific projects still reside with state governments, which means each state must approve its portion of proposed transmission lines crossing multiple state lines.

Federal lands require additional permits from the federal agency with oversight. A recent example of the complexity such projects involves the Northern Pass proposal to bring 1,090 MWs of hydroelectric power from Canada to Massachusetts (through New Hampshire).[106] First, the U.S. Department of Energy (DOE) had to approve importing the electricity. Then, because proposed routes crossed national forest and state park lands in New Hampshire, the U.S. Department of Interior, which manages national forests, approved the plan.

Residents and conservation groups, however, raised concerns about the environmental impact.[107] To try to accommodate concerns, the utilities suggested downgrading the voltage and burying some of the transmission line. The project also involved eminent domain issues (when the line crossed private lands). A robust campaign in New Hampshire led the state's Site Evaluation Committee to not approve the project. Ultimately, the project was abandoned after ten years and $300 million in expenses.[108]

Clearly, regulatory and social hurdles, coupled with the huge cost of infrastructure projects, remain an issue for the electric industry. As a result, it becomes more and more difficult for utilities to keep up with the strains of new demands on the BPS.

Increasing Renewable Resources

At the end of the twentieth century, the electrical power industry in the U.S. generated nearly all electricity at some ten thousand utility-scale electric generation facilities. Those plants converted primary energy (largely in fossil fuels) into electricity on a fairly uniform and consistent basis. Consumption (i.e., load), however, varied during the day and across seasons. Nearly all generation facilities were *dispatchable* and not subject to an intermittent fuel source, meaning the operator could bring them on line or take them off line to match demand and keep the grid stable.

In the twenty-first century, the electric industry evolved away from legacy, dispatchable fossil fuel generation facilities towards numerous, dispersed generation facilities, most of which convert primary energy in *sunlight* and *wind* into electricity. Solar facilities can produce electrical power only when the sun shines, and wind turbines only when the wind blows. Also, the rate (kilowatts) of electricity production by solar and wind depends on how bright the sun shines or how hard the wind blows.[109]

Significant renewable generation installations also create challenges to operating a grid originally designed for large baseload power plants. To increase the output and predictability of weather-dependent, intermittent wind and solar generators, utilities must invest in advanced weather forecasting tools that integrate load data, generation, and cutting-edge weather prediction. This work demonstrates the kind of advanced tools needed to ensure the lights stay on as more intermittent resources connect to the grid.

In addition to the challenge of intermittency, transmission grid operators do not have information on the *capability* of a lot of installed solar generation. Unlike large solar farms, a lot of solar generation connects at the distribution level and, therefore, remains "invisible" to grid operators in real time because it sits *behind the meter*. Grid operators can monitor only *net* load at the meter. Without real-time data on behind-the-meter generation, operators cannot know what will happen in a disturbance. If generation suddenly trips offline, contingency analysis simulations may under-predict resulting load levels. This has resulted in recent changes to interconnection standards requiring updated *ride-through* and voltage control capabilities, even for smaller, inverter-based installations (like solar), to keep those resources online through minor grid fluctuations.

Generation "Ride Through"

Owners of generation facilities protect their equipment from damage by calibrating their protection systems to automatically disconnect from the grid during system disturbances. "Ride through" refers to allowing the generator to remain connected to the grid during small or transient voltage or frequency fluctuations.

Another challenge for grid operators stemming from the proliferation of solar projects is the changing load demand curve. Increases in renewable generation have drastically affected the timing of daily load peaks. Historically, load peaked in the mid-to-late afternoon with the commercial energy demand peak. Now, in locations with significant solar generation penetration, load peaks after sundown. This situation happens most intensely on sunny summer days, as shown in Figure 12-4. Loads traditionally ramped up in the morning, peaked in mid-afternoon, and dropped gradually; now, on a sunny day, midday loads drop significantly due to generation from solar panels and rise to a daily peak after the sun sets. The load will also fluctuate during the day depending on cloud cover.

The dramatic ramp in demand at sunset challenges grid operators to quickly increase electricity production to replace the generation previously supplied by solar panels. The challenge is to ensure enough generation to meet demand without being overly conservative and paying for unused generation. Figure 12-1 also shows the importance of predicting the daily load based on the weather forecast.

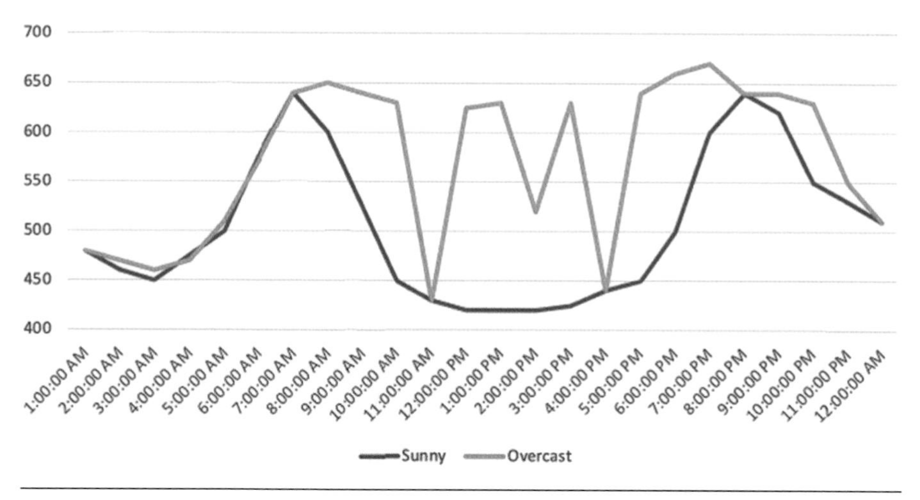

Figure 12-4. Demand Curve on Summer Sunny vs. Spotty Overcast Days.

Dispatchable vs. Nondispatchable Generation

Electricity consumption in the twenty-first century continues to vary during the day and across seasons. At the same time, the world strives to change the energy mix to address climate goals. For example, to meet the world's climate change goals, the amount of wind and solar generation must increase to 85% by 2050.[110] Connecting renewable energy sources (RES) to the grid is not as simple as it may seem, and their effectiveness depends entirely on weather conditions. As a result, Reliability Coordinators and Balancing Authorities (those responsible for planning and managing the electric grid) consider RES *unstable* and their operation, without an advanced management system, can cause serious grid imbalances. Figure 12-5 depicts dispatchable and nondispatchable generation in relation to the grid and end-use consumers.

> **Dispatchable Generation**
> Generation for which the precise output can be controlled with certainty that, barring unexpected mechanical or electrical failure, will produce an expected amount.

Grid stability still depends on balancing inputs into the grid with withdrawals from it. The nondispatchable nature of renewable generation points to the need for vast amounts of electric storage to balance grid input with grid output.

Figure 12-6 from the U.S. Energy Information Administration (EIA) shows total energy (not just electrical energy) consumed in the U.S. during 2020. Replacing the petroleum, natural gas, and coal (which in 2020 accounted for 78% of U.S.

consumption) with energy generated from RES means renewables must increase by a factor of 6.5 times if energy consumption remains constant – a staggering thought.

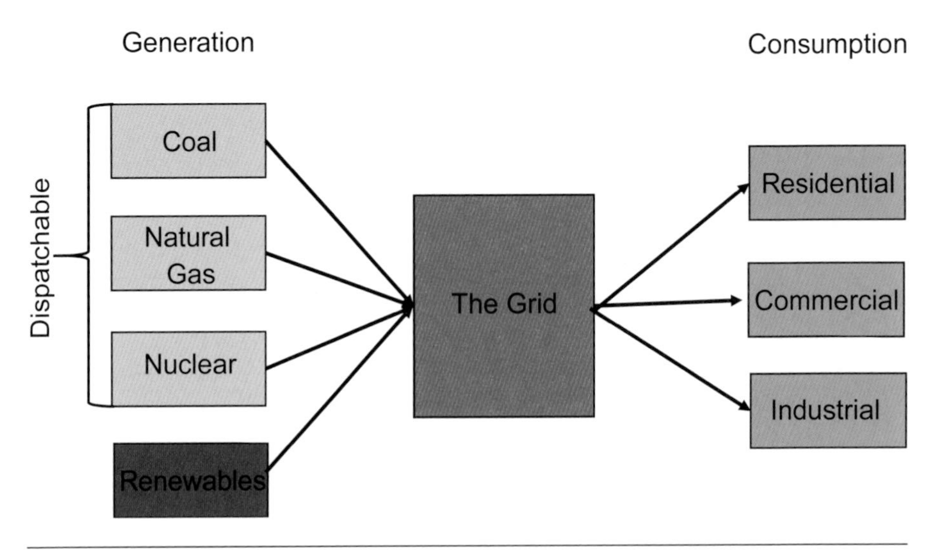

Figure 12-5. Dispatchable vs. Nondispatchable Generation.

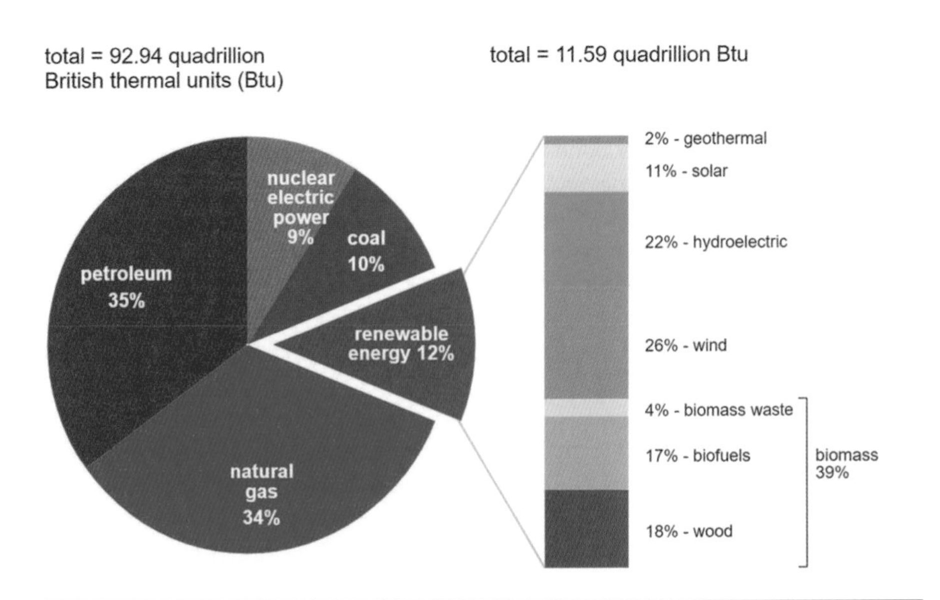

Figure 12-6. U.S. Primary Energy Consumption by Energy Source, 2020

Taking a closer look at Figure 12-6, solar and wind make up a bit more than one-third of the 12% renewables in the U.S. If all the electricity consumed in the U.S. came from wind and solar, those facilities would have to increase by more than 13 times. Converting that number to a percent means wind and solar facilities must increase by more than 1,300%.

Microgrids

Microgrids are a group of interconnected loads and distributed energy resources (DERs) in a clearly defined electrical boundary, acting as a single controllable entity with respect to the traditional (macro) grid. Most microgrids can connect and disconnect from the grid to operate in *connected* or *island* mode.

Macrogrids connect homes, businesses, and other buildings to central power generation sources. That interconnectedness means that when part of the macrogrid goes down, it affects everyone connected to that grid. Microgrids can help address that situation by operating separately when needed.

Distributed generators, batteries, and/or renewable resources like solar panels can power a microgrid. Depending on the fuel and how it is managed, microgrids may run indefinitely in island mode. A microgrid connects to the macrogrid and maintains voltage at the same level as the macrogrid unless a problem occurs or some other reason to disconnect arises. A switch typically separates the microgrid from the macrogrid automatically or manually. Thereafter, the microgrid functions as an *island*, taking the place of the traditional grid during emergency conditions and connecting to a local resource too small or unreliable for the traditional grid.

Microgrids provide consumer(s) energy independence. They can also serve as an electricity source for areas too far from the traditional grid or for which it would cost too much to connect to the traditional grid. Microgrids, however, face the difficulty of addressing control and protection because all ancillary services to stabilize the system must come from within the microgrid's infrastructure. Key features of microgrids include:

- operation in island mode or connected to the macrogrid;
- appearance to the macrogrid as a single controlled entity;
- combination of interconnected loads and electricity generation sources;
- provision of varied levels of power quality and reliability for consumers; and
- ability to accommodate total system energy requirements.

Figure 12-7 depicts a sample microgrid.

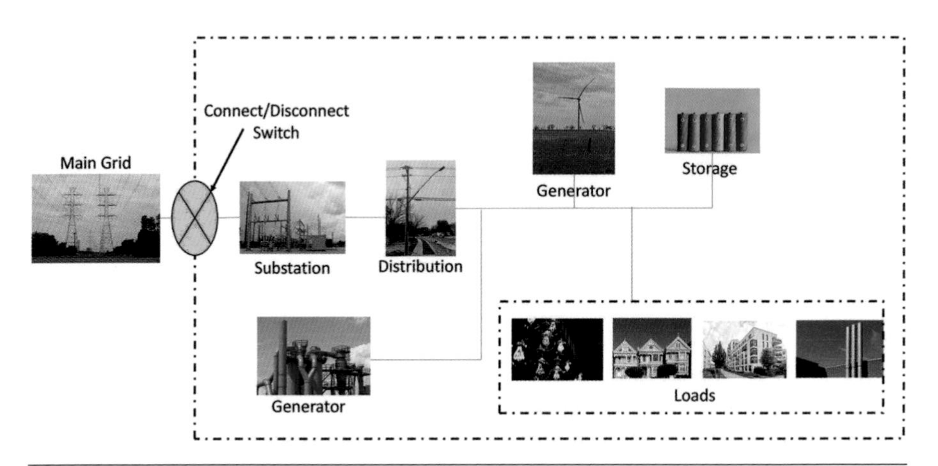

Figure 12-7. Microgrid.

Microgrid Types

Several types of microgrids exist to serve specific communities or customers.

Campus/Institutional Microgrids

This type of microgrid aggregates existing onsite generation to support numerous loads in a small geographical area by combining onsite generation with several load centers co-located in a campus or institutional setting (e.g., an industrial park).[111] The most common size of this type of microgrid is from 4 MW to 40+ MW.

Community Microgrids

Community microgrids serve many customers through, for example, houses with renewable sources to meet their demand (e.g., rooftop solar panels) plus those of their neighbors in the same community. The community microgrid may also have energy storage connected to the renewable sources on either the AC or DC side of the power inverter (a feature not specific to microgrids).

Remote Off-grid Microgrids

These microgrids never connect to the macrogrid and always operate in island mode due to economics or geographic location. They are usually placed in areas far from transmission and distribution infrastructure. Several independent microgrids may supply large remote areas. Although these microgrids are designed for self-sufficiency, intermittent renewable generation and its unpredictability can cause electricity shortfalls, leading to unacceptable voltage or frequency deviations. To remedy that situation, operators may connect the microgrid to a neighboring microgrid to exchange electricity and improve voltage and frequency control.

Military Base Microgrids

This type of microgrid focuses on physical and cybersecurity for military facilities to provide reliable power without needing the macrogrid.

Commercial and Industrial Microgrids

These microgrids are becoming more popular for electricity supply security and reliability due to the fact that many manufacturing processes cannot easily withstand an interruption in electricity supply because it would cause significant revenue loss or long restart time (e.g., a large computer chip fabrication plant). Industrial microgrids often include combined heat and power generation supplied by waste processing and can include storage capability.

Advantages

The primary advantage of microgrids is they can operate as a single collective load and customers benefit from power quality and enhanced reliability versus relying on only the macrogrid. Distributed power production using smaller generating systems (like combined heat and power or small-scale renewables) can produce energy efficiency and environmental advantages over large, central generation stations. Additionally, the microgrid concept focuses on creating a design for local power delivery meeting exact customer needs by efficiently and economically integrating customers with distribution and generation at a local level.[112]

Microgrids can also improve reliability due to redundant distribution, smart switches, automation, local power generation, and the ability to island from the macrogrid, thus either eliminating or greatly minimizing blackouts and disturbances. A microgrid's improved reliability can reduce costs due to power outages, brownouts, and poor power quality as well as generate revenue by allowing the sale of excess electricity to the macrogrid. Microgrids can also supply ancillary services such as demand response, real-time price response, day-ahead price response, voltage support, capacity support, spinning reserves, etc.

Disadvantages

Because of the nascent status of technologies to implement microgrids, some disadvantages exist. For example, technology challenges limit economies of scale. Also, electricity storage types and capabilities serve as a weak link for microgrid operations because their limited duration means they cannot serve throughout a lengthy emergency. Due to the early state of microgrids, we have limited information on the true total cost of operating them. A 2011 MIT study indicated that, of the 160 microgrid projects encompassing 1.2 GW of installed distributed generation (at that time), the majority were demonstrations and research pilots. Consequently, the study considered microgrids expensive because they require new, advanced electronics and sophisticated coordination among different customers or areas still in their infancy. The MIT study stated (among other things):

> It is our sense that in most situations, the cost of configuring an area as a microgrid does not justify the reliability benefits, which may be achieved through other means, such as backup generators. Despite the challenges, microgrids have the potential to bring new control flexibility to the distribution system and thus will continue to receive much academic interest.[113]

Energy Storage

Solving the grid stability issues associated with RES requires developing dispatchable energy storage systems – discussed in Chapter 8. These storage systems receive electricity from nondispatchable wind and solar, store it as chemical energy, and convert it back to electrical energy when needed. While there are several storage solutions in development, most grid-scale storage is currently battery storage. According to the EIA, large-scale battery (electrochemical) storage capacity in the U.S. grew from fifty-nine megawatts (MWs) in 2010 to 869 MWs at the end of 2018.[114]

Electrochemical Storage Safety

One of the biggest issues facing energy storage is safety. One incident in 2019, at the McMicken Battery Storage facility in Arizona, sent nine first responders to the hospital and destroyed the facility. Late in the afternoon, the fire department received reports of smoke from the building housing that energy storage system. Hazardous Material units and first responders arrived shortly thereafter. Several hours after receiving the reports of smoke, firefighters opened a door to the facility, and the site suffered a catastrophic failure. In July 2020, Arizona Public Service issued a report. As it turns out, the event was not a fire but an extensive *cascading thermal runaway* event caused by an internal cell failure. The fire suppression system operated correctly but was designed to put out fires in ordinary combustible materials and could not prevent or stop cascading thermal runaway. As a result, thermal runaway continued through every cell and module in one rack of the facility due to heat transfer and the lack of sufficient thermal barriers between battery cells. The uncontrolled cascading produced a large amount of flammable gases sufficient to create a flammable atmosphere in the container. When firefighters opened the door, the flammable gases contacted a heat source or spark and exploded.[115]

Additionally, South Korea has experienced more than thirty fires and explosion incidents at battery storage facilities.[116] Consequently, authorities revised regulations in an attempt to improve safety. Nonetheless, these events create a considerable amount of concern regarding the safety of such facilities.

Electrochemical Storage Capacity

The electric industry measures electrochemical storage in two ways: *power capacity* and *energy capacity*. Power capacity is the maximum amount of output possible *in any instant* (measured as MWs), determined by the inverter size. Electrochemical sources, by design, can maintain power output for only a period of time before needing to recharge. The *energy capacity* of an electrochemical storage system equals the total amount of energy it can store or discharge *for a particular duration,* measured as megawatt-hours (MWh).

Figure 12-8 depicts how a storage facility can complement renewable energy resources.

Battery Duration

A battery's *duration* is the length of time it can maintain power output at a particular discharge rate (usually expressed in hours).

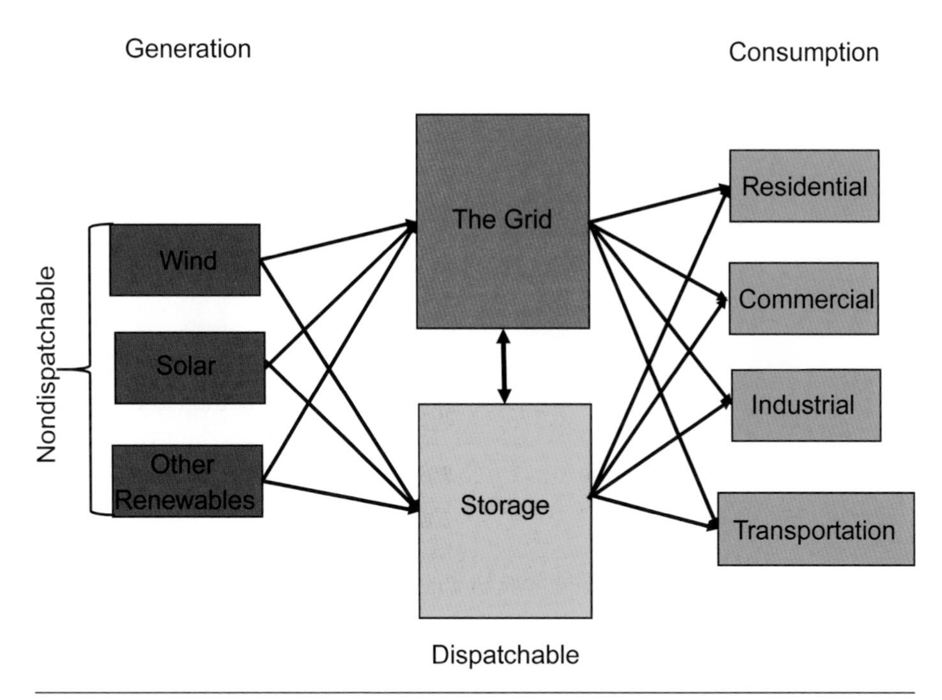

Figure 12- 8. Storage Facility Supplementing a Grid with Nondispatchable Generation Resources.

Storage Installations in the U.S.

Hydroelectric power (mechanical energy storage) accounts for most (97%) of the large-scale energy storage capacity in the U.S. But, *new* large-scale[117] energy storage resources since 2003 have been almost entirely electrochemical.[118] The first large-scale electrochemical storage installation in the U.S. still in service in 2018 went on line in 2003. Between 2011 and 2018, companies installed 810 MW of power capacity from large-scale electrochemical storage. Approximately 73% of large-scale electrochemical storage power capacity and 70% of energy capacity in the U.S. is installed in areas covered by independent system operators (ISOs) or regional transmission organizations (RTOs).[119]

Electrochemical Storage Installations in Other Countries

Other parts of the world have also seen increases in large-scale electrochemical facilities. Figure 12-9 depicts global investment in grid-scale electrochemical storage.

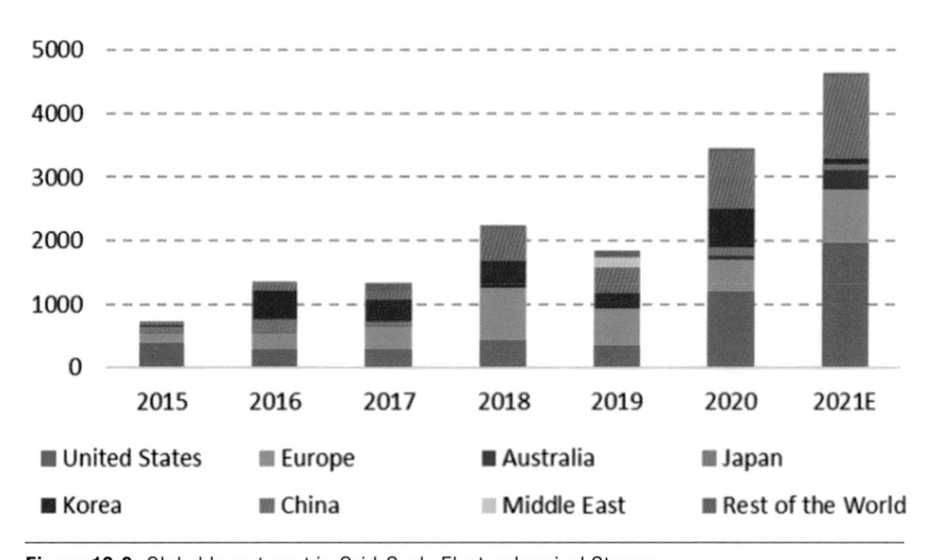

Figure 12-9. Global Investment in Grid-Scale Electrochemical Storage.

The following are some examples of large-scale electrochemical facilities around the world.

United Kingdom: In June 2021, the United Kingdom saw its first grid-scale electrochemical storage system directly connected to the electricity grid in Oxford, England.[120]The system is part of a project integrating energy storage, electric vehicle (EV) charging, low-carbon heating, and smart energy management technologies.

The U.K. also has a 50 MW electrochemical storage facility (with a 50 MWh capacity) consisting of 150,000 lithium-ion battery cells in Hertfordshire, England. The facility stores excess power from solar panels.[121] In Glassenbury, Kent, and Cleator, Cumbria, the U.K. has electrochemical facilities of 40 MW and 10 MW, respectively.[122]

Saudi Arabia: Claiming to be the world's largest, an electrochemical storage facility will begin construction in Saudi Arabia as part of a 28,000-square-kilometer development with fifty hotels powered entirely by wind and solar energy. The complex will rely on a 1,000 MWh electrochemical storage facility when the wind and solar generation do not generate power.[123]

Australia: The State of Victoria recently commissioned a 300 MW electrochemical facility near Geelong with a capacity of 450 MWh. The facility puts electricity onto the grid when it becomes unstable (to help prevent blackouts).[124]

Previously, Australia's largest electrochemical storage power station was the Hornsdale Power Reserve near a wind farm,[125] which provides a total of 189 MWh of storage capable of discharging at 150 MW into the power grid and provides grid stability.[126] If the wind suddenly decreases or other network issues arise, the system operator can rely on the stored energy to bridge the gap until starting slower, conventional generators.

The state of New South Wales (NSW) has a twenty-year road map for its electricity infrastructure to include energy storage such as pumped hydro and on-demand supply, including gas and electrochemical storage. The government hopes to add 2 GW of stored energy and plans to hold auctions to provide companies a guaranteed minimum floor price for the energy produced on land allocated by the plan.[127]

Canada: In Ontario, electrochemical storage with 53 MWh capacity and 13 MW of power went into service in 2016. The Independent Electricity System Operator (IESO) awarded the contract for the facility, which uses the energy storage to provide fast grid services primarily for voltage and reactive power control.[128]

Japan: Japan has two large-scale electrochemical storage facilities— Buzen Substation and at the Rokkasho Wind Development project. The Buzen project uses sodium-sulphur technology and has 50 MW output and 300 MWh capacity and is part of a pilot project to balance supply and demand using energy storage systems.[129]

The Rokkasho Wind Development project also uses sodium sulphur and has an output of 34 MW and a capacity of 245 MWh. The facility charges the electrochemical storage at night when demand is low and uses the stored electricity along with electricity from windmills during the day to provide a steady supply of electricity to the grid even when power production drops due to low wind speed.[130]

South Korea: South Korea has 56 MW of energy storage capacity consisting of two lithium-nickel-manganese-cobalt oxide energy storage systems (a 24 MW system with 9 MWh capacity and a 16 MW system with 6 MWh capacity) and a 16 MW/5 MWh lithium titanate oxide system. It uses the storage for frequency regulation of the grid.[131]

Germany: Since 2016, Germany has had an electrochemical storage unit in the town of Lünen with a total capacity of 13 MWh that uses second-life batteries from electric vehicles. The owner markets the output in the German electricity balancing sector.[132]

In Schwerin, the electricity supplier operates electrochemical storage consisting of 25,600 lithium manganese cells with 5 MWh of capacity and an output of 5 MW to offset short-term power fluctuations.[133]

In Braderup (Schleswig-Holstein, Germany), they have a system consisting of lithium-ion electrochemical storage (2 MW power and 2 MWh capacity) and a vanadium flow electrochemical storage facility (330 kW power and 1 MWh storage capacity). The system connects to the local community wind park (which has 18 MW of installed capacity).[134]

Portugal: Graciosa Island has a 3.2 MWh lithium-ion storage facility. The island uses a 1 MW solar energy plant and a 4.5 MW wind farm for almost all its electricity needs. The former diesel power plant serves as a backup system when the solar and wind plants cannot satisfy customer needs.[135]

Overall, electrochemical storage facilities continue to proliferate at a brisk pace, and regulated and unregulated utilities strive to integrate intermittent, non-dispatchable generation resources into the legacy electric grid.

Electric Vehicles

Electric vehicles (EVs) are still relatively uncommon but continue growing in popularity. A recent Bloomberg New Energy Finance study shows EVs currently make up 3% of car sales worldwide (5.6 million EVs). The study projects that, by 2025, EVs will reach 10% of passenger vehicle sales and continue growing (to 28% in 2030 and 58% in 2040).[136] The owner of an EV with a daily commute of 25 miles (40km) needs approximately 6–8 kWh of electricity to recharge the battery (about one-third of the electricity currently used in the average U.S. home in a day).[137]

Figure 12-10 shows actual and forecast global energy usage by primary fuel from 2010 to 2050. The brown area – petroleum and other liquids – are primarily used by vehicles. Converting the entire liquid fuel vehicle to EVs would result in a massive increase in electrical demand.

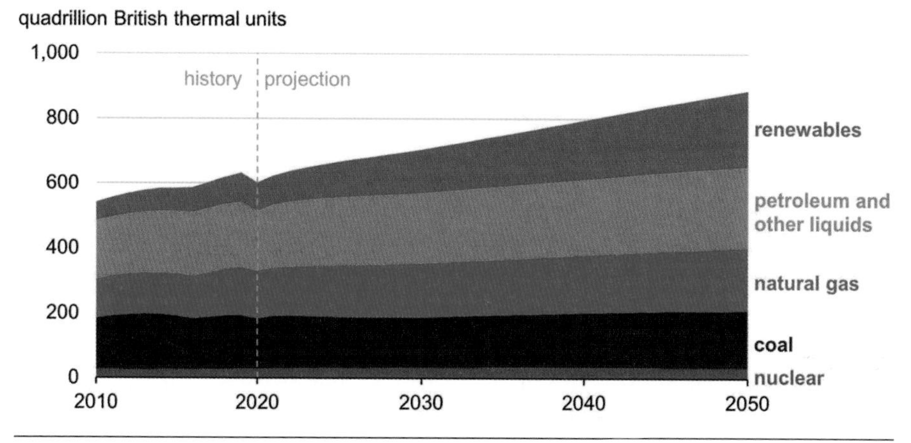

Figure 12-10. Global Primary Energy Consumption by Energy Source.

EV Charging

Drivers are used to adding from 350 to 550 miles of fuel to their cars in less than ten minutes. EV charging is different. There are three levels of EV chargers:

- Level 1 – operates at 120V and adds three to five miles of range per hour
- Level 2 – operates at 208 to 240V and adds twelve to twenty miles of range per hour
- Level 3 – operates at 400 to 900V and adds three to twenty miles of range per minute[138]

Level 2 chargers are the most popular for home use as they balance charging time with cost. Level 3 chargers, sometimes called *fast chargers*, consume significant amounts of electricity but allow drivers to recharge in time frames closer to internal combustion engines. That surge in electricity demand can create problems for the stability of the electric grid.

An EV with a 60 kWh battery that charges 80% in 6 hours (using a *Level 2* charger as opposed to a *DC fast charger* described previously), consumes 8 kW/hour.[139] Thus, the habits and daily schedules of EV owners will likely cause peak demand at certain times of day (presumably, most EV owners will charge their vehicles beginning at 5:00–7:00 p.m. when they return home from work), which may lead to electricity network congestion or voltages dropping below acceptable levels.[140]

Each distribution line in a network has a related impedance that causes the voltage at each house in the network to decrease the further it is from the distribution transformer (as described in another chapter, voltage is like the electrical *pressure* in the network). As distribution lines draw more current due to EV charging stations, the decrease in voltage becomes more dramatic. If voltage in some houses drops below acceptable limits (especially those far from the transformer), appliances may fail or suffer.[141]

Additionally, distribution networks are generally three-phase (i.e., three lines carry the current, each out of phase with the others by one-third of a cycle), but most houses connect to only one phase. If a disparate number of homes with EV charging stations connect to the same phase, that phase could get out of balance with the others and cause a loss of network efficiency. Mass EV charging could also affect overall network power quality by distorting the shape of the frequency waveform carrying the current. Modeling and simulations show these impacts can occur at fairly low rates of EV ownership.[142]

According to Forbes, by 2035, EVs will make up 37% of cars in Germany.[143] Based on the status of the German power grid, Forbes predicts widespread blackouts from EVs as early as 2032.[144] For the U.K., brownouts from EVs could start once they make up 25-30% of cars on the road (if no significant action is taken), estimated to occur in the early 2030s.[145] To address these situations and prevent widespread service interruptions, grid operators must spend significant sums of money to reinforce electric power grids. By the time EVs equal 50% of the cars and trucks in Germany, grid operators would have had to spend an estimated €11 billion to prevent service interruptions.[146]

One way to address these concerns is a *smart charging* option offering incentives to EV owners to charge vehicles during off-peak hours or to coordinate charging with other local owners. This approach, however, needs charging stations and their meters to have sophisticated two-way communications so grid operators can control them remotely. EV owners will likely need monetary incentives to encourage this behavior (and penalties to ensure compliance). Another approach would use decentralized, local energy storage at the charging point (like a microgrid, described above) to avoid overloading the grid during peak use times.

Another way to address these concerns is to increase use of communication technologies in the distribution grid. These grids (with increased communication and control capabilities) are called *smart grids* that create a more sustainable and reliable electrical grid by better integrating resources. The interaction of EVs and smart grids creates a relationship among communication, electricity, and transportation networks requiring a holistic and interdisciplinary approach to the challenge.

The faster smart charging becomes standardized and required for all EV chargepoints, the better grid operators can manage large-scale EV adoption. Whether such solutions will eventually solve these problems remains unclear, but these approaches could ease the stress on electric grids and provide an opportunity to make better decisions on infrastructure changes and additions.

The electric power grid will continue to change and evolve. New technologies will make renewable energy sources and electric vehicles more viable.

Conclusion

The electrical power industry faces significant challenges, including:

- Increased demand and the requirement for significant new infrastructure
- Aging infrastructure
- The NIMBY phenomenon
- Nondispatchable generation
- Intermittent generation
- Energy storage safety
- Microgrid integration
- Integrating widespread use of electric vehicles

Addressing those challenges requires the industry to develop novel approaches to siting facilities, forecasting weather, monitoring equipment, and controlling facilities. The electrical power industry has risen to the occasion consistently in the past, and there is no reason to believe it will not do so now. Stay tuned – as the expression goes, *the future is bright.*

Summary

- The primary challenge of the electrical power industry is balancing the disparate and often conflicting needs and wants of its stakeholders.
- Demand forecasts vary, but they all show electrical demand rising into the foreseeable future, requiring new generation, transmission, distribution, and storage resources.
- The increasing demand for electricity to satisfy only existing customers will require massive infrastructure investments in generation, transmission, and distribution facilities.
- The switch from carbon-based energy to "cleaner" energy, much of which is forecast to be electricity, will increase the required investments even more.
- On top of those investments is the investment required for storing electricity due to the intermittent nature of renewable generation resources.
- In addition to adding new infrastructure, the electrical power industry must maintain its existing, aging infrastructure.
- The reliability of the legacy BPS is declining and will continue to do so absent large investments in reliability.
- Solving the grid stability issues associated with RES requires developing dispatchable energy storage systems.
- The electric power grid will continue to change and evolve. New technologies will make renewable energy sources and electric vehicles more viable.

Glossary

Access control list (ACL): Specifies users or system processes granted access to computer system objects and the allowed operations on given objects. Each entry in a typical ACL specifies a subject and an operation.

Adequacy: The ability of the electric system to supply the aggregate electrical demand and energy requirements of end-use customers at all times, taking into account scheduled and reasonably expected unscheduled outages of system elements.

Administrative and general expenses: Expenses of an electric utility relating to the operation of its corporate offices and administrative affairs, as contrasted with expenses incurred for specialized functions. Examples include salaries, office supplies, advertising, and other general expenses.

Aggregator: An entity that combines the loads of multiple end-use customers to negotiate the purchase or transmission of electricity and other related services for these customers.

Alternating current (AC): An electric current that reverses direction at regular intervals and a magnitude that varies continuously in a sinusoidal manner.

Ampere: The unit of measurement of electrical current produced in a circuit by 1 volt acting through a resistance of 1 ohm.

Ancillary services: Services to support the transmission of electricity from generation sites to customer loads that include regulation, spinning reserve, non-spinning reserve, replacement reserve, and voltage support.

Apparent power: The product of the voltage (in volts) and the current (in amperes). It comprises both active and reactive power measured in "volt-amperes" and is often expressed in "kilovolt-amperes" (kVA) or "megavolt-amperes" (MVA).

Automatic circuit reclosers (ACRs): A class of switchgear designed for overhead electricity distribution networks to detect and interrupt momentary faults.

Auxiliary generator: A generator at an electric plant providing power to operate the plant equipment, including related demands such as plant lighting, during periods when the electric plant is not operating and power is unavailable from the grid. An auxiliary generator can serve as a blackstart generator to start the main central station generators.

Auxiliary power units (APUs): Small gas turbines designed for auxiliary power of larger machines.

Available but not needed capability: Net capability of main generating units that are operable but not considered necessary to carry load and cannot be connected to load within thirty minutes.

Average revenue per kilowatt-hour: The average revenue per kilowatt-hour of electricity, calculated by dividing the total monthly revenue by the corresponding total monthly sales.

Avoided cost: The cost to an electric utility of electric power the utility would generate itself or purchase from another source if it did not purchase the power from a qualifying facility (QF). FERC regulations implemented the Public Utility Regulatory Policies Act of 1978 (PURPA), which requires utilities to purchase electricity from QFs at a price at or below their avoided costs.

Balancing Authority (BA): The responsible entity that integrates resource plans ahead of time, maintains load-interchange-generation balance within a Balancing Authority Area, and supports Interconnection frequency in real time.

Balancing Authority Area (BAA): The geographic area controlled by a BA that supports interconnection frequency in real time.

Base bill: A charge calculated by taking the rate from the appropriate electric rate schedule and applying it to the level of consumption.

Base load: The minimum amount of electric power delivered or required over a given period of time at a steady rate.

Base load capacity: The generating equipment normally operated to serve base loads on an around-the-clock basis.

Base load plant: A plant normally operated to take all or part of the minimum load of a system that produces electricity at an essentially constant rate and that runs continuously. These units are operated to maximize system mechanical and thermal efficiency and minimize system operating costs.

Bilateral agreement: A written agreement signed by two parties that specifies the terms of an agreement.

Bilateral energy transaction: A transaction between two willing parties to enter into a physical or financial agreement to trade energy commodities.

Billing period: The time between meter readings.

Biofuels: Liquid fuels and blending components produced from biomass (plant) feedstocks used primarily for transportation.

Biomass: Organic nonfossil material of biological origin constituting a renewable energy source.

Biomass resources: Any plant-derived organic matter available on a renewable basis, including dedicated energy crops and trees, agricultural food and feed crops, agricultural crop wastes and residues, wood wastes and residues, aquatic plants, animal wastes, municipal wastes, and other waste materials.

Biomass technologies. Technologies that convert renewable biomass fuels into electricity (and heat) using boilers, gasifiers, turbines, generators, and fuel cells.

Blackstart: The procedure to recover from a total or partial shutdown of a transmission system that caused an extensive loss of supply. Blackstart involves isolated power stations starting individually and gradually reconnecting to each other to form an interconnected system again.

Boiler: A device for generating steam for power, processing, heating, or hot water. Heat from an external combustion source is transmitted to a fluid contained within the tubes found in the boiler shell and delivered to an end use at a desired pressure, temperature, and quality.

Boiler fuel: An energy source to produce heat transferred to the boiler vessel to generate steam or hot water.

British thermal unit (BTU): A standard unit for measuring the quantity of heat energy equal to the quantity of heat required to raise the temperature of one pound of water by 1°F.

Bulk power transactions: The wholesale sale and purchase of electricity for many different aspects of electric utility operations, from maintaining load to reducing costs.

Bundled utility service: A means of operation whereby energy, transmission, and distribution services, as well as ancillary and retail services, are provided by one entity.

Bus: An electrical conductor serving as a common connection for two or more electrical circuits.

California power exchange: A state-chartered, nonprofit corporation which provides day-ahead and hour-ahead markets for energy and ancillary services in accordance with the power exchange tariff. The power exchange is a scheduling coordinator and is independent of both the independent system operator and all other market participants.

Capability: The maximum load a generating unit or other electrical apparatus can carry under specified conditions for a given period of time without exceeding approved limits of temperature and stress.

Capacity: The amount of electric power a generator, turbine, transformer, transmission circuit, or system is capable of delivering.

Capacity factor (nameplate): The ratio of the electrical energy a generating unit could have produced based on its availability during a period of time compared to the electrical energy the unit could have produced at continuous full power operation during the same period, expressed as a percentage.

Capacity factor (operating): The ratio of the electrical energy actually produced by a generating unit during a period of time compared to the electrical energy the unit could have produced at continuous full power operation during the same period expressed as a percentage.

Class rate schedule: An electric rate schedule applicable to one or more specified classes of service, groups of businesses, or customer uses (e.g., residential, commercial, industrial).

Circuit: A conductor or a system of conductors through which electric current is transmitted.

Coal bed degasification: The removal of methane or coal bed gas from a coal mine before or during mining.

Coal gasification: The process of converting coal into gas. The basic process involves crushing coal to a powder, then heating it in the presence of steam and oxygen to produce a gas. The gas is then refined to reduce sulfur and other impurities. The gas can be used as a fuel or processed further and concentrated into chemical or liquid fuel.

Cogeneration: The production of electrical energy and another form of useful energy (such as heat or steam) through the sequential use of energy during a separate, usually industrial, process.

Cogeneration system: A system using a common energy source to produce electricity and steam for other uses, resulting in increased fuel efficiency.

Cogenerator: A generating facility that produces electricity and another form of useful thermal energy (such as heat or steam) for industrial, commercial, heating, or cooling purposes. To receive status as a QF under PURPA, the facility must produce electric energy and "another form of useful thermal energy through the sequential use of energy" and meet certain ownership, operating, and efficiency criteria established by the FERC. (Code of Federal Regulations, Title 18, Part 292) www.govinfo.gov/help/cfr

Coincidental demand: The sum of two or more demands occurring at the same time.

Coincidental peak load: The sum of two or more peak loads occurring at the same time.

Coulomb (C): The amount of electrical charge contained in approximately 6,240,000,000,000,000,000 electrons or, more practically, the amount of electricity transported in 1 second by a current of 1 amp.

Combined cycle: An electric generating technology producing electricity from otherwise lost waste heat exiting from one or more combustion turbines. The exiting heat goes to a conventional boiler or heat recovery steam generator (HRSG) for use by a steam turbine to produce electricity, increasing the electric generating unit's efficiency.

Combined-cycle unit: An electric generating unit with one or more combustion turbines and one or more boilers, with a portion of the required energy input to the boiler(s) provided by the exhaust gas of the combustion turbine(s).

Combined heat and power (CHP) plant: A plant designed to produce heat and electricity from a single heat source. Note: This term replaces the term "cogenerator" used by EIA in the past. CHP better describes the facilities because some plants do not produce heat and power in a sequential fashion and, as a result, do not meet the legal definition of cogeneration specified in PURPA.

Combined pumped-storage plant: A pumped-storage hydroelectric power plant using pumped water and natural stream flow to produce electricity.

Commercial sector: An energy-consuming sector consisting of service-providing facilities and equipment of businesses and other private and public organizations, such as religious, social, or fraternal groups. Common uses of energy associated with this sector include space heating, water heating, air-conditioning, lighting, refrigeration, cooking, and running a wide variety of other equipment.

Competitive bidding. A procedure used in many jurisdictions to select suppliers of electric capacity and energy. Under competitive bidding, a central entity solicits bids from prospective power generators to meet current or future electricity demand.

Competitive transition charge: A nonbypassable charge levied on each customer of a distribution utility, including those served under contracts with nonutility suppliers, to recover the utility's "stranded costs" from a transition to competition in the electric market.

Concentrating solar power: A solar electric generation technology that uses reflective materials, such as mirrors, to concentrate solar energy converted into electricity. The facility collects sunlight focused with mirrors to create a high-intensity power source. The heat source produces steam or mechanical power to run a generator to create electricity.

Congestion: When insufficient transfer capacity is available to simultaneously implement all the preferred schedules for electricity transmission.

Connected load: The sum of continuous ratings or capacities for a system, part of a system, or a customer's electric power-consuming apparatus.

Connection: The physical connection of transmission lines, transformers, switch gear, etc., between two electric systems, permitting the transfer of electric energy in one or both directions.

Conservation: A reduction in energy consumption corresponding with a reduction in service demand. Service demand can include buildings-sector end uses such as lighting, refrigeration, and heating; industrial processes; or vehicle transportation. Unlike energy efficiency, typically a technological measure, conservation better associates with behavior. Examples of conservation include adjusting the thermostat to reduce the output of a heating unit, using occupancy sensors to turn off lights or appliances, and carpooling.

Contract price: The delivery price determined when a contract is signed.

Contract receipts: Purchases based on a negotiated agreement over a given time period.

Control area: An electric power system or combination of electric power systems controlled by a single entity. A common automatic control scheme applies to the systems to match the load in the electric power system to the output of the generators within the electric power system, plus capacity and energy obtained from entities outside the electric power system. The control area maintains scheduled interchange with other control areas, maintains the frequency of the electric power system within prescribed limits, and provides sufficient generating capacity to maintain operating reserves.

Cooperative electric utility (Co-op): An electric utility owned by and operated for the benefit of those using its service. The co-op will generate, transmit, and/or distribute electric energy to a specified area not served by another utility. Such ventures are generally exempt from federal income tax laws. The Rural Utilities Service (prior Rural Electrification Administration) of the U.S. Department of Agriculture initially financed most electric cooperatives (although co-ops exist in more jurisdictions than just the U.S.).

Cost-based rates: A ratemaking concept used to design and develop rate schedules to ensure recovery of only the cost of providing the service.

Cost-of-service (COS) regulation: A traditional electric utility regulation under which a utility sets its rates based on the cost of providing service to customers pl;us the right to earn a limited profit approved by a regulatory body.

Current: Transmission of electricity in an electrical conductor. The strength or rate of movement of the electricity measured in amperes.

Customer choice: The right of customers to purchase energy from a supplier other than their legacy supplier or from more than one seller in a retail market.

Day-ahead schedule: A schedule prepared by a scheduling coordinator or the independent system operator the day before the real-time operations day. This schedule indicates the levels of generation and demand scheduled for each settlement period that trading day.

Demand: The rate at which electric energy is delivered to or by a system or part of a system at a given instant or averaged over a designated time period (*See* peak demand).

Demand bid: A bid into the power exchange indicating a quantity of energy or an ancillary service an eligible customer is willing to purchase and, if relevant, the maximum price the customer will pay.

Demand charge: Portion of the consumer's bill for electric service based on the consumer's maximum electric usage and calculated based on the billing demand charges per the applicable rate schedule.

Demand charge credit: Compensation received by the buyer when the seller cannot meet the contract's delivery terms.

Demand indicator: A measure of the number of energy-consuming units or amount of service or output for which energy inputs are required.

Demand interval: The time period during which transmission of electricity is measured (usually in 15-, 30-, or 60-minute increments).

Demand metered: Metered demand during a billing period.

Demand response programs: Incentive-based programs encouraging electric power customers to temporarily reduce their demand for power in exchange for a reduction in their electricity bills. Some demand response programs allow electric power system operators to directly reduce load, while in others, customers retain control. Customer-controlled reductions in demand may involve actions such as curtailing load, operating on-site generation, or shifting electricity use to another time period. Demand response programs are one type of demand-side management, which also covers broad, less immediate, programs such as promoting energy-efficient equipment in residential and commercial sectors.

Demand-side management (DSM): A utility action reducing or curtailing end-use equipment or processes. DSM is often used to reduce customer load during peak demand and/or in times of supply constraint. DSM includes focused, deep, and immediate programs such as the brief curtailment of energy-intensive processes used by a utility's most demanding industrial customers and broad, shallow, and less immediate programs such as promoting energy-efficient equipment in residential and commercial sectors.

Derate: A decrease in the available capacity of an electric generating unit, commonly due to:

- a system or equipment modification;
- environmental, operational, or reliability considerations.

Causes include high cooling water temperatures, equipment degradation, and historical performance during peak demand periods. In this context, a derate is typically temporary and due to transient conditions.The term can also refer to discounting a portion of a generating unit's capacity for planning purposes.

Deregulation: The elimination of some or all regulations from a previously regulated industry or sector of an industry.

Direct access: The ability of a retail customer to purchase electricity or other energy directly from a supplier other than the traditional supplier.

Direct control load management: The magnitude of customer demand an operator can interrupt at the time of the seasonal peak load by interrupting power supply to individual appliances or equipment on customer premises. This type of control usually reduces residential customer demand.

Direct current (DC): An electric current of constant direction with a magnitude that does not vary or varies only slightly, as in a battery. (Compare with alternating current.)

Direct electricity demand control: The utility installs a radio-controlled device on customer equipment. During periods of particularly heavy electricity use, the utility will send a signal to buildings in its service territory with this device and turn off the equipment for a certain period.

Direct use: Use of electricity that is: (1) self-generated; (2) produced by either the same entity that consumes it or an affiliate; and (3) used in direct support of a service or industrial process located in the same facility or group of facilities housing the generating equipment.

Dispatchability: The ability of a generating unit to increase or decrease generation or be brought online or shut down at the request of a system operator.

Dispatchable electrical power: Electric generation that can be turned on and off when and as needed.

Dispatching: The operating control of an integrated electric system involving operations such as: (1) assigning load to specific generating stations and other sources of supply to effect the most economical supply, as the total or the significant area loads rise or fall; (2) the control of operations and maintenance of high-voltage lines, substations, and equipment; (3) the operation of principal tie-lines and switching; (4) the scheduling of energy transactions with connecting electric utilities.

Distribution: The delivery of energy to end-use customers.

Distribution provider: Entity that provides and operates the wires between the transmission system and the end-use customer. For end-use customers served at transmission voltages, the transmission owner also serves as the Distribution Provider. Thus, specific voltage does not define the Distribution Provider but, rather, the entity performing the distribution function at any voltage.

Distribution system: The portion of the transmission and facilities of an electric system dedicated to delivering electric energy to end-use customers.

Divestiture: The stripping off of one function from the others by selling (spinning off) or in some other way changing the ownership of assets related to a function. Stripping off is most commonly associated with spinning off generation assets so the shareholders that own the transmission and distribution assets no longer own them.

Electric generation industry: Stationary and mobile generating units connected to the electric power grid that can generate electricity. The electric generation industry includes the electric power sector (utility generators and Independent Power Producers) and industrial and commercial power generators, including CHP producers, but excludes units at single-family dwellings.

Electric industry re-regulation: The design and implementation of regulations applied after restructuring the electric power industry. Re-regulation applies to entities that continue to exhibit characteristics of a natural monopoly and may employ the same or different regulatory practices as those used before restructuring.

Electric industry restructuring: The process of replacing a monopolistic system of vertically integrated electric utilities with competing sellers, allowing individual retail customers to choose their supplier but still receive delivery over the electric power lines of the local utility. It includes reconfiguration of vertically integrated electric utilities.

Electric plant (physical): A facility containing prime movers, electric generators, and auxiliary equipment to convert mechanical, chemical, and/or fission energy into electric energy.

Electric rate schedule: A statement of the electric rate and the terms and conditions governing its application, including attendant contract terms and conditions accepted or approved by a regulatory body with appropriate oversight authority.

Electricity: A form of energy characterized by the presence and motion of elementary charged particles generated by friction, induction, or chemical change.

Electricity broker: An entity that arranges the sale and purchase of electric energy, transmission of electricity, and/or other related services between buyers and sellers but does not take title to any of the products sold.

Electricity congestion: A condition that occurs when insufficient transmission capacity exists to transmit all the electricity needed to serve load at a particular point in time.

Electricity demand: The requirement for electric energy as an input to provide products and/or services.

Electricity demand bid: A bid into a power exchange indicating a quantity of energy or an ancillary service an eligible customer is willing to purchase and, if relevant, the maximum price the customer will pay.

Electricity generation: The process of producing electric energy or the amount of electric energy produced by transforming other forms of energy; commonly expressed in kilowatt-hours (kWh) or megawatt-hours (MWh).

Electricity-only plant: A plant designed to produce electricity only. *See* also Combined heat and power (CHP) plant.

Electricity paid by household: The household paid the electric utility company directly for all household uses of electricity (such as water heating, space heating, air-conditioning, cooking, lighting, and operating appliances). Bills paid by a third party are not counted as "paid by household."

Electric plant (physical): A facility containing prime movers, electric generators, and auxiliary equipment to convert mechanical, chemical, and/or fission energy into electric energy.

Electric power: The rate at which electric energy is transferred. Electric power is measured by capacity and commonly expressed in megawatts (MWs).

Electric power grid: A system of synchronized power providers and consumers connected by transmission and distribution lines and operated by one or more control centers. In the continental United States, the electric power grid consists of three systems: the Eastern Interconnect, the Western Interconnect, and the Texas Interconnect. In Alaska and Hawaii, several systems encompass areas smaller than the state (e.g., the interconnect serving Anchorage, Fairbanks, and the Kenai Peninsula; and individual islands).

Electric power plant: *See* Electric plant (physical).

Electric power system: A grouping of individual electric power equipment/facilities.

Electric rate: The price for a specified amount and type of electricity by class of service in an electric rate schedule or sales contract.

Electric rate schedule: A statement of the electric rate and terms and conditions governing its application, including attendant contract terms and conditions accepted or approved by a regulatory body with appropriate oversight authority.

Electric utility: An entity or instrumentality aligned with distribution facilities to deliver electric energy for use by end users. Included are IOUs, municipal and state utilities, federal electric utilities, and co-ops and includes some tariff-based entities corporately aligned with companies that own distribution facilities.

Electric utility sector: Privately and publicly owned establishments that generate, transmit, distribute, or sell electricity that meet the definition of an electric utility; does not include nonutility power producers.

Electricity sales: The amount of electricity sold in a given time period; usually grouped by classes of service, such as residential, commercial, industrial, and other. "Other" sales include sales for public street and highway lighting and other sales to public authorities, sales to railroads and railways, and interdepartmental sales.

Encryption: A method to convert data into secret code to hide the information's true meaning. The science of encrypting and decrypting information is called cryptography.

Energy charge: That portion of the charge for electric service based upon the electric energy (kWh) consumed.

Energy conservation features: Building shell conservation features, HVAC conservation features, lighting conservation features, and other conservation features incorporated by a building; does not include DSM program participation.

Energy deliveries: Energy generated by one electric utility system and delivered to another system.

Energy efficiency: A ratio of service provided to energy input (e.g., lumens to watts, in the case of light bulbs). Services provided can include buildings-sector end uses such as lighting, refrigeration, and heating; industrial processes; or vehicle transportation. Unlike conservation, which involves some reduction of service, energy efficiency provides energy reductions without sacrifice of service. May also refer to the use of technology to reduce the energy needed for a given purpose or service (e.g., replacing incandescent bulbs with light-emitting diode (LED) bulbs).

Energy efficiency program: Program aimed at reducing the electrical energy used by specific end-use devices and systems, typically without affecting the services provided. These programs reduce overall electricity consumption, often without explicit consideration for the timing of program-induced savings. Such savings are generally achieved by substituting technologically more advanced equipment to produce the same level of end-use services (e.g. lighting, heating, motor drive) with less electricity. Examples include high-efficiency appliances, efficient lighting programs, high-efficiency heating, ventilating, and air-conditioning (HVAC) systems or control modifications, efficient building design, advanced electric motor drives, and heat recovery systems.

Energy Information Administration (EIA): An independent agency in the United States Department of Energy (DOE) that develops surveys, collects energy data, and does analytical and modeling analyses of energy issues. The agency must satisfy the requests of Congress, other areas within the DOE, FERC, the executive branch, and its own independent needs, as well as assist the general public or other interest groups, without taking a policy position.

Energy intensity: A ratio of energy consumption to another metric, typically national gross domestic product in the case of a country's energy intensity. Sector-specific intensities may refer to energy consumption per household, per unit of commercial floorspace, per dollar value of industrial shipment, or another metric indicative of a sector. Improvements in energy intensity include energy efficiency and conservation, as well as structural factors not related to technology or behavior.

Energy management and control system (EMCS): An energy conservation feature using mini and microcomputers, instrumentation, control equipment, and software to manage a building's use of energy for heating, ventilation, air-conditioning, lighting, and/or business-related processes. These systems can also manage fire control, safety, and security. EMCS does not include time-clock thermostats.

Energy marketer: An entity that arranges bulk power transactions for end users. An energy marketer's main functions are to determine the best overall fuel choice(s) for customers and to deliver that fuel to the customer.

Energy Policy Act of 1992 (EPACT): Act creating a new class of power generators, exempt wholesale generators, exempt from the provisions of the Public Holding Company Act of 1935 and granted FERC the authority to order and condition access by eligible parties to the interconnected transmission grid.

Energy Policy Act of 2005 (EPAct 2005): Signed into law August 8, 2005. The first effort of the U.S. government to address U.S. energy policy after EPACT. Intended to encourage energy efficiency and conservation, promote alternative and renewable energy sources, reduce dependence on foreign sources of energy, increase domestic production, modernize the electricity grid, and encourage expansion of nuclear energy. In the area of solar and energy-efficiency measures, EPAct 2005 created a number of tax credit opportunities that include residential solar photovoltaic and hot water heating systems, high-efficiency new homes, improvements to existing homes including high-efficiency air conditioners and equipment, residential fuel cell systems, fuel cell and microturbines used in businesses, and tax deductions for highly efficient commercial buildings.

Energy receipts: Energy brought into a site from another location.

Energy service provider: An entity that provides service to a retail or end-use customer.

Energy source: Any substance or natural phenomenon that can be consumed or transformed to supply heat or power. Examples include petroleum, coal, natural gas, nuclear, biomass, electricity, wind, sunlight, geothermal, water movement, and hydrogen in fuel cells.

Exempt wholesale generator (EWG): Wholesale generators created under the EPACT exempt from certain financial and legal restrictions stipulated in the Public Utilities Holding Company Act of 1935.

Facility: A location or site containing prime movers, electric generators, and/or equipment for converting mechanical, chemical, and/or nuclear energy into electric energy. A facility may contain more than one generator of either the same or different prime mover type. For a cogenerator, the facility includes the industrial or commercial process.

Facilities charge: An amount paid by the customer, in a lump sum or periodically, as reimbursement for facilities furnished. The charge may include operation and maintenance as well as fixed costs.

Federal electric utilities: An electric utility classification applying to utilities that are agencies of the federal government involved in generating and/or transmitting electricity. Most of the electricity generated by federal electric utilities is sold at wholesale prices to local government-owned and cooperatively owned utilities and IOUs. These government agencies consist of: the Army Corps of Engineers and the Bureau of Reclamation, which generate electricity at federally owned hydroelectric projects. Four power marketing administrations sell the relatively low-cost power on a preferential basis to local government-owned and cooperatively owned utilities (Bonneville Power Administration (BPA), Western Area Power Administration (WAPA), Southeastern Power Administration (SEPA) and Southwestern Power Administration (SWPA)). Also includes the Tennessee Valley Authority (TVA), which produces and transmits electricity in the Tennessee Valley region.

Federal Energy Regulatory Commisison (FERC): Successor to the Federal Power Commission.

Federal Power Act: Enacted in 1920 and amended in 1935, the law consists of three parts. The first part incorporated the Federal Water Power Act, administered by the former Federal Power Commission, whose activities were confined almost entirely to licensing nonfederal hydroelectric projects. Parts II and III extended the Act's jurisdiction to include regulating the interstate transmission of electrical energy and rates for its sale as wholesale in interstate commerce. The FERC now administers this law.

Federal Power Commission (FPC): The predecessor agency of the FERC, created on June 10, 1920, and originally charged with regulating the electric power and natural gas industries. Abolished on September 30, 1977, when the DOE was created; its functions were divided between the DOE and FERC.

Firewall: Part of a computer system or network designed to block unauthorized access while permitting outward communication.

Firm power: Power or power-producing capacity intended to be available at all times during the period covered by a guaranteed commitment to deliver, even under adverse conditions.

Flue-gas particulate collector: Equipment used to remove fly ash from the combustion gases of a boiler plant before discharge to the atmosphere. Particulate collectors include electrostatic precipitators, mechanical collectors (cyclones), fabric filters (baghouses), and wet scrubbers.

Fluidized-bed combustion: A method of burning particulate fuel, such as coal, in which the amount of air required for combustion far exceeds that found in conventional burners. The fuel particles are continually fed into a bed of mineral ash in the proportions of 1 part fuel to 200 parts ash, while a flow of air passes up through the bed, causing it to act like a turbulent fluid.

Forced outage: The shutdown of electrical equipment for emergency reasons or a condition in which the equipment is unavailable due to unanticipated breakdown.

Fossil fuel: An energy source formed in the Earth's crust from decayed organic material. The common fossil fuels are petroleum, coal, and natural gas.

Fossil fuel electric plant: An electric generation plant using a fossil fuel as its source of energy.

Fuel: Any material substance consumed to supply heat or power. Includes petroleum, coal, and natural gas (the fossil fuels) and other consumable materials such as uranium, biomass, and hydrogen.

Fuel cell: Device capable of generating an electrical current by converting the chemical energy of a fuel directly into electrical energy. Fuel cells differ from conventional electrical cells in that the cell does not contain active materials such as fuel and oxygen, which are supplied from outside. It does not contain an intermediate heat cycle like most other electrical generation techniques.

Fuel expenses: Costs for the fuel used to produce steam or to drive another prime mover to generate electricity. Other associated expenses include unloading the shipped fuel and all handling of the fuel up to the point where it enters the first bunker, hopper, bucket, tank, or holder in the boiler-house structure.

Full forced outage: The net capability of main generating units unavailable for emergency reasons.

Futures market: A trade center for quoting prices on contracts for the delivery of a specified quantity of a commodity at a specified time and place in the future.

Gas: A nonsolid, nonliquid combustible energy source that includes natural gas, coke-oven gas, blast-furnace gas, and refinery gas.

Gas turbine: A rotating engine that extracts energy from a flow of combustion gas. It has an upstream compressor coupled to a downstream turbine, and a combustion chamber in between. Gas turbine may also refer to just the turbine element. Energy is added to the gas stream in the combustor, where air is mixed with fuel and ignited. Combustion increases the temperature and volume of the gas flow. This is directed through a nozzle over the turbine's blades, spinning the turbine and powering the compressor. Energy is extracted in the form of shaft power, compressed air, or thrust.

Gas turbine plant: A plant in which the prime mover is a gas turbine.

Generating unit: Any combination of physically connected generators, reactors, boilers, combustion turbines, and other prime movers operated together to produce electric power.

Generation: The process of producing electric energy by transforming other forms of energy; also, the amount of electric energy produced, expressed in kilowatt-hours.

Generation company: An entity that owns or operates generating plants. The generation company may own the generation plants or interact with the short-term market on behalf of plant owners.

Generator capacity: The maximum output, commonly expressed in megawatts (MWs), that generating equipment can supply, adjusted for ambient conditions.

Generator nameplate capacity (installed): The maximum rated output of a generator, prime mover, or other electric power production equipment under specific conditions designated by the manufacturer. Installed generator nameplate capacity is commonly expressed in megawatts (MWs) and usually indicated on a nameplate physically attached to the generator.

Geothermal energy: Hot water or steam extracted from geothermal reservoirs in the Earth's crust, used for geothermal heat pumps, water heating, or electricity generation.

Geothermal plant: A plant in which the prime mover is a steam turbine driven either by steam produced from hot water or by natural steam from heat found in rock.

Gigawatt (GW): One billion watts or one thousand megawatts.

Gigawatt-hour (GWh): One billion watt-hours.

Greenhouse effect: The result of water vapor, carbon dioxide, and other atmospheric gases trapping radiant (infrared) energy and keeping the earth's surface warmer than it would otherwise be. Greenhouse gases within the lower levels of the atmosphere trap this radiation, which would otherwise escape into space, and subsequent re-radiation of some of this energy back to the Earth maintains higher surface temperatures than would occur if the gases were absent.

Gross generation: The total amount of electric energy produced by generating units and measured at the generating terminal in kilowatt-hours (kWh) or megawatt-hours (MWh).

Heat content: The amount of heat energy available by the transformation or use of a specified physical unit of an energy form (e.g., a ton of coal, a barrel of oil, a kilowatt-hour of electricity, a cubic foot of natural gas, or a pound of steam) commonly expressed in British thermal units (Btu). Note: Heat content of combustible energy forms can be expressed in terms of either gross heat content (higher or upper heating value) or net heat content (lower heating value), depending on whether the available heat energy includes or excludes the energy used to vaporize water (contained in the original energy form or created during the combustion process). The EIA typically uses gross heat content values.

Heat rate: The total amount of energy required to produce one kilowatt-hour (kWh) of electricity by an electric generator or power plant. The input rate required for generating unit power, also described as the ratio of thermal inputs to electrical output. The lower the heat rate, the higher the efficiency.

Hedging contracts: Contracts that establish future prices and quantities of electricity independent of the short-term market. Derivatives may be used for this purpose.

Hourly nonfirm transmission service: Point-to-point transmission scheduled and paid for on an as-available basis and subject to interruption.

Hydroelectric power: The use of flowing water to produce electrical energy.

Hydroelectric pumped storage: Process for storing electric energy in the form of water pumped from a lower reservoir into an upper reservoir during off-peak periods and subsequently released through hydroelectric turbines into the lower reservoir for generation during peak-demand periods to serve electrical system loads or provide electric power during system emergencies.

Hydrogen: A colorless, odorless, highly flammable gaseous element; the lightest of all gases and the most abundant element in the universe; occurring chiefly in combination with oxygen in water and also in acids, bases, alcohols, petroleum, and other hydrocarbons.

Impedance: The opposition to power flow in an AC circuit and any device that introduces such opposition in the form of resistance, reactance, or both. Measured as the ratio of voltage to current, where a sinusoidal voltage and current of the same frequency are used for the measurement; measured in ohms.

Incremental energy costs: The additional cost of producing and/or transmitting electric energy above some previously determined base cost.

Independent power producer (IPP): A legal entity or instrumentality that owns or operates facilities to generate electricity that is not an electric utility.

Independent system operator (ISO): An independent entity established to coordinate regional transmission in a nondiscriminatory manner and ensure the reliability of the electric system.

Instantaneous peak demand: The maximum demand at the instant of greatest load.

Interchange: Energy transfers that cross Balancing Authority boundaries.

Interchange authority: The responsible entity that authorizes implementation of valid and balanced Interchange Schedules between Balancing Authority Areas and ensures communication of Interchange information for reliability purposes.

Interchange schedule: A schedule between balancing authorities agreed upon in the interchange confirmation process and available for reliability assessments.

Interchange transaction: An agreement to transfer energy from a seller to a buyer that crosses one or more Balancing Authority Area boundaries.

Interdepartmental service: Amounts charged by the electric department (typically a municipal utility) at tariff or other specified rates for electricity supplied by it to other utility departments.

Integrated gasification combined-cycle technology: Coal, water, and oxygen are fed to a gasifier that produces syngas. This medium-BTU gas is cleaned (particulates and sulfur compounds removed) and fed to a gas turbine. The hot exhaust of the gas turbine and heat recovered from the gasification process go through a heat-recovery steam generator (HRSG) to produce steam to drive a steam turbine to produce electricity.

Integrated resource planning (IRP): A process by which an electric utility plans for its future resource needs. Key characteristics include a long-term forecast of power needs and a comprehensive evaluation of all resource options (supply- and demand-side).

Inter-Control Center Communications Protocol (ICCP) or Telecontrol Application Service Element (TASE.2): An international standard for communications between control centers in the electrical power sector.

Interconnected system: A system consisting of two or more individual power systems normally operating with connecting tie-lines.

Interconnection: Two or more electric systems having a common transmission line that permits a flow of energy between them. The physical connection of the electric power transmission facilities allows for the sale or exchange of energy. The term is also used to describe one of the three common transmission systems in the U.S.: Eastern Interconnection, Western Interconnection, and Texas Interconnection.

Interconnection Reliability Operating Limit (IROL): A System Operating Limit that, if violated, could lead to instability, uncontrolled separation, or cascading outages that adversely impact the reliability of the BES.

Intermediate load: The range from base load to a point between base load and peak. This point may be the midpoint, a percent of the peak load, or the load over a specified time period.

Intermittent electric generator: An electric generating plant with output controlled by the natural variability of the energy resource rather than dispatched based on system requirements. Intermittent output usually results from the direct, nonstored conversion of naturally occurring energy sources such as solar, wind, or free-flowing rivers (run-of-river hydroelectricity).

Internal combustion engine: An engine with one or more cylinders in which the process of combustion takes place, converting energy released from the rapid burning of a fuel-air mixture into mechanical energy. Diesel or gas-fired engines are the principal types used in electric plants.

Internal combustion plant: A plant in which the prime mover is an internal combustion engine usually operated during periods of high demand for electricity.

Interruptible gas: Gas sold to customers with a provision allowing curtailment or cessation of service at the discretion of the distributing company under certain circumstances, as specified in the service contract.

Interruptible load: A Demand-Side Management category representing the consumer load that, in accordance with contractual arrangements, can be interrupted by the action of the consumer at the direct request of the system operator; usually involves large-volume commercial and industrial consumers. Interruptible Load does not include Direct Load Control.

Interruptible load or interruptible demand (electric): Demand that the end-use customer makes available to its Load-Serving Entity through an agreement for curtailment.

Interruptible power: Power and usually the associated energy made available by one utility to another. This transaction is subject to curtailment or cessation of delivery by the supplier in accordance with a prior agreement with the other party or under specified conditions.

Interruptible rate: A special electricity or natural gas arrangement under which, in return for lower rates, the customer must either reduce energy demand on short notice or allow the electric or natural gas utility to temporarily cut off the energy supply the utility can maintain service for higher-priority users. This interruption or reduction in demand typically occurs during periods of high demand for the energy (summer for electricity and winter for natural gas).

Investor-owned utility (IOU): A privately owned electric utility whose stock is publicly traded; typically regulated and authorized to achieve an allowed rate of return.

Joule (J): The meter-kilogram-second unit of work or energy, equal to the work done by a force of 1 newton (N) when its point of application moves through a distance of 1 meter in the direction of the force; equivalent to 107 ergs and 1 watt-second.

Jurisdictional utilities: Utilities regulated by public laws.

Kilovolt-ampere (kVA): A unit of apparent power, equal to 1,000 VA; the mathematical product of the volts and amperes in an electrical circuit.

Kilowatt (kW): One thousand watts.

Kilowatt-hour (kWh): A measure of electricity defined as a unit of work or energy measured as 1 kilowatt (1,000 watts) of power expended for 1 hour. One kWh is equivalent to 3,412 Btu.

Line pack: The volume of gas in a pipeline necessary for the line to remain "full" while accepting new molecules at receipt points and delivering molecules from the line at delivery points.

Load: An end-use device or customer that receives power from the electric system.

Load control program: A program in which the utility offers a lower rate in return for permission to turn off equipment for short periods of time by remote control allowing the utility to reduce peak demand.

Load curve: The relationship of power supplied to the time of occurrence that illustrates the varying magnitude of the load during the period covered.

Load diversity: The difference between the peak of coincident and noncoincident demands of two or more individual loads.

Load factor: The ratio of the average load to peak load during a specified time interval.

Load following: Regulation of the power output of electric generators within a prescribed area in response to changes in system frequency, tie-line loading, or the relation of these to each other to maintain the scheduled system frequency and/or established interchange with other areas within predetermined limits.

Load leveling: Any load control technique that dampens the cyclical daily load flows and increases base load generation. Peak load pricing and time-of-day charges are two techniques to reduce peak load and maximize efficient generation of electricity.

Load loss (3 hours): Any significant incident on an electric utility system resulting in a continuous outage of three hours or longer to more than 50,000 customers or to more than one half of the total customers served immediately prior to the incident, whichever is less.

Load-serving entity: Secures energy and transmission service (and related interconnect operations services) to serve the electrical demand and energy requirements of its end-use customers.

Load shedding: Intentional action to reduce firm customer load to maintain the continuity of service of the bulk electric power system. The routine use of load control equipment to reduce firm customer load is not considered load shedding.

Local distribution company (LDC): A legal entity engaged primarily in the retail sale and/or delivery of natural gas through a distribution system that includes main lines (pipelines designed to carry large volumes of gas, usually located under roads or other major right-of-ways) and laterals (pipelines of smaller diameter to connect an end user to the main line). Since the restructuring of the gas industry, the sale of gas and/or delivery arrangements may be handled by other agents, such as producers, brokers, and marketers referred to as non-LDC.

Manufacturing sector: An energy-consuming subsector of the industrial customer class, consisting of all facilities and equipment engaged in the mechanical, physical, chemical, or electronic transformation of materials, substances, or components into new products. Assembly of component parts of products is included, except for that included in construction.

Marginal cost: The increase in total cost of production as a result of producing one more unit of output. Because certain overhead costs are fixed, the marginal cost is almost aways less than the total per-unit cost of production averaged over all units produced. In the case of electrical energy, marginal cost typically means the cost of the last MW procured to meet customer load.

Market price contract: A contract in which the price of a product is not specifically determined at the time of contract signing but is based on the prevailing market price at the time of delivery. May include a floor price, that is, a lower limit on the eventual settled price. The floor price and the method of price escalation generally are determined when the contract is signed. May also include a price ceiling or a discount from the agreed-upon market price reference.

Market clearing price: The price at which supply equals demand for any electrical energy market.

Market-based pricing: Prices determined in an open market system of supply and demand in which prices are set solely by agreement as to what buyers will pay and sellers will accept. Such prices could recover less or more than full costs, depending upon what the buyers and sellers see as their relevant opportunities and risks.

Maximum demand: The greatest of all demands of the load that has occurred within a specified period of time.

Megawatt (MW): One million watts of electricity.

Megawatt-hour (MWh): One thousand kilowatt-hours or 1 million watt-hours.

Merchant facilities: High-risk, high-profit facilities that operate, at least partially, at the whims of the market. IPPs are an example of merchant facilities.

Metered data: End-use data obtained through the direct measurement of the total energy consumed for specific uses within the individual household. Individual appliances can be submetered by connecting recording meters directly to individual appliances.

Metered peak demand: The presence of a device to measure the maximum rate of electricity consumption per unit of time, allowing electric utilities to bill customers for maximum consumption along with total consumption.

Methane: A simple organic compound with a molecular weight of 16.043 g/mol, with a tetrahedral structure with four hydrogen atoms bound to one carbon atom, denoted as CH_4. It is a colorless, flammable, odorless hydrocarbon gas that is the major component of natural gas and an important source of hydrogen in various industrial processes.

Microgrid: A local energy grid with control capability, which means it can disconnect from the traditional grid and operate autonomously.

Municipal electric utility: A class of electric utility owned by the municipality in which it operates. Municipal electric utilities are financed through municipal bonds and are self-regulated.

National Association of Regulatory Utility Commissioners (NARUC): An affiliation of the public service commissioners to promote the uniform treatment of members of the railroad, public utilities, and public service commissions of the fifty United States, the District of Columbia, the Commonwealth of Puerto Rico, and the territory of the Virgin Islands.

National Electric Light Association (NELA): An early organization that governed the activities of investor-owned electric utilities and the forerunner of the Edison Electric Institute (EEI).

National Rural Electric Cooperative Association (NRECA): A national organization dedicated to representing the interests of cooperative electric utilities and the consumers they serve. Members come from the forty-six states with electric distribution cooperatives.

Native load: The end-use customers a Load-Serving Entity is obligated to serve.

Natural Gas Policy Act of 1978 (NGPA): A framework for the regulation of most facets of the natural gas industry, signed into law on November 9, 1978.

Net actual interchange: The algebraic sum of all metered interchange over all interconnections between two physically adjacent BAAs.

Net energy for load: Net BAA generation, plus energy received from other BAAs, less energy delivered to BAAs through interchange. It includes BAA losses but excludes energy required for storage at energy storage facilities.

Net generation: The amount of gross generation less the electrical energy consumed at the generating station(s) for station service. Note: Electricity required for pumping at pumped-storage plants is regarded as electricity for station service and deducted from gross generation.

Net interstate flow of electricity: The difference between the sum of electricity sales and losses within a state and the total amount of electricity generated within that state. A positive number indicates more electricity (including associated losses) came into the state than went out of the state during the year; conversely, a negative number indicates more electricity (including associated losses) went out of the state than came into the state.

Net summer capacity: The maximum output, commonly expressed in megawatts (MW), that generating equipment can supply to system load, as demonstrated by a multi-hour test, at the time of summer peak demand (period of June 1 through September 30) that reflects a change in capacity due to electricity use for station service and ambient weather conditions.

Net winter capacity: The maximum output, commonly expressed in megawatts (MW), that generating equipment can supply to system load, as demonstrated by a multi-hour test, at the time of peak winter demand (period of December 1 through February 28) that reflects a change in capacity due to electricity use for station service.

Noncoincidental peak load: The sum of two or more peak loads on individual systems that do not occur in the same time interval. Meaningful only when considering loads within a limited period of time, such as a day, week, month, a heating or cooling season, and usually for not more than one year.

Nonfirm power: Power or power-producing capacity supplied or available under a commitment having limited or no assured availability.

Nonrenewable fuels: Fuels not easily made or renewed; typically oil, natural gas, and coal.

Nonspinning reserve: The generating capacity not currently running but capable of connecting to the bus and serving load within a specified (relatively short) time.

Nonutility generation: Electric generation by end users, or small power producers under PURPA, to supply electric power for industrial, commercial, and military operations or sales to electric utilities.

Nonutility power producer: A legal entity or instrumentality that owns or operates facilities for electric generation that is not an electric utility. Nonutility power producers include qualifying cogenerators, qualifying small power producers, and other nonutility generators (including IPPs). Nonutility power producers have no franchised service area and do not file forms listed in the Code of Federal Regulations, Title 18, Part 141.

North American Electric Reliability Corporation (NERC): A nonprofit corporation formed as the successor to the North American Electric Reliability Council; authorized by FERC to serve as the Electric Reliability Organization pursuant to EPAct 2005 to develop and maintain mandatory reliability standards for the bulk power system, with the fundamental goal of maintaining and improving the reliability of that system. NERC consists of regional reliability entities covering the interconnected power regions of the contiguous United States, Canada, and Mexico.

North American Energy Standards Board (NAESB): An electric industry forum to develop and promote standards for a marketplace for wholesale and retail natural gas and electricity.

Nuclear fuel: Fissionable materials enriched to such a composition that, when placed in a nuclear reactor, will support a self-sustaining fission chain reaction producing heat in a controlled manner for process use.

Nuclear power: Electricity generated by the use of the thermal energy released from the fission of nuclear fuel in a reactor.

Nuclear reactor: An apparatus in which a nuclear fission chain reaction can be initiated, controlled, and sustained at a specific rate. A reactor includes fuel (fissionable material), moderating material to control the rate of fission, a heavy-walled pressure vessel to house reactor components, shielding to protect personnel, a system to conduct heat away from the reactor, and instrumentation for monitoring and controlling the reactor's systems.

Nuclear Regulatory Commission (NRC): Licenses operators of nuclear power plants. Reactor operators are authorized to control equipment that affects the power of the reactor in a nuclear power plant.

Ocean energy: Draws on the energy of ocean waves, tides, or on the thermal energy (heat) stored in the ocean. Ocean energy technologies include wave energy, tidal energy, and ocean thermal energy conversion systems.

Ocean energy systems: Energy conversion technologies that harness ocean energy.

Ocean thermal energy conversion (OTEC). The process or technologies for producing energy by harnessing the temperature differences (thermal gradients) between ocean surface waters and that of ocean depths. Warm surface water is pumped through an evaporator containing a working fluid in a closed Rankine-cycle system. The vaporized fluid drives a turbine/generator.

Off-peak. Period of relatively low system load. These periods often occur in daily, weekly, and seasonal patterns and differ for each individual electric utility

Ohm: A measure of the electrical resistance of a material equal to the resistance of a circuit in which the potential difference of one volt produces a current of one ampere.

On-peak: Periods of relatively high system load. These periods often occur in daily, weekly, and seasonal patterns and differ for each individual electric utility.

Open access: A regulatory mandate to allow others to use a utility's transmission and distribution facilities to move bulk power from one point to another on a nondiscriminatory basis for a cost-based fee.

Open access – FERC: FERC Order No. 888 requires public utilities to provide nondiscriminatory transmission service over their transmission facilities to third parties to move bulk power from one point to another on a nondiscriminatory basis for a cost-based fee. Order 890 expanded Open Access to cover the methodology for calculating available transmission transfer capability; improvements that opened a coordinated transmission planning processes; standardization of energy and generation imbalance charges; and other reforms regarding the designation and undesignation of transmission network resources.

Open access transmission tariff: Transmission tariff accepted by the FERC requiring the Transmission Service Provider to furnish all shippers with nondiscriminating service comparable to that provided by Transmission Owners to themselves.

Outage: The period during which a facility is out of service.

Peak demand: The maximum load during a specified period of time; also known as "peak load."

Peak load plant: A plant usually housing old, low-efficiency steam units, gas turbines, diesels, or pumped-storage hydroelectric equipment normally used only during the peak-load periods.

Peaking capacity: Capacity of generating equipment normally reserved for operation during the hours of peak loads. Some generating equipment may be operated at certain times as peaking capacity and at other times to serve loads on an around-the-clock basis.

Photovoltaic cell (PVC): An electronic device consisting of layers of semiconductor materials fabricated to form a junction (adjacent layers of materials with different electronic characteristics) and electrical contacts. Capable of converting incident light directly into electricity (DC).

Photovoltaic module: An integrated assembly of interconnected cells capable of producing voltage when exposed to radiant energy, especially light designed to deliver a selected level of working voltage and current at its output terminals, packaged for protection against environmental degradation and suited for incorporating in photovoltaic power systems.

Planned generator: Proposed electric generating equipment at an existing or planned facility or site.

Planning authority: The responsible entity that coordinates and integrates transmission facilities and service plans, resource plans, and protection systems.

Plant: Either a synonym for an industrial establishment or a generating facility, or a term used to refer to a particular process in an establishment.

Plant use: Electric energy used in operating a plant and includes energy required for pumping at pump-storage plants.

Power: The rate of producing, transferring, or using energy, most commonly associated with electricity, measured in watts and often expressed in kilowatts (kW) or megawatts (mW). Also known as "real" or "active" power.

Power distributors and dispatchers: Also called load dispatchers or systems operators, these workers control transmission of electricity through transmission lines to end users. They monitor and operate current converters, voltage transformers, and circuit breakers. Dispatchers also monitor other distribution equipment. They may call control room operators to start or stop boilers and generators to bring production into balance with needs. They also handle emergencies, such as transformer or transmission line failures, and they route current around affected areas.

Power exchange: An entity providing a competitive spot market for electric power through auctions of generation and demand bids.

Power factor: The ratio of real power (kilowatts) to apparent power (kilovolt-amperes) for any given load and time.

Power marketers: Business entities engaged in buying and selling electricity. Power marketers do not usually own generating or transmission facilities. Unlike brokers, they take ownership of the electricity and are involved in interstate trade.

Power plant operators: Operators who control and monitor boilers, turbines, generators, and auxiliary equipment in power generating plants, and monitor instruments to maintain voltage and regulate electricity flows from the plant. When power requirements change, they start or stop generators and connect or disconnect them from circuits. They often use computers to record switching operations and loads on generators, lines, and transformers. Operators may also use computers to prepare reports of unusual incidents, malfunctioning equipment, or maintenance performed during their shifts.

Power pool: An association of two or more interconnected electric systems having an agreement to coordinate operations and plan for improved reliability and efficiencies.

Power production plant: All the land and land rights, structures and improvements, boiler or reactor vessel equipment, engines and engine-driven generators, turbo generator units, accessory electric equipment, and miscellaneous power plant equipment grouped together for an individual facility.

Price: The amount of money or consideration-in-kind for which a product or service is bought, sold, or offered for sale.

Primary energy source: A fuel to create a secondary energy source (e.g., coal, natural gas).

Prime mover: The engine, turbine, waterwheel, or similar machine driving an electric generator or, for reporting purposes, a device that converts energy to electricity directly (e.g., photovoltaic solar and fuel cells).

Profit: The income remaining after all business expenses are paid.

Public Utility Regulatory Policies Act of 1978 (PURPA): Statute requiring states to implement utility conservation programs and create special markets for cogenerators and small producers who meet certain standards, including the requirement that states set the prices and quantities of power that the utilities must buy from such facilities.

Pumped-storage hydroelectric plant: A plant that usually generates electric energy during peak load periods by using water previously pumped into an elevated storage reservoir during off-peak periods when excess generating capacity is available. When additional generating capacity is needed, the water can be released from the reservoir through a conduit to turbine generators located in a power plant at a lower level.

Purchased power: Power purchased or available for purchase.

Purchased power adjustment: A clause in a rate schedule providing for adjustments to the bill when energy from another electric system is acquired and its cost varies from a specified unit base amount.

Pure pumped-storage hydroelectric plant: A plant that produces power only from water previously pumped to an upper reservoir.

Qualifying facility (QF): A cogeneration or small power production facility meeting certain ownership, operating, and efficiency criteria established by the FERC pursuant to PURPA.

Rates: The authorized charges per unit or level of consumption for a specified time period for any of the classes of utility services provided to a customer.

Rate base: The value of property upon which a utility can earn a specified rate of return as established by a regulatory authority. The rate base generally represents the value of property used by the utility in providing service and may be calculated by any one or a combination of the following accounting methods: fair value, prudent investment, reproduction cost, or original cost.

Ratemaking authority: A utility commission's legal authority to fix, modify, approve, or disapprove rates as determined by the powers given by a state or federal legislature.

Reactive power: The portion of electricity that establishes and sustains the electric and magnetic fields of AC equipment. Reactive power must be supplied to most types of magnetic equipment, such as motors and transformers. Reactive power is provided by generators, synchronous condensers, or electrostatic equipment, such as capacitors, and directly influences electric system voltage. It is a derived value equal to the vector difference between the apparent power and the real power. It is usually expressed as kilovolt-amperes reactive (KVAR) or megavolt-amperes reactive (MVAR).

Real Power: The component of electric power that performs work, typically measured in kilowatts (kW) or megawatts(MW) and sometimes referred to as "active power." The terms "real" or "active" are often used to modify the base term "power" to differentiate it from "reactive power" and "apparent power."

Receipts:

- Deliveries of fuel to an electric plant
- Purchases of fuel
- All revenues received by an exporter for the reported quantity exported

Regional Transmission Group (RTG): A utility industry concept that the Federal Energy Regulatory Commission (FERC) embraced for the certification of voluntary groups that would be responsible for transmission planning and use on a regional basis.

Regulation: The governmental function of controlling or directing economic entities, through the process of rulemaking and adjudication.

Reliability: The adequacy and reliability of the interconnected BES. Adequacy means the ability of the electric system to supply the aggregate electrical demand and energy requirements of customers at all times, taking into account scheduled and reasonably expected unscheduled outages of system elements. Security means the ability of the electric system to withstand sudden disturbances such as electric short circuits or unanticipated loss of system elements.

Reliability coordinator (RC): The highest level of authority responsible for the reliable operation of the BES, with the wide area view of the BES and the operating tools, processes, and procedures, including the authority to prevent or mitigate emergency operating situations. The RC has a purview broad enough to enable calculating Interconnection Reliability Operating Limits, which it may base on the operating parameters of transmission systems beyond any single Transmission Operator's vision.

Reliability Standards: As defined in EPAct of 2005, a requirement, approved by the FERC to provide for reliable operation of the BPS. The term includes requirements for operating existing BPS facilities, including cybersecurity protection, and the design of planned additions or modifications to such facilities to provide for reliable operation of the BPS, but the term does not include any requirement to enlarge such facilities or to construct new transmission capacity or generation capacity.

Renewable energy resources: Naturally replenishing but flow-limited electric energy resources; virtually inexhaustible in duration but limited in the amount of energy available per unit of time. Renewable energy resources include biomass, hydro, geothermal, solar, wind, ocean thermal, wave action, and tidal action.

Re-regulation: The design and implementation of regulatory practices applied to the remaining regulated entities after restructuring of the electric industry. The remaining regulated entities continue to exhibit characteristics of a natural monopoly, where imperfections in the market prevent the realization of more competitive results and where, in light of other policy considerations, competitive results are unsatisfactory in one or more respects. Regulation could employ the same or different regulatory practices as those used before restructuring.

Reserve generating capacity: Amount of generating capacity available to meet peak or abnormally high demands for power or to generate power during scheduled or unscheduled outages.

Reserve margin (operating): The amount of unused available capability of an electric power system (at peak load for a utility system) as a percentage of total capability.

Residential consumers: Consumers using a commodity for residential uses in single and multifamily dwellings and mobile homes.

Restricted-universe census: This is the complete enumeration of data from a specifically defined subset of entities including, for example, those that exceed a given level of sales or generator nameplate capacity.

Restructuring: The process of replacing a monopoly system of electric utilities with competing sellers, allowing individual retail customers to choose their electricity supplier while receiving delivery over the power lines of the local utility. It includes reconfiguration of vertically integrated electric utilities.

Retail sales (electric): Sales made directly to the customer consuming the energy product.

Retail wheeling: The process of moving electric power from a point of generation across third-party-owned transmission and distribution systems to a retail customer.

Revenue: The total amount of money received by an entity from sales of its products or services; gains from the sales or exchanges of assets, interest, and dividends earned on investments; and other increases in the owner's equity, except those arising from capital adjustments.

Right-of-way: A corridor of land on which electric lines may be located. The Transmission Owner may own the land in fee, own an easement, or have certain franchise, prescription, or license rights to construct and maintain lines.

Running and quick-start capability: The net capability of generating units serving load or having quick-start capability. In general, quick-start capability refers to generating units that can be available within a 30-minute period.

Rural Electrification Administration (REA): A lending agency of the U.S. Department of Agriculture that makes self-liquidating loans to qualified borrowers to finance electric and telephone service to rural areas. The REA finances the construction and operation of generating plants, electric transmission and distribution lines, or systems for furnishing initial and continued electric services to persons in rural areas not receiving central station service.

Sales for resale: A type of wholesale sale covering energy supplied to other electric utilities, cooperatives, municipalities, and federal and state electric agencies for resale to ultimate consumers.

SCADA: Abbreviation for supervisory control and data acquisition system to acquire process variables and issue commands to control a system.

Scheduled outage: The shutdown of equipment for inspection or maintenance, in accordance with an advance schedule.

Scheduling coordinators: Entities certified by the FERC to act on behalf of generators, supply aggregators (wholesale marketers), retailers, and customers to schedule electricity deliveries.

Seasonal rates: Electric rate whereby an electric utility provides service to consumers at different rates depending on the season. The electric rate schedule usually takes into account demand, based on weather and other factors.

Secondary energy source: Energy created from a primary energy source (e.g., electricity).

Small power producer (SPP): Under PURPA, a small power production facility (or small power producer) generates electricity using waste or renewable (biomass, conventional hydroelectric, wind and solar, and geothermal) energy as a primary energy source. Fossil fuels can be used, but renewable resource must provide at least 75 percent of total energy input. (*See* Code of Federal Regulations, Title 18, Part 292.)

Spark spread: A measurement of the difference between the price a generator can obtain from selling one megawatt hour (MWh) of electricity and the cost of the natural gas needed to generate the MWh of electricity. Spark spread measures potential profit for generating electricity on a particular day.

Spinning reserve: That reserve generating capacity running at zero load and synchronized to the electric system.

Solar energy: The radiant energy of the Sun, which can be converted into other forms of energy such as heat or electricity.

Spot purchases: A single shipment of fuel or volumes of fuel purchased for delivery within a specified time period. Spot purchases are often made by a user to fulfill a certain portion of energy requirements, to meet unanticipated energy needs, or to take advantage of low fuel prices.

Stability: The property of a system or element by virtue of which its output will ultimately attain a steady state. The amount of power that can be transferred from one machine to another following a disturbance. The stability of a power system is its ability to develop restoring forces equal to or greater than the disturbing forces so as to maintain a state of equilibrium.

Static reactive power devices: Maintain steady-state voltage levels. An example is switched capacitor banks. Static reactive power compensation can be thought of as keeping reactive power available in generators by supplying the reactive power consumption needs of heavily loaded transmission lines.

Station service: The electric energy produced by a generation resource used within the facility to power the lights, motors, control systems, and other auxiliary electrical loads necessary for the facility to operate.

Standby facility: A facility supporting a utility system and generally running under no-load and available to replace or supplement a facility normally in service.

Standby service: Support service available to supplement a customer, utility system, or another utility if a schedule or an agreement authorizes the transaction. The service is not regularly used.

Steam electric power plant (conventional): A plant in which the prime mover is a steam turbine to drive the turbine in a boiler using fossil fuels.

Stranded benefits: Benefits associated with regulated retail electric service which may be at risk under open market retail competition. Examples include conservation programs, fuel diversity, reliability of supply, and tax revenues based on utility revenues.

Stranded costs: A public utility's existing infrastructure investments that became redundant after substantial changes in regulatory or market conditions. An incumbent electric power utility will have made substantial investments over the years and will carry debt associated with those investments. Public Utility Commissions typically allow legacy utilities to recover their stranded costs during the transition to a competitive electric market.

Substation: Facility equipment that switches, changes, or regulates electric voltage.

Surplus energy: Energy generated beyond the immediate needs of the producing system that may be supplied by spinning reserves and sold on an interruptible basis.

Switching station: Facility equipment used to tie together two or more electric circuits through switches selectively arranged to permit a circuit to be disconnected or to change the electric connection between the circuits.

Switchyard: A substation associated with an electric generation facility connecting the generator(s) to the transmission grid.

System: Physically connected generation, transmission, and distribution facilities operated as an integrated unit under one central management or operating supervision.

System operating limit (SOL): The value (such as MW, MVAR, amperes, frequency, or volts) that satisfies the most limiting of the prescribed operating criteria for a specified system configuration to ensure operation within acceptable reliability criteria. SOLs are based upon certain operating criteria including, but not limited to:

- facility ratings (applicable pre- and postcontingency equipment ratings or facility ratings);
- transient stability ratings (applicable pre- and postcontingency stability limits);
- voltage stability ratings (applicable pre- and postcontingency voltage stability);
- system voltage limits (applicable pre- and postcontingency voltage limits).

System operator: An individual at a control center who monitors and controls the electric system in real time.

Transmission Control Protocol/Internet Protocol (TCP/IP): A set of rules governing the connection of computer systems to the internet.

Telemetering: The process by which measurable electrical quantities from substations and generating stations are instantaneously transmitted to the control center and by which operating commands from the control center are transmitted to the substations and generating stations.

Terawatt-hour: One trillion watt hours.

Thermal rating: The maximum amount of electrical current a transmission line or electrical facility can conduct over a specified time period before it sustains permanent damage by overheating or before it sags to the point it violates public safety or regulatory requirements.

Tie-line (electric): A circuit connecting two BAAs. Also, describes circuits within an individual electrical system.

Tidal energy: An ocean energy technology involving erecting a dam across the opening to a tidal basin. The dam includes a sluice opened to allow the tide to flow into the basin; the sluice is then closed and, as the sea level drops, traditional hydropower technologies can generate electricity from the elevated water in the basin.

Topology: The form taken by a network of interconnections of circuit components. Different specific values or ratings of the components are considered the same topology.

Transformer: An electrical device for changing AC voltage.

Transmission (1): Movement or transfer of electric energy over an interconnected group of lines and associated equipment between points of supply and points at which it is transformed for delivery to consumers or other electric systems. Transmission is considered to end when the energy is transformed for distribution to the consumer.

Transmission (2): An interconnected group of lines and associated equipment to move or transfer electric energy between points of supply and points at which it is transformed for delivery to customers or other electric systems.

Transmission and distribution loss: Electric energy lost due to transmission and distribution of electricity. Much of the loss is thermal in nature.

Transmission constraint: A limitation on one or more transmission elements that may be reached during normal or contingency system operations.

Transmission line: A system of structures, wires, insulators and associated hardware that transmits electric energy from one point to another in an electric power system. Lines are operated at relatively high voltages varying from 69 kV up to 765 kV and are capable of transmitting large quantities of electricity over long distances.

Transmission operator: The entity responsible for the reliability of its localized transmission system and operating or directing the operation of the transmission facilities.

Transmission owner: The entity that owns and maintains transmission facilities.

Transmission service provider: The entity that administers the transmission tariff and provides transmission service to transmission customers under applicable transmission service agreements.

Transmission system: An interconnected group of electric transmission lines and associated equipment to move or transfer electric energy in bulk between points of supply and points at which it is transformed for delivery over the distribution system lines to consumers or to other electric systems.

Transmitting utility: A regulated entity that owns and may construct and maintain wires used to transmit wholesale power. It may or may not handle the power dispatch and coordination functions. It is regulated to provide nondiscriminatory connections, comparable service, and cost recovery.

Turbine: A machine for generating rotary mechanical power from the energy of a stream of fluid (such as water, steam, or hot gas). Turbines convert the kinetic energy of fluids to mechanical energy through the principles of impulse and reaction or a mixture of the two.

Ultimate customer (or end-use customer): A customer that purchases electricity for its own use and not for resale.

Unbundling: Separating vertically integrated monopoly functions into their component parts to separate service offerings.

Uprate: An increase in available electric generating unit power capacity due to a system or equipment modification. An uprate is typically a permanent increase in the capacity of a unit.

Utility distribution companies: The entities that continue providing regulated services for the distribution of electricity to customers and serve customers who do not choose direct access. Regardless of where a consumer chooses to purchase power, the customer's current utility, also known as the utility distribution company, will deliver the power to the consumer.

Vertical integration: The combination within a firm or business enterprise of one or more stages of production or distribution. In the electric industry, it refers to the historical arrangement whereby a utility owns its own generating plants, transmission system, and distribution lines to provide all aspects of electric service.

Volt (V): The volt is the International System of Units (SI) measure of electric potential or electromotive force. A potential of 1 V appears across a resistance of 1 ohm when a current of 1 ampere flows through that resistance.

Voltage: The difference in electrical potential between any two conductors or between a conductor and ground.

Voltage reduction: Any intentional reduction of system voltage by 3% or greater for reasons of maintaining the continuity of service of the bulk electric power supply system.

Volt-ampere reactive (VAR): A reactive load, typically inductive from electric motors, that causes more current to flow in the distribution network than is actually consumed by the load. This requires excess capability on the generation side and causes greater power losses in the distribution network.

Water turbine: A turbine that uses water pressure to rotate its blades; the primary types are the Pelton wheel, for high heads (pressure); the Francis turbine, for low to medium heads; and the Kaplan, for a wide range of heads. Primarily used to power an electric generator.

Watt (W): The unit of electrical power equal to 1 ampere under a pressure of 1 volt. A watt is equal to 1/746 horsepower.

Watt-hour (Wh): The electrical energy unit of measure equal to one watt of power supplied to, or taken from, an electric circuit steadily for one hour.

Wheeling service: The movement of electricity from one system to another over transmission facilities of interconnecting systems. Wheeling service contracts can be established between two or more systems.

Whitelisting: The process of adding a specific computer domain to an approved list so traffic from that IP address is not blocked.

Wholesale competition: A system whereby a distributor of power would have the option to buy its power from a variety of power producers, and the power producers would be able to compete to sell their power to a variety of distribution companies.

Wholesale power market: The purchase and sale of electricity from generators to resellers (who sell to retail customers), along with the ancillary services to maintain reliability and power quality at the transmission level.

Wholesale sales: Energy supplied to other electric utilities, cooperatives, municipals, and federal and state electric agencies for resale to ultimate consumers.

Wholesale transmission services: The transmission of electric energy sold, or to be sold, in the wholesale electric power market.

Wind energy: Kinetic energy present in wind motion that can be converted to mechanical energy for driving pumps, mills, and electric power generators.

Wind farm (wind generating station): A group of wind turbines interconnected to a common utility system through a system of transformers, distribution lines, and (usually) one substation. Operation, control, and maintenance functions are often centralized through a network of computerized monitoring systems, supplemented by visual inspection. In Europe, this is called a generating station.

Wind turbines: Turbines that capture the wind's energy with two or three propeller-like blades mounted on a rotor to generate electricity. The turbines sit high on the top of the towers, taking advantage of the stronger and less turbulent wind at 100 ft (30 m) or more above ground. A blade acts much like an airplane wing. When the wind blows, a pocket of low-pressure air forms on the downwind side of the blade. The low-pressure air pocket pulls the blade toward it, causing the rotor to turn. This is called lift. The force of the lift is actually much stronger than the wind's force against the front side of the blade, called drag. The combination of lift and drag causes the rotor to spin, like a propeller, and the turning shaft spins a generator to make electricity. Wind turbines can be used as stand-alone applications or connected to a utility power grid or even combined with a photovoltaic (solar cell) system. Stand-alone turbines are typically used for water pumping or communications. However, homeowners and farmers in windy areas may also use turbines to generate electricity. For utility-scale sources of wind energy, a large number of turbines are usually built close together to form a wind farm.

Wires charge: A broad term referring to fees levied on power suppliers or their customers for the use of the transmission or distribution wires.

Common units of measure

Kilo (k) = 1,000 (one thousand)
Mega (M) = 1,000,000 (one million)
Giga (G) = 1,000,000,000 (one billion)
Tera (T) = 1,000,000,000,000 (one trillion)

Ampere: The unit of measurement of electrical **current** produced in a circuit by 1 volt acting through a resistance of 1 Ohm.

Kiloampere (kA) = 1,000 (one thousand) amperes
Mega-amperes (MA) = 1,000,000 (one million) amperes
Giga-amperes (GA) = 1,000,000,000 (one billion) amperes
Tera-ampere (TA) = 1,000,000,000,000 (one trillion) amperes

Volt (V): The unit of measurement of electrical **potential** required to drive 1 ampere of current against 1 ohm resistance.

Kilovolt (kV) = 1,000 (one thousand) volts
Megavolt (MV) = 1,000,000 (one million) volts
Gigavolt (GV) = 1,000,000,000 (one billion) volts
Teravolt (TV) = 1,000,000,000,000 (one trillion) volts

Watt (W): The unit of electrical **powe**r equal to 1 ampere under a pressure of 1 volt.

Kilowatt (kW) = 1,000 (one thousand) watts
Megawatt (MW) = 1,000,000 (one million) watts
Gigawatt (GW) = 1,000,000,000 (one billion) watts
Terawatt (TW) = 1,000,000,000,000 (one trillion) watts

Watt-hour (Wh): The unit of electrical **energy** equal to 1 watt of power supplied to, or taken from, an electric circuit steadily for 1 hour.

Kilowatt-hours (kWh) = 1,000 (one thousand) watt-hours
Megawatt-hours (MWh) = 1,000,000 (one million) watt-hours
Gigawatt-hours (GWh) = 1,000,000,000 (one billion) watt-hours
Terawatt-hours (TWh) = 1,000,000,000,000 (one trillion) watt-hours

Notes

1. Extracted March 17, 2021 from, https://www.merriam-webster.com/dictionary/electricity

2. Extracted March 17, 2021 from, https://wtamu.edu/~cbaird/sq/2014/02/19/what-is-the-speed-of-electricity/

3. Extracted June 10, 2021 from, https://www.merriam-webster.com/dictionary/magnetism

4. Miller, E. M, & Keith, David W., Joule, Climatic Impacts of Wind Power, December 19, 2018.

5. This situation caused the blackouts experienced in Texas (and much of the southern U.S.) in February 2021.

6. BP, plc. Statistical Review of World Energy 2020, 69th edition; p. 60.

7. Consider this story from The Gambia. One Saturday afternoon, the hospital lights came on (powered by a local generator), which was not typical. The lights generally did not come on after 2:00 p.m. on weekends but were needed for an emergency caesarian section. When the doctor removed the infant from its mother, the technician suctioned its nose and mouth for more than twenty-five minutes. Ultimately, the caregivers gave up because the child had died. The doctor stated the baby suffocated in utero because the hospital did not have enough power to use the ultrasound machine for every pregnancy. If they had had sufficient power, he would have detected the problem earlier and could have saved the baby. *The Moral Case for Fossil Fuels*, Alex Epstein, Penguin Group (2014). Most people do not realize the current worldwide standard of living – higher than ever in history – is almost directly attributable to the widespread use of fossil fuels to generate electricity. That electricity powers hospitals, schools, nursing homes, fire and police stations, banks, homes, computers, to charge phones, to cook meals, etc..

8. Extracted March 17, 2021 from, https://ourworldindata.org/energy; https://www.forbes.com/sites/jamesellsmoor/2019/05/23/sdg-7-at-current-rate-2030-renewable-energy-goals-will-be-missed/?sh=731506fd3f0b

9. Extracted April 12, 2021 from, https://www.universetoday.com/82402/who-discovered-electricity/#:~:text=Electricity%20is%20a%20form%20of,lightning%20and%20electricity%2C%20nothing%20more.

10. Yıldız İlhami, Craig MacEachern, *Comprehensive Energy Systems* (2018).

11. Gilbert, William, On the Loadstone and Magnetic Bodies. A translation by P. Fleury Mottelay. New York, John Wiley & Sons.

12. Extracted April 20, 2021 from, https://www.ieee-pes.org/images/files/pdf/Chinas_Power_Industry_Milestones_IEEE_PES-20081010.pdf

13. Extracted April 20, 2021 from, https://new.siemens.com/global/en/company/about/history/company/1865-1896.html

14. Extracted April 20, 2021 from, https://www.zum.de/whkmla/sp/0809/kyungmook/km2.html

15. Smil, Vaclav (2005), *Creating the Twentieth Century: Technical Innovations of 1867–1914 and Their Lasting Impac*t. Oxford University Press. p. 62. ISBN 0-19-516874-7.

16. Extracted April 13, 2021 from, https://www.hydropower.org/iha/discover-history-of-hydropower

17. *Ibid.*

18. *Ibid.*

19. Extracted December 29, 2021 from, https://portlandgeneral.com/about/rec-fish/willamette-river/willamette-falls-and-sullivan-plant-history

20. Extracted April 13, 2021 from, https://edisontechcenter.org/LauffenFrankfurt.html

21. Extracted June 12, 2021 from, https://dr.library.brocku.ca/handle/10464/13120

22. Extracted June 12, 2021 from, https://npgallery.nps.gov/NRHP/SearchResults?view=list

23. Extracted April 20, 2021 from, https://www.power-technology.com/features/oldest-geothermal-plant-larderello/

24. Extracted June 23,2021 from, https://en.wikipedia.org/wiki/Larderello

25. Extracted April 14, 2021 from, https://www.power-technology.com/features/oldest-geothermal-plant-larderello/

26. Extracted July 7, 2021 from, https://www.energy.gov/ne/articles/9-notable-facts-about-world-s-first-nuclear-power-plant-ebr-i

27. Extracted April 20, 2021 from, https://www.statista.com/statistics/267158/number-of-nuclear-reactors-in-operation-by-country/#:~:text=Currently%2C%20there%20are%20440%20nuclear,from%20nuclear%20sources%20in%202018.

28. Extracted April 20, 2021 from, https://www.eia.gov/energyexplained/geothermal/use-of-geothermal-energy.php

29. Nalbandian, H. "Performance and Risks of Advanced Pulverized-Coal Plants," *Energeia*, Vol. 20, No. 1, 2009, pages 1–6.

30. Ansolobehere, S., J. Beer, J. Deutch, A.D. Ellerman, J. Friedman, H. Herzog, H. Jacoby, P. Joskow, G. McRae, R. Lester, E. Moniz and E. Steinfeld "The Future of Coal – Options for a Carbon-Constrained World," MIT study on the future of coal, Massachusetts Institute of Technology. 2007, ISBN 978-0-615-14092-6.

31. Extracted June 12, 2021 from, https://www.eia.gov/electricity/annual/pdf/epa.pdf "Electric Power Annual 2017", October 2018, Revised December 2018, U.S. Energy Information Administration, https://www.eia.gov/electricity/annual/pdf/epa.pdf

32. Extracted June 12, 2021 from, https://www.energy.gov/fe/how-gas-turbine-power-plants-work

"How Gas Turbine Power Plants Work," Office of Fossil Energy,

33. Extracted March 17, 2021, from, https://www.ipieca.org/resources/energy-efficiency-solutions/power-and-heat-generation/combined-cycle-gas-turbines/#:~:text=The%20overall%20electrical%20efficiency%20of,cycle%20application%20of%20around%2033%25.

34. Extracted May 26, 2021, from, https://www.nsenergybusiness.com/features/top-countries-wind-energy-capacity/#:~:text=China%20boasts%20the%20world's%20largest,more%20than%20any%20other%20country.

35. Extracted May 26, 2021, from, https://www.eia.gov/energyexplained/wind/types-of-wind-turbines.php

36. Extracted May 27, 2021 from, https://www.pnas.org/content/107/42/17899

37. Extracted March 17, 2021 from, https://www.eia.gov/todayinenergy/detail.php?id=37972

38. Extracted April 2, 2022 from, https://www.eia.gov/tools/glossary/index.php?id=Capacity_factor#:~:text=Capacity%20factor%3A%20The%20ratio%20of,Top

39. Extracted March 17, 2021 from, https://www.eia.gov/electricity/monthly/current_month/epm.pdf

40. Extracted October 21, 2021 from, https://csrc.nist.gov/glossary/term/intelligent_electronic_device

41. Extracted October 24, 2021 from, https://www.eia.gov/energyexplained/electricity/delivery-to-consumers.php

42. Extracted November 16, 2021 from, https://www.eia.gov/tools/faqs/faq.php?id=65&t=2

43. Extracted March 17, 2021 from, https://www.eia.gov/tools/faqs/faq.php?id=108&t=3

44. Extracted March 17, 2021 from, https://library.municode.com/tx/austin/codes/utilities_criteria_manual

45. Extracted September 1, 2021 from, https://csrc.nist.gov/glossary/term/supervisory_control_and_data_acquisition

46. The North American Electric Reliability Corporation (NERC), the organization given authority to promulgate and enforce Reliability Standards for the North American Bulk Electric System (BES), has issued a litany of Critical Infrastructure Protection Reliability Standards addressing cybersecurity for the BES.

47. Microsoft Windows™ and Windows™ used throughout this publication are trademarks of Microsoft Corporation.

48. Extracted March 17, 2021 from, https://isssource.com/classic-hacker-case-maroochy-shire/ extracted April 9, 2022.

49. Extracted March 17, 2021 from, https://www.bloomberg.com/news/articles/2021-06-04/hackers-breached-colonial-pipeline-using-compromised-password extracted April 9, 2022.

50. Midcontinent Independent System Operator, Inc. (MISO); Saskatchewan Power Corporation; Southwest Power Pool, Inc. (SPP); Hydro-Quebec TransEnergie; ISO-NE; New Brunswick Power Corporation; New York Independent System Operator (NYISO); Ontario IESO; PJM Interconnection, LLC; Florida Reliability Coordinating Council, Inc. (FRCC); Southern Company Services, Inc. – Trans; TVA; VACAR South; Electric Reliability Council of Texas, Inc. (ERCOT); and California Independent System Operator (CAISO).

51. Chapter 6 contains a detailed discussion of utility controls systems.

52. Extracted March 17, 2021 from, https://www.ercot.com/calendar/event?id=1613577826960, and accessing Key Documents, 2.2 Revised ERCOT Presentation, slide 13.

53. *Ibid.* at Slide 11.

54. *Ibid.* at Slide 12.

55. *Ibid.* at Slide 15.

56. Extracted March 17, 2021 from, https://www.energy.gov/eere/water/pumped-storage-hydropower; accessed February 14, 2022.

57. *Ibid.*

58. Extracted March 17, 2021 from, https://en.wikipedia.org/wiki/Energy_density

59. Extracted June 12, 2021 from, https://cleantechnica.com/2022/01/04/largest-pumped-hydro-facility-in-world-turns-on-in-china/

60. Extracted June 12, 2021 from, https://heindl-energy.com/technical-concept/basic-concept

61. A 317 MW CAES facility using an underground salt dome was recently proposed in Anderson County, Texas. http://www.apexcaes.com/project; accessed February 15, 2022.

62. Techno-economic Performance Evaluation of Compressed Air Energy Storage in the Pacific Northwest, US Department of energy, February 2013.

63. Extracted June 12, 2021 from, https://sandia.gov/ess-ssl/gesdb/public/projects.html#526; https://sandia.gov/ess-ssl/gesdb/public/projects.html; https://www.atlantica.com/web/en/company-overview/our-assets/asset/solana/; https://www.energy.gov/lpo/solana.

64. Extracted June 12, 2021 from, https://highviewpower.com/news_announcement/highview-power-breaks-ground-on-250mwh-cryobattery-long-duration-energy-storage-facility/.

65. Extracted June 12, 2021 from, https://www.eia.gov/electricity/data/eia860. U.S. DOE 2020 Form EIA-860 Data - Schedule 3, Detailed power generation data with previous form data (EIA-860A/860B) representing all U.S. power generation including storage, as distinct from the U.S. DOE Global Energy Storage Database (https://sandia.gov/ess-ssl/gesdb/public/projects.html) which depicts only global storage; accessed October 2, 2021.

66. Extracted April 17, 2022 from, https://www.iea.org/reports/the-future-of-hydrogen.

67. Extracted April 17, 2022 from, https://www.eia.gov/dnav/ng/TblDefs/ng_enr_nprod_tbldef2.asp

68. North American Electrical Reliability Corporation Frequently Asked Questions, August 2013.

69. *Ibid.*

70. *Ibid.*

71. Extracted June 12, 2021 from, https://www.nerc.com/pa/Stand/Glossary%20of%20Terms/Glossary_of_Terms.pdf

72. NERC Sanction Guidelines, Appendix 4B to NERC Rules of Procedure, Section 3.2.1.1.

73. NERC Sanction Guidelines, Appendix 4B to NERC Rules of Procedure, Section 3.2.1.2.

74. Extracted March 20, 2022 from, https://www.nerc.com/pa/rrm/ea/ERO_EAP_Documents%20DL/ERO_EAP_v3.1.pdf

75. *Crown corporations* serve as the principal entities in provinces with vertically integrated markets. Newfoundland and Labrador Power is a Crown corporation with huge hydroelectric operations on the Churchill River. Ontario Power Generation is a Crown corporation with approximately 50% of the generating capacity in Ontario. The other major participant in the Ontario market is Hydro One, which owns and operates the transmission grid and the largest distribution network. British Columbia (BC) Hydro, SaskPower, ManitobaHydro and Hydro-Québec are all Crown corporations and control virtually all electricity provision services in their provinces. Each owns and operates most of the province's generation, transmission, and distribution assets.

76. If a generating facility uses flammable substances (gas or oil) in amounts above established limits, its operation may become subject to a license requirement. The Federal Service for Environmental, Technological and Nuclear Supervision issues such licenses.

77. *National Association for the Advancement of Colored People v. FPC*, 520 F.2d 432, 438 (D.C. Cir. 1975), aff'd, 425 U.S. 662 (1976).

78. Extracted March 20, 2022 from, https://www.ferc.gov/industries-data/electric/power-sales-and-markets/electric-competition

79. Extracted March 20, 2022 from, (https://texasmonitor.org/city-of-austin-one-of-many-sucked-into-biomass-plant-money-pit/; https://www.kxan.com/investigations/austins-biomass-power-plant-sat-idle-during-texas-winter-energy-crisis/#:~:text=AUSTIN%20%28KXAN%29

%20%E2%80%93%20Austin%20Energy%E2%80%99s%20biomass%20power%20plant,the%20worst%20winter%20energy%20crises%20in%20state%20history; https://www.bizjournals.com/austin/news/2021/06/25/austin-biomass-power-plant-idled.html).This is true for government-operated utilities. In Austin, Texas, for example, the City Council mandated the municipal utility – Austin Energy (AE) – buy more and more *green* power even though that mandate required the utility to enter many contracts for wind energy at prices significantly higher than those for traditional fossil fuel plants. AE even entered into a contract with a developer of a biomass plant at a total cost of approximately one *billion* dollars. The city ended up buying out that biomass contract in 2019 for $460 million. Despite the staggering cost, the plant produced hardly any electricity.

80. Many countries continue to rely on regulated markets or government control of generation, transmission, and distribution. This text does not discuss that approach in great detail. In regulated markets, utilities comply with electricity rates set by public agencies. Regulated markets are often considered monopolies because they limit consumer choice. They do, however, provide benefits such as stable prices and long-term certainty. Regulatory bodies analyze utility financial data to determine if its rates are fair, just, and reasonable based on long-standing regulatory jurisprudence. Such rates contain two main elements: cost recovery and earning a reasonable return on investment.

81. Extracted March 20, 2022 from, https://www.electricchoice.com/map-deregulated-energy-markets/.

82. Extracted March 20, 2022 from, https://aemo.com.au/-/media/Files/Electricity/NEM/National-Electricity-Market-Fact-Sheet.pdf.

83. Nakajima, Daiki Japan's Energy Market Reform - Full Retail Choice In Electricity Market, Japan External Trade Organization. New York.

84. Extracted March 20, 2022 from, https://eepublicdownloads.entsoe.eu/clean-documents/Publications/Statistics/Factsheet/entsoe_sfs2015_web.pdf).

85. Extracted March 20, 2022 from, https://simsee.org/simsee/biblioteca/ElectricityMarketsInRussiaUSandEU.pdf.M. Oksanen, R. Karjalainen, S. Viljainen and D. Kuleshov, *Electricity Markets in Russia, the US, and Europe*.

86. Some people blame ERCOT's *energy only* market for the situation in February 2021 (jokingly referred to by Texans as *snowpocalypse*) when ERCOT depleted all available generation and, consequently, had to enforce blackouts lasting several days.

87. Extracted March 20, 2022 from, https://www.gao.gov/assets/gao-18-131.pdf. The New England ISO, Midcontinent ISO, New York ISO, and PJM markets in the eastern U.S. all employ a capacity market. Power plant owners earn revenue in capacity auctions for making a *capacity commitment* – an agreement to make available their power plants to meet customers' electricity needs during a specific delivery period, if needed. To participate in the auction, power generators offer to commit capacity (in MW) at a specified price. The market operator administers the auction by selecting offers and establishing a final price. Capacity commitments count toward each electricity supplier's *resource adequacy requirement*, and each electricity supplier pays part of the total cost of capacity commitments, generally in proportion to their customers' share of the region's total electricity needs (i.e., load ratio share). According to the FERC, regions with capacity markets paid more than $51 billion between 2013 and 2016 for capacity commitments. *See Report to Congressional Committees – ELECTRICITY MARKETS Four Regions Use Capacity Markets to Help Ensure Adequate Resources, but FERC Has Not Fully Assessed Their Performance.*

88. Section 39.051 of the Texas Utilities Code requires electric utilities to separate their customer energy services activities (otherwise widely available in the competitive market) from their regulated utility activities. Additionally, it requires each electric utility to separate its

business activities from one another into: (1) a power generation company; (2) a retail electric provider; and (3) a T&D utility.

Extracted March 20, 2022 from, (https://www.dfat.gov.au/sites/default/files/deregulation-of-the-energy-industry-australian-experience.pdf).When Australia deregulated its electric market, it also required vertical separation of generation, T&D, and retail. *See* Felix Karmel, *Deregulation and Reform of the Electricity Industry in Australia* (February 1, 2018) at p. 6.

89. Extracted March 20, 2022 from, https://www.npr.org/2021/02/22/970074424/why-some-texas-residents-now-face-huge-electricity-bills and (https://thehill.com/policy/energy-environment/539693-texas-households-face-massive-electricity-bills-some-as-high-as-17k). During the Texas severe weather event in February 2021, customers with variable-rate contracts for electricity faced *huge* electric bills when wholesale electric prices spiked. Variable-rate contracts transfer market risk to the customer because the retailer simply passes on to the customer whatever the retailer paid for the wholesale electricity. It can be a good product during times of low (or declining) prices, but when the grid operator experiences generation shortages (as ERCOT did during the freezing temperatures), wholesale prices can soar and retail electric providers pass on those costs to customers with variable-rate contracts. *See, Why Some Texas Residents Now Face Huge Electricity Bills,* NPR, Feb. 22, 2021; Brooke Seipel, *Texas households face massive electricity bills, some as high as $17K, after winter storm,* Feb. 19, 2021.

90. Texas Administrative Code, Title 16, Part 2, Chapter 25, Sub-Chapter K.

91. ExtractedMarch20,2022from,https://www.electric.coop/electric-cooperative-fact-sheet

92. Extracted March 20, 2022 from, https://www.rescoop.eu/about-us

93. *Ibid.*

94. National Electrification Administration Reform Act of 2013, Republic Act (R.A.) No. 10531, Section 2.

95. An IOU's service territory is often referred as a *certificated* area because the utility receives a *certificate of convenience and necessity,* defined as a certificate issued by an agency granting a company authority to operate a public service.

96. In some cases, one holding company owns multiple IOUs with centralized functions. Each IOU is run as a separate company for regulatory purposes.

97. The REP usually includes a profit component in the usage charge and adds an *administration* fee or *customer charge* to cover overhead expenses.

98. Extracted March 20, 2022 from, https://www.marketwatch.com/story/that-kind-of-bill-sets-us-way-back-some-texans-face-massive-electric-bills-after-severe-weather-why-its-happening-and-what-they-can-do-about-it-11614118493; and https://www.wbrz.com/news/winter-storm-uri-could-affect-your-utility-billThis arrangement led to many shocking electric bills in the wake of Winter Storm Uri, which hit the U.S. with record cold temperatures in February 2021. The storm affected natural gas, wind, and solar generation. For example, natural gas prices increased to record highs. Electric generators paid exorbitant prices for fuel to operate generation facilities while demand for electricity soared. Ultimately, consumers felt the impact of those high fuel costs through *fuel cost adjustments.*

99. Extracted December 24, 2021 from, https://www.asce.org/advocacy/infrastructure/failure-to-act-reports, American Society of Civil Engineers (ASCE), *Failure to Act: Electric Infrastructure Investment Gaps in a Rapidly Changing Environment,* 2020 (12/24/21).

100. *Ibid.*

101. ASCE, *supra* Note i.

102. Extracted March 20, 2022 from, https://www.energy.gov/sites/prod/files/oeprod/DocumentsandMedia/adequacy_report_01-09-09.pdf.

U.S. Department of Energy Electric Advisory Committee report, *Keeping the Lights On in a New World*, January 2009.

103. *Ibid.*

104. Extracted December 24, 2021 from, https://www.linkedin.com/pulse/top-10-challenges-electric-distribution-utilities-steven-collier , Collier, Steve, *Top 10 Challenges for Electric Distribution Utilities*, June 30, 2015.

105. Extracted December 24, 2021 from, https://iopscience.iop.org/article/10.1088/1748-9326/ab875d/pdf Carley, Konisky, Atiq, and Land, *Energy infrastructure, NIMBYism, and public opinion: a systematic literature review of three decades of empirical survey literature*, August 25, 2020, Environmental Research Letters.

106. Extracted December 24, 2021 from, https://www.nhpr.org/nh-news/2019-07-19/in-unanimous-vote-n-h-supreme-court-upholds-northern-pass-denial Ropeik, Annie, *In Unanimous Vote, N.H. Supreme Court Upholds Northern Pass Denial*, N.H. Public Radio, July 19, 2019.

107. *Ibid.*

108. *Ibid.*

109. Extracted December 24, 2021 from, https://www.ucsusa.org/resources/environmental-impacts-wind-power, https://www.ucsusa.org/resources/environmental-impacts-solar-power, and https://www.nrel.gov/state-local-tribal/blog/posts/life-cycle-assessment-and-photovoltaic-pv-recycling-designing-a-more-sustainable-energy-system.html. Another challenge, as mentioned above, is where to build such facilities. Wind and solar "farms" need vast open spaces and consume huge tracts of land. The NIMBY phenomenon mentioned earlier affects this issue. Environmental issues also become a concern. *See, e.g.*, Union of Concerned Scientists, *Environmental Impacts of Wind Power*, March 5, 2013, (12/24/21), which claims wind farms use between 30 and 141 acres of land *per megawatt*. On the same day, the same organization published a paper titled, *Environmental Impacts of Solar Power* which states photovoltaic (PV) projects use between 3.5 and 10 acres per megawatt and concentrating solar thermal plants (CSPs) use between 4 and 16.5 acres per megawatt. That same report states PV cell manufacturing uses numerous hazardous materials, including hydrochloric acid, sulfuric acid, nitric acid, hydrogen fluoride, 1,1,1-trichloroethane and acetone. PV cells also contain a number of toxic materials, including gallium, arsenide, copper-indium-gallium-diselenide and cadmium-telluride. The National Renewable Energy Laboratory estimates the useful life of PV systems as 25-40 years). How to dispose of those toxic materials at that time remains a concern.

110. *Ibid.*

111. Extracted December 20, 2021 from, https://microgridknowledge.com/campus-microgrids-higher-education/

112. *Ibid.*

113. *Ibid.*

114. Extracted December 19, 2021 from, https://www.eia.gov/analysis/studies/electricity/batterystorage/pdf/battery_storage.pdf

115. Extracted January 31, 2022 from, https://www.aps.com/-/media/APS/APSCOM-PDFs/About/Our-Company/Newsroom/McMickenFinalTechnicalReport.ashx?la=en&hash=50335FB5098D9858BFD276C40FA54FCE

116. Extracted January 31, 2022 from, https://www.sciencedirect.com/science/article/abs/pii/S0950423021001686.

117. Installations greater than 1 MW are considered "large scale."

118. *Ibid.*

119. The authors discuss ISOs and RTOs in the chapter discussing markets.

120. Extracted December 19, 2021 from, https://eandt.theiet.org/content/articles/2021/06/battery-storage-system-connected-to-transmission-grid/

121. Extracted December 19. 2021 from, https://www.businessgreen.com/news/3035576/uks-largest-grid-battery-storage-facility-completed-in-hertfordshire and https://www.edie.net/news/8/UK-s-largest-battery-storage-facility-comes-online-in-Hertfordshire/

122. Extracted December 19. 2021 from, https://theenergyst.com/vlc-energy-connects-50mw-battery-storage-portfolio/

123. Extracted December 19. 2021 from, https://eandt.theiet.org/content/articles/2020/11/largest-battery-storage-facility-to-power-luxury-hotel-resort-in-saudi-arabia/.

124. Extracted December 20. 2021 from, https://www.abc.net.au/news/2020-11-05/new-tesla-battery-for-moorabool-victoria/12851698.

125. Extracted December 21. 2021 from, https://hornsdalepowerreserve.com.au/.

126. Extracted December 21. 2021 from, https://www.abc.net.au/news/2020-09-02/tesla-battery-expanded-as-sa-energy-minister-lauds-benefits/12622382 (12/20/21).

127. Extracted December 21. 2021 from, https://eandt.theiet.org/content/articles/2020/11/australia-s-new-south-wales-announces-a-32bn-renewables-push/.

128. Extracted December 21. 2021 from, https://www.solarserver.de/2016/01/20/leclanche-soll-eines-der-weltgroessten-energiespeicher-systeme-mit-13-mw-53-mwh-nach-ontario-liefern/

129. Extracted December 21. 2021 from, https://electronics360.globalspec.com/article/6402/mitsubishi-installs-50mw-energy-storage-system-to-japanese-power-company

130. Extracted December 21. 2021 from, https://www.yokogawa.com/us/library/resources/references/stardom-and-fa-m3-ensure-smooth-supply-of-power-to-grid-by-wind-farm-equipped-with-large-capacity-nas-batteries/

131. Extracted December 21. 2021 from, https://www.ee-news.ch/de/erneuerbare/article/33162/kokam-liefert-56-mw-fur-speicherprojekt-zur-frequenzregulierung

132. Extracted December 21. 2021 from, https://www.greencarcongress.com/2015/11/20151104-daimler.html

133. Extracted December 21. 2021 from, https://www.wemag.com/aktuelles-presse

134. Extracted December 21. 2021 from, https://ees-ev.de/in-braderup-steht-der-stromspeicher-der-zukunft/

135. Extracted December 21. 2021 from, https://www.sunwindenergy.com/photovoltaics-wind-energy/recharge-invests-hybrid-renewable-energy-storage-project-graciosa

136. Extracted December 21. 2021 from, https://policyadvice.net/insurance/insights/electric-car-statistics/

137. Extracted December 21, 2021 from, https://theconversation.com/australias-electricity-grid-can-easily-support-electric-cars-if-we-get-smart-115294.

138. Extracted April 3, 2022 from, https://www.forbes.com/wheels/advice/ev-charging-levels/

139. Extracted December 24. 2021 from, https://getelectricvehicle.com/impact-of-ev-on-power-grid/_Krishnan, Sibi. *Impact of EV on power utility grid: Positive or Negative*?

140. A study based on an area in Melbourne, Australia, indicates EV penetration of only 10% could lead to network failures; *supra* Note 41.

141. *Ibid.*

142. *Ibid.*

143. Extracted December 21. 2021 from, https://www.forbes.com/sites/oliverwyman/2019/05/15/as-more-evs-hit-the-road-blackouts-become-likely/?sh=5359a5a5dc30 (12/21/21).

144. *Ibid.*

145. *Ibid.*

146. *Ibid.*

Figure References

Figure 1-4. Source: U.S. EIA.

Figure 1-6. Source: Graphs extracted from https://www.eia.gov/todayinenergy/detail.php?id=41433 U.S. Energy Information Administration, International Energy Outlook 2019 Reference case.

Figure 1-22. Source: Produced by BP, plc. Statistical Review of World Energy 2020, 69th edition; p. 60.

Figure 1-23. Source: Produced by Power Knowledge & Development from the U. S. Energy Information Administration (EIA) data.

Figure 1-24. Source: Produced by Power Knowledge & Development from the U. S. Energy Information Administration (EIA) data.

Figure 1-25. Source: IEA (2020), World Energy Outlook 2020, IEA, Paris. https://www.iea.org/reports/world-energy-outlook-2020.

Figure 2-1. Source: Public Domain. Created: before 1886 date QS:P,+188 6-00-00 T00:00:00Z/7,P1326,+1886-00-00T00:00:00Z/9.

Figure 2-3. Source: Wikimedia Commons, the free media repository. 6 Nov 2020, 05:07 UTC. 12 Apr 2021, 21:35, https://commons.wikimedia.org/w/index.php?title=File:Jablochkoff%27s_candle.jpg&oldid=510497268.

Figure 2-4. Source: By Newton Henry Black – Newton Henry Black, Harvey N. Davis (1913) Practical Physics, The MacMillan Co., USA, p. 242, fig. 200, Public Domain, https://commons.wikimedia.org/w/index.php?curid=73846.

Figure 2-5. Source: Public Domain, https://commons.wikimedia.org/w/index.php?curid=3261751.

Figure 2-6. Source: U.S. Department of Energy.

Figure 2-7. Source: Public domain – Project Gutenberg e-text 17167.png.

Figure 2-9. Source: Extracted June 12, 2021 from, https://npgallery.nps.gov/NRHP/SearchResults?view=list

Figure 2-10. Source: GNU Free Documentation License.

Figure 2-1. Source: Public Domain Robert W. Righter (1996) Wind Energy in America: A History, University of Oklahoma Press, page 44. Retrieved on 27 December 2008. ISBN: 0806128127. Picture is in the public domain.

Figure 3-1. Source: Produced by Power Knowledge & Development from data contained in bp Statistical Review of World Energy 2021.

Figure 3-2. Source: Produced by Power Knowledge & Development from data contained in bp Statistical Review of World Energy 2021.

Figure 3-4. Source: Courtesy Mechanical Dynamics & Analysis (MD&A).

Figure 3-6. Source: Courtesy Mechanical Dynamics & Analysis (MD&A).

Figure 3-9. Source: Courtesy Mechanical Dynamics & Analysis (MD&A).

Figure 3-11. Source: Produced by Paul Meier

Figure 3-13. Source: Extracted May 4, 2021, from https://www.nrc.gov/reactors/pwrs.html

Figure 3-14. Source: Courtesy STP Nuclear Operating Company.

Figure 3-15. Source: Produced by Paul Meier

Figure 3-17. Source: Extracted May 26, 2021 from https://commons.wikimedia.org/wiki/File:Parabolic_trough_at_Harper_Lake_in_California.jpg and used under the Creative Commons Attribution-Share Alike 3.0 Unported license. No modifications were made to the original picture.

Figure 3-18. Source: Extracted May 26, 2021, from https://www.eia.gov/energyexplained/hydropower. Source: Tennessee Valley Authority (public domain).

Figure 3-19. Source: Photo by Theo Miesner.

Figure 3-20. Source: Extracted May 26, 2021, from https://www.energy.gov/eere/wind/inside-wind-turbine

Figure 3-22. Source: https://science.nasa.gov/science-news/science-at-nasa/2002/solarcells

Figure 3-23. Source: Photo by Theo Miesner.

Figure 3-25. Source: https://www.energy.gov/eere/geothermal/downloads/enhanced-geothermal-system-egs-fact-sheet

Figure 3-26. Source: Extracted May 27, 2021, from https://www.energy.gov/ne/articles/what-generation-capacity

Figure 4-1. Source: Extracted May 27, 2021, from https://www.energy.gov/ne/articles/what-generation-capacity

Figure 5-2. Source: https://www.energy.gov/oe/activities/technology-development/grid-modernization-and-smart-grid/role-microgrids-helping.

Figure 5-6. Source: Used under terms of the GNU Free Documentation License.

Figure 5-8. Source: https://commons.wikimedia.org/wiki/File:Diapositiva14.PNG.

Figure 6-1. Source: Courtesy of ERCOT.

Figure 7-6. Source: Courtesy ERCOT.

Figure 7-8. Source: Screen capture from www.ercot.com.

Figure 7-10. Source: Screen capture from www.austinenergy.com.

Figure 8-2. Source: Prepared by Tom Miesner from various sources.

Figure 8-4. Source: Derived from U.S. Energy Information Administration Form EIA-860 data which was extracted March 17, 2021 from, https://www.eia.gov/electricity/data/eia860

Figure 8-5. Source: Environmental and Energy Study Institute.

Figure 8-6. Source: Sandia National Laboratories.

Figure 8-7. Source: Courtesy Ameren Corporation.

Figure 8-8. Source: U.S. DOE Office of Energy Efficiency and Renewable Energy. Extracted June 12, 2021 from, https://www.energy.gov/eere/water/pumped-storage-hydropower

Figure 8-10. Source: Used under the terms of the GNU Free Documentation License available at https://commons.wikimedia.org/wiki/Commons:GNU_Free_Documentation_License.

Figure 8-12 Source: Abengoa media kit.

Figure 8-14. Source: Picture by Tom Ortman.

Figure 9-1. Source: Statistical Review of World Energy 2021 © BP p.l.c. 2021.

Figure 9-2. Source: U.S. Energy Information Administration Monthly Data Review. Table 4.3 April 2021.

Figure 9-8. Source: Compiled by Pipeline Knowledge, LLC from CIA Fact Book data.

Figure 9-9. Source: Courtesy ConocoPhillips.

Figure 9-17. Source: EIA.

Figure 9-25. Source: U.S. Energy Information Administration Natural Gas Monthly February 13, 2020.

Figure 10-2. Source: https://www.pinterest.com/margeschifini/oldies-but-goodies/.

Figure 10-3. Source: Taken by author at NERC offices in Atlanta, Georgia.

Figure 10-4. Source NERC ERO Enterprise Guide for Internal Controls Version 2, September 2017 (page iii) on www.nerc.com.

Figure 10-6. Source: Screen capture from www.nerc.com.

Figure 10-7. Source: Screen capture from www.nerc.com.

Figure 10-8. Source: Screen capture from www.nerc.com.

Figure 10-9. Source: Screen capture from www.wecc.org.

Figure 10-10. Source: Screen capture from www.nerc.com.

Figure 11-2. Source: Produced from data extracted from the U.S. EIA at https://www.eia.gov/electricity/wholesale/#history.

Figure 11-3. Source: U.S. Energy Information Administration.

Figure 11-4. Source: U.S. Energy Information Administration.

Figure 11-5. Source: Courtesy Stephen Elliott Consulting, LLC.

Figure 11-6. Source: Courtesy Stephen Elliott Consulting, LLC.

Figure 11-7. Source: Courtesy Stephen Elliott Consulting, LLC.

Figure 11-8. Source: Courtesy Stephen Elliott Consulting, LLC.

Figure 11-9. Source: U.S. EIA extracted March 22, 2022 from https://www.eia.gov/todayinenergy/detail.php?id=42915.

Figure 12-2. Source: Source: IEA (2020), World Energy Outlook 2020, IEA, Paris https://www.iea.org/reports/world-energy-outlook-2020.

Figure 12-3. Source: Produced by BP, plc. Statistical Review of World Energy 2020, 69th edition; p. 60.

Figure 12-6. Source: U.S Energy Information Administration, Monthly Energy Review, Table 1.3 and 10.1, April 2021, preliminary data

Figure 12-9. Source: IEA World Energy Investment 2021.

Figure 12-10. Source: U.S Energy Information Administration, *International Energy Outlook 2021* Reference case

Index

A

B

C

D

E

F

G

H

I

J

K

L

S

T

U

V

W